MW01069921

OCCUPATIONAL MATHEMATICS

Gerald E. Gannon, Ed.D.

Professor
California State University
Fullerton, California

Willard R. Daggett, Ed.D.

Director
Division of Occupational Education Programs
New York State Education Department
Albany, New York

CINCINNATI WEST CHICAGO, IL DALLAS PELHAM MANOR, NY LIVERMORE, CA

ISBN: 0-538-13900-5

Library of Congress Catalog Card Number: 84-71052

1 2 3 4 5 6 7 8 9 K 9 8 7 6 5

Printed in the United States of America

Cover illustration:
Logique Mathematique, Logique Symbolique

Contents

PART II. MATHEMATICS AND YOUR LIFE

PART III. MATHEMATICS, PRODUCTS, AND TRAVEL

PART V. DATA AND INFORMATION PROCESSING

APPENDICES

Preface

Occupational Mathematics is designed to impart problem-solving skills. The concept is straightforward: Mathematics is a tool for the needs of everyday living and for the work place. Practical uses for mathematics go far beyond the manipulation of numbers. Mathematics is a thought-based discipline.

Thus, this book concentrates on helping students learn the skills and processes for solving problems. They do this by thinking their way through problems through the creation of *algorithms* before any numbers or calculations are applied. Students even are told that, with teacher permission, some of the problems in this book may be worked with the aid of a calculator.

This is a distinctive feature of this text: If students know what calculator keys to press, in which sequence, they have effectively solved the problem. Manipulation of numbers simply implements and bears out that solution.

Another special feature of this book lies in the practicality of the situations and problems that students encounter. All problems within the book relate directly to occupational or consumer situations. The objective is to pose problems with which students can relate. Students can then identify with the need to solve the problems and are encouraged to use the orderly, algorithmic tools that are unfolded for them.

The emphasis on algorithmic methods has another advantage: Algorithm design is a technique for computer-related problem solving. Computer programs, for the most part, start as algorithms. Thus, as students master the content of this book, they also enhance their computer literacy and build skills that will be transferable into the the world of data processing.

CONTENT ORGANIZATION

Occupational Mathematics has five parts and 15 units.

Mathematics and You

PART I deals with mathematics as a special form, or language, of communication.

Unit 1 introduces the concept of mathematics as a language. Students are given comparisons between statements in English and in mathematical notation. In keeping with the language concept, algorithms are introduced as the sentences and paragraphs of mathematics. Students are shown how to write algorithms as step-by-step procedures for solving problems. The use of algorithms serves to establish mathematics as a problem-solving language. The idea of problem solving is central to the content of this text. Calculators and computers are introduced in this unit as tools to help people solve problems.

Unit 2 acquaints students with mathematical methods of presenting information. These methods include tables, maps, drawings, pie charts, bar charts, scatter graphs, and line graphs. The concept of mathematics as a problem-solving language is reinforced. Students are shown that solving a problem develops information. Information, in turn, is the basis for informed decisions. The techniques covered in this unit are introduced as methods for presenting numerical information in easy-to-understand formats.

Unit 3 deals with fractions and decimals. The use of fractions and decimals is brought into practical focus through calculations of gross pay. In keeping with the occupational nature of the text, students are shown how to verify gross pay on a paycheck. Emphasis is placed on the relationships between fractional and decimal values, and conversion skills are stressed. As a practical illustration, students are shown why fractions are converted to decimal equivalents to simplify payroll calculations.

Unit 4 explains the earned-hours method of calculating gross pay when overtime is involved. The logic of this approach is illustrated through the implementation of algorithms on hand-held calculators. Percentages are introduced as decimal values expressed as parts of 100. Conversions from decimals to percentages are stressed. Students are shown how percentage calculations are used to determine the gross pay of many workers. Commissions, bonuses, shift differentials, and cost-of-living raises are presented as examples of payroll calculations frequently based on percentages. The writing of algorithms for solving payroll problems is a central theme in this unit.

Mathematics and Your Life

PART II presents a variety of mathematical functions related to everyday living; leisure and sports; banking, interest, and retirement planning; retail cashiering; and health care subjects, such as exercise and diet.

Unit 5 utilizes everyday living experiences to introduce ratios, proportions, and equations. Students are shown how to solve problems through the use of proportions and equations. Exercises include shopping and other consumer purchasing decisions based on unit pricing. Rounding and basic estimating skills also are covered. The emphasis is on practicality, such as estimating fuel requirements for a trip.

Unit 6 discusses the significant role that mathematics plays in leisure activities. Emphasis is placed on sports and games. Students are shown how mathematics is used to compare and evaluate sports performances, in both individual and team competitions. Team standings, batting averages, and pitchers' earned run averages are covered as examples of the mathematical grading of sports activities. Percentages are calculated extensively in this unit. Handicapping also is introduced as a method of equalizing competition between individuals or groups of varying skill levels. Scoring systems used for bowling and golf, the most common human competitions involving handicapping, are covered thoroughly.

Unit 7 provides an easily understood but meaningful introduction to the concepts of profit and interest. The description of profit as the difference between revenue and cost helps students understand basic business concepts, which are expanded upon in the following unit. Interest is defined as the charge a lender makes to a borrower for the use of money. Simple interest is discussed in terms of an annual rate. Students learn that a finance charge often involves fees or other costs in addition to interest. Compounding of interest occupies a major portion of this unit. Students are shown that growth of savings accounts is accelerated with more frequent compounding.

Unit 8 introduces basic mathematical functions involved in entry-level business experiences. Concepts discussed include making change, transaction accumulation, and computation of sales tax. A discussion of merchandizing introduces students to the concepts of handling and quantity as major determining factors in the pricing of merchandise. Students are introduced to the basic responsibilities of a cashier, including the cashier's role as a banker. Factors affecting merchandise markup decisions are discussed. These factors include sales volume, costs of handling, prevailing market prices, and demand for merchandise.

Unit 9 illustrates the importance of mathematics in measuring and maintaining physical well-being. Students are shown how mathematics is used in monitoring such health indicators as pulse rate. Proper aerobic exercise rates are discussed, as are the mathematical calculations used to determine them. Other calculations involving the body include those used to determine proper dosages of medications and an individual's caloric needs based upon activity levels. Graphs and charts that present statistical information are described as valuable uses of mathematics in the health-care field.

Mathematics, Products, and Travel

PART III covers the use of mathematics in working with machines, in operating a manufacturing business, and in the storage and transportation of inventory.

Unit 10 deals with energy, power, and work as performed by machines. Students are shown how to measure mechanical energy, or work, in foot-pounds. Horsepower is presented as a measurement of mechanical power, and methods for determining horsepower requirements are discussed. Transmission of power through belts and gears is covered. Also explained is the role of idler gears in some systems. Another major topic is mechanical advantage, including the transfer of force through levers and gears. Torque is introduced as a turning or twisting force, and torque multiplication is illustrated as a form of mechanical advantage. Basic principles of hydraulics also are covered.

Unit 11 introduces students to the world of manufacturing through product arithmetic. The point is made that the arithmetic involved in operating a manufacturing business is similar for organizations of any size. A pie chart is used to show the numerical parts of a business. The entire pie represents the gross income of the business. Costs and profit represent slices of the pie. Mathematics is applied to business through equations that summarize key values. These equations are used to measure and evaluate such areas as profit, operating costs, and labor costs. Business costs are categorized as materials, labor, overhead, and distribution, including sales. Labor costs are shown to include fringe benefits and unproductive labor factors. Students are shown how adjustments to one or more items affect entire equations. Thus, mathematical calculations are presented as basic ingredients of business decisions. Presentation of information also is emphasized as an important tool in business problem solving and decision making.

Unit 12 deals with inventory problems. Students are introduced to mathematical methods for controlling inventory levels. An inventory equation is introduced as a method of solving problems and making decisions involving restocking of inventory. Factors in the inventory equation are the cost of inventory, labor, overhead, and the cost of money borrowed to purchase inventory. Also discussed are such factors as safety stocks and lead time for deliveries. Graphs are emphasized as tools for presenting information on sales performance and inventory levels. As inventory decision-making tools, graphs can be used to trigger orders. Storing and moving inventory are other major topics discussed in this unit. Concepts presented include relay trucking systems and hub-based air transportation.

Mathematics and Your Environment

PART IV discusses the role of mathematics in the design and construction of shelter and in the delivery of life support services through infrastructure systems.

Unit 13 presents basic mathematical concepts used in the construction of residential structures. Shapes, sizes, and slopes are discussed. Students are shown how to calculate spatial relationships, such as the slope, or pitch, of a gable roof. Methods are introduced for estimating amounts of materials needed for structures. Area calculations are presented as basic mathematical steps in making decisions about room capacities and flooring needs. Electrical requirements and usage are calculated, based on the relationships among volts, amperes, and watts. The concept of degree days is illustrated as a mathematical tool for calculating fuel requirements for indoor heating.

Unit 14 explains infrastructure systems and illustrates how mathematics is used in providing necessary life support services for an area. Students are shown how to calculate volumes, both in terms of spaces to be filled and materials to be used or stored. Large measurements, such as the capacity of a water reservoir in acre-feet, are discussed. Other basic calculations covered in this unit include measuring cut firewood in terms of cords. Freight capacities based on weight and volume also are discussed, along with mathematical methods for calculating profit or loss on freight operations. Containerization of cargo is presented as a major cost-saving device for the transportation of freight.

Data and Information Processing

PART V deals with the use of computers to increase people's problem-solving capacities.

Unit 15 shows how mathematics is used to establish the value of a computer system. Computers are defined as tools used by people to process data and to produce useful information. Information processing systems are described, and computer logic is explained as a method of making mathematical comparisons. The ability of computers to support greatly increased business volumes is emphasized. Tangible and intangible benefits of computers are discussed. The four basic steps of an information processing cycle—input, processing, output, and storage—also are discussed in terms of realistic business applications. The use of mathematics for scheduling computer jobs completes this unit.

P A R T

Mathematics and You

IN THIS PART

Have you ever thought about "speaking" mathematics? That is exactly what you do when you solve a problem with numbers. The first unit in this textbook describes mathematics as a special language for solving problems. Some of the special features of the language of mathematics are presented in the second unit. The methods presented make problems easy to understand—and to solve.

When you become a full-time worker, you will be paid for your services. Mathematics will be used to determine how much you are paid. You, in turn, will use mathematics to make sure you receive the correct amount on your paycheck. The special mathematical procedures used to compute your pay are covered in Units 3 and 4.

1 Speaking and Using Mathematics

YOUR LEARNING JOB

When you have completed the exercises and assignments for this unit, you should be able to:

- ☐ **Explain why mathematics is considered a language.**
- ☐ **Describe how a problem stated in English can be translated into mathematical terms.**
- ☐ **Explain the functions of algorithms in solving problems.**
- ☐ **Describe the roles of calculators and computers in mathematical problem solving.**

WARM-UP

A. Find the sum for each problem.

1. 4,345 + 332 =
2. 9,182 + 712 + 123 =
3. 2,001 + 6,999 + 3,987 + 123 =
4. 34,847 + 29,132 + 876 + 3,928 + 89,912 + 4,294 =
5. 1,987,121 + 398,654 + 3,875,153 + 56,234,987 + 3,456 =

B. Find the difference in each problem.

1. 734 − 529 =
2. 5,435 − 5,193 =
3. 89,286 − 27,205 =
4. 938,001 − 38,999 =
5. 63,254,512 − 49,768,624 =

C. Find the product in each problem.

1. 7,435 × 321 =
2. 55,319 × 5,834 =
3. 18,708 × 2,321 =
4. 1,429,438 × 4,385 =
5. 234,919 × 2,043 =

D. Find the quotient in each problem.

1. 1,265 ÷ 55 =
2. 17,765 ÷ 95 =
3. 883,272 ÷ 298 =
4. 1,427,166 ÷ 963 =
5. 11,985,444 ÷ 3,462 =

MATHEMATICS AS A LANGUAGE

Mathematics is a language. As a language, mathematics has special features that can provide a valuable tool for you. A knowledge of mathematics can help in your future as a student and as a successful worker.

Mathematics, as a language, must be able to do certain things. One of these is to *express* and *communicate ideas* and *images.* This is really what a language does.

To express an idea or image is to state it clearly. When language is used, the person who is talking or writing is expressing. The listener or reader should be able to understand, or *comprehend,* the message. When messages have been expressed and comprehended, communication has taken place. That is, communication requires a complete process of sending and receiving messages.

Languages communicate about ideas and images. An idea or image is a basic unit of information. To illustrate, consider an idea expressed in a language you know and use, English:

All people are created free and equal.

This is an idea because it states a belief. English can also be used to describe an image:

The green leaves stand out against the blue sky.

For some kinds of communication, mathematics is a better language than English. For example, consider the statement:

The value of two added to the value of two produces a sum of four.

This is a message that is easier to state in the language of mathematics:

$$2 + 2 = 4$$

Using mathematics, the statement is shorter, more precise, and easier to understand. Even though the form is different, the statement still delivers the same message. This example adds to the description of mathematics as a language: Mathematics is a language with special capabilities for presenting and dealing with numbers. This text helps you to understand and use the language of mathematics. The skills you gain in the use of mathematics are directed toward your future work experiences and career.

Examples: English to Math

Below are some statements written in English, followed by the same statements written in the language of mathematics:

The Carlsons are planning to visit some friends in another state. The driving distance is 800 miles. They plan to drive 450 miles the first day and 350 miles the second.

$$450 + 350 = 800$$

You have to work 20 math problems for homework that is due in two days. You have a rehearsal for a school play tomorrow. Thus, you need to finish more of your homework today. If you can finish 15 problems today, you'll have to do only five problems tomorrow.

$$20 - 15 = 5$$

Heidi plans to sew curtains for her two bedroom windows. Each window will require 4 yards of material. She will have to buy 8 yards of material.

$$2 \times 4 = 8$$

Miguel and his family are planning a trip to the seashore. By driving along a certain route, Miguel figures he can average 40 miles per hour for the trip. The distance from Miguel's home to the beach is 80 miles. The trip will take two hours.

$$80 \div 40 = 2$$

EXERCISE 1-1

Read the English statement of each problem. Then, write the same statement in the language of mathematics.

1. Saul and Pilar are planning to race their bicycles. The race will involve four laps around the track at the local high school. The distance around the track is 440 yards. The total distance of the race will be 1,760 yards.

2. Amanda and Benigo are planning a party to celebrate the Fourth of July. They have invited 15 of their friends. Each person will be served one sandwich. Amanda and Benigo will need 17 sandwiches.

3. Your English class is having a spelling bee. There are 12 contestants. You are to find 15 words for each contestant. You will have to locate 180 words.

4. Your math class is having a test. There are 10 questions on the test and 25 students in the class. The teacher will have to correct 250 problems.

5. Leroy lives 2 miles from school. Leroy's father drives him to school five days a week. Leroy walks home from school. Leroy walks 10 miles a week.

6. Bob works for a company that makes candy bars. Bob's job is to pack 12 candy bars in every box. Bob then places 12 boxes in a shipping carton. Each carton shipped contains 144 candy bars.

7. Roman, a truck driver, is assigned a 3,000-mile trip from New York to Los Angeles. Roman spends 60 hours in actual driving time. His average speed is 50 miles per hour.

8. Monica loves to cook and wants to bake a cake that can feed 20 people. The recipe says it takes one egg for each two persons. Monica needs 10 eggs.

ALGORITHMS: THE SENTENCES AND PARAGRAPHS OF MATHEMATICS

When you use the English language, you express ideas and images in specific forms. For English, these forms include words, sentences, and paragraphs. Mathematics, on the other hand, is a problem-solving language. To deal with problems in mathematics, it is necessary to establish a series of steps to be followed. When all of the identified steps have been completed, the last step delivers a solution to the problem.

For example, when you buy food in a supermarket, each item is added to the total of other items purchased. The total keeps changing with each purchase entered in the register. The steps are illustrated in Figure 1-1. New purchases are entered and new totals developed repeatedly until your order is complete. Thus, in the language of mathematics, the equivalents of sentences and paragraphs are statements describing solutions of problems. These statements are organized in the form of lists of steps. When the steps are followed one after another to completion, the problem is solved. Such a list, or sequence, of steps is called an *algorithm*. Specifically, an algorithm is a series of steps followed in sequence to produce a known result.

1. Enter price for 1 can of peas.
2. Enter price for 1 quart of milk.
3. Total purchases.
4. Enter price for 1 candy bar.
5. Total purchases.

Figure 1-1. Supermarket purchases are entered in order in this algorithm.

You can express many of the tasks you perform each day in the form of algorithms. These tasks include getting dressed, taking a shower or a bath, and going to school. Each of these processes can be expressed as an algorithm by describing it in a step-by-step sequence. One example is a routine you may follow for starting off the day. That routine, expressed as an algorithm, might be as shown in Figure 1-2.

Another example of an algorithm would be the sequence of steps involved in filling a car's fuel tank. At a self-serve pump, the process might consist of the steps listed in Figure 1-3.

1. Turn off the alarm.
2. Get out of bed.
3. Make the bed.
4. Shower and get dressed.
5. Eat breakfast.
6. Brush teeth.
7. Pack lunch and book bag.
8. Leave for school.

Figure 1-2. Algorithm shows a routine for starting the day.

1. Turn off engine.
2. Remove filler cap.
3. Turn pump on.
4. Insert pump nozzle in filler neck.
5. Pump desired amount of fuel into tank.
6. Return pump nozzle to pump and turn pump off.
7. Replace filler cap.
8. Pay attendant.

Figure 1-3. Step-by-step procedures for filling a vehicle's fuel tank are listed in this algorithm.

EXERCISE 1-2

Write algorithms for the following activities.

1. Putting on a pair of shoes with laces.
2. Making a sandwich.
3. Brushing your teeth.
4. Washing a car.
5. Adding a column of whole numbers.

ALGORITHMS FOR SOLVING PROBLEMS

Algorithms are tools for you. An algorithm enables you to get a problem set in your mind. Once you have defined a problem in an algorithm, you are ready to apply mathematics for its solution.

Algorithms can be written to describe step-by-step procedures for everyday activities. PHOTO BY CLIFF CREAGER

To illustrate this step-by-step procedure, the following examples are presented in both descriptive (English) and numerical (number) forms.

> Suppose you were to purchase three items in a store. The prices of the three items are $4.50, $2.75, and $1.85. Suppose also that you had to pay a 6 percent sales tax on the items. How would you figure your change from a $20 bill?

There are several steps to follow in calculating the answer. The key to finding the correct answer lies in stating the problem accurately. For example:

> First, I add the costs of the three items. Then, I multiply that total by 6 percent to find the sales tax. Next, I add the sales tax to the total price of the purchases. Finally, I subtract that total from $20 to determine my change.

The preceding statement is a clear summary of the mathematical problem and the steps required to solve it. The problem is stated as an algorithm in Figure 1-4. In this example, the descriptive list

Descriptive	Numerical
1. List prices of all purchases.	$ 4.50 2.75 1.85
2. Add prices.	9.10
3. Calculate 6 percent of total.	0.06 × 9.10
4. Write result (tax).	0.55
5. Add tax to total price.	9.10 + 0.55
6. Write result.	9.65
7. Subtract new total from $20.	20.00 − 9.65
8. Difference is change from $20	$10.35

Figure 1-4. Algorithm shows steps for totaling purchases and adding sales tax.

is the algorithm. The numeric entries, as they would be made into a cash register, are the working, or *execution,* of the problem. An algorithm describes a method of finding a solution. Using mathematics executes the solution by actually working the problem. Here is another example of solving problems with algorithms:

> You decide to treat your friends to the movies, using some of the money ($10) you received as a birthday gift. Three of your friends accept your invitation. Admission to the theater is $2.25. You want to know two things. First, do you have enough money to pay for the tickets? If you do, how much money will you have left for refreshments?

Your first step is to state your problem clearly and accurately in your natural language. Understanding the situation will help you to write an appropriate algorithm.

> I have $10 that I want to use to take Frank, Yoko, and Maria to the movies. I also want to know whether I will have any money left over for refreshments after buying the tickets. To start with, I know I will need four tickets. I have to multiply this value by the price of the tickets. This will tell me how much the tickets cost. Then I can compare the ticket cost with my $10. This will give me my answer.

Figure 1-5 shows this problem's algorithm and execution.

Descriptive	Numerical
1. List the price of one admission.	$ 2.25
2. Multiply by four.	2.25 × 4
3. Write the total.	9.00
4. Compare the total with $10.	10.00 [>] (is greater than) 9.00
5. If price is higher than $10, cancel invitation.	
6. If price is lower than $10, subtract.	10.00 − 9.00 = 1.00
7. Difference is amount available.	$ 1.00

Figure 1-5. Algorithm shows steps in calculating the cost of taking a group to the movies.

EXERCISE 1-3

Solve each problem in the same manner as the examples above. First, state the problem in English. Next, write an algorithm to describe the solution. Finally, perform the mathematical calculations to produce the solution.

1. Jerry wants to buy four cassettes. The music store has the cassettes he wants on sale for $4.25 each. The sales tax of 5 percent comes to 85 cents. Jerry has $20 to spend. Can he buy all four cassettes? If so, how much change will he receive? If not, how much more money will he need?

2. For part-time work mowing lawns, you earn $6 for each lawn. In a normal week, you mow 12 lawns. Your goal is to earn enough to buy a portable stereo that costs $360. How long will it take to earn $360?

3. Hoshimoto and three of his friends plan to go to Yellowstone National Park for a 14-day vacation. They estimate car expenses of 22 cents per mile for 3,700 miles. The four friends plan to share these costs equally. In addition, meals and motel rooms are expected to average $40 dollars per day, per person. How much money will each person need?

4. Huck and Jim are traveling down the Mississippi River. On the first day, they travel 50 miles. On the second day, they

travel 10 miles. On the third day, they travel 72 miles. On the fourth day, they travel 38 miles. Huck and Jim plan on traveling 170 miles. How much farther must they travel after the fourth day?

5. Jose has to be in Chicago Sunday night. Chicago is 1,560 miles away. Jose would like to limit his driving to 400 miles per day. He wants to leave home as late as possible. On what day should he plan to leave on his trip?

TOOLS FOR USE IN MATHEMATICS AND PROBLEM SOLVING

The examples and exercises in this unit involve simple mathematical calculations. You probably figured some of them in your head. The others probably required only quick pencil-and-paper calculations.

Many mathematical problems are either longer or more complex. Remember, mathematics is a language. When people want to communicate lengthy or complicated messages, they use communication devices, or tools. A business report, for example, would be typed or printed. The president of a large company would not write a long government report in pen and ink.

Similarly, mathematical tools are used to execute lengthy or complex mathematical problems. These tools perform calculations at a much faster rate than the human mind. There are two major types of mathematical tools: calculators and computers.

Calculators

People have been using mathematical tools throughout recorded history. The earliest known calculating device is the *abacus,* developed in China thousands of years ago. An abacus consists of counters and rods. The counters are manipulated by hand. Thus, the abacus is a *manual,* or hand-operated, *calculator.* A calculator is a device that performs mathematical computations. The operator of an abacus performs calculations by sliding the counters along the rods.

The first *mechanical* tool for performing calculations was the *adding machine.* Mechanical means that calculations are performed by a machine. An American, William Burroughs, developed the

The abacus is the earliest known calculating device. However, it is still used as a mathematical aid by many people. PHOTO BY LISA SCHWABER-BARZILAY

first commercially successful adding machine in 1888. The adding machine could do more than simply add numbers, of course. Adding machines also could perform other calculations: subtraction, multiplication, and division.

Early in this century, adding machines and mechanical calculators were designed to use electric power. The use of electricity greatly increased the speed of calculations.

After World War II, *electronic processing* devices were developed to serve the growing needs of business and industry. Electronic processing means that calculations are performed without mechanical motion. Electronic devices perform calculations by applying and routing electric current. Electronic processing produced two major advantages: greatly increased speed and reduced size.

Electronics made possible the development of the small, highly efficient devices we now call calculators. Calculators come in many sizes—some small enough to be held in the palm of your hand. Data entries and the results of calculations are displayed electronically. Some calculators also print entries and results on paper, similar to an adding machine.

Calculators operate electronically to perform calculations rapidly on human command.

Calculators perform one function at a time in response to keyboard entries. All functions of a calculator are responses to human inputs.

Computers

A *computer* is a device that processes data automatically to combine or change those data for new meaning. Remember, data are raw facts and figures that are meaningless in and of themselves. *Data processing* describes the activities that are applied to transform data into useful *information*. Information, in this sense, consists of data that have taken on meaning through processing. For example, data items might include height, weight, color of hair, color of eyes, and other details. Information items from these data items would be names of people: Sally Wong, James Jones, etc. To qualify as a computer, a device or set of equipment must have other specific capabilities. A computer must be able to:

- Accept data items as *inputs*
- Process data to produce information
- Deliver results in usable form—known as *outputs*
- Store data, information, and sets of instructions known as *programs* for later reference and use.

Thus, the basic parts of a computer system, illustrated in Figure 1-6, are input, processing, output, and *storage*. Computer storage

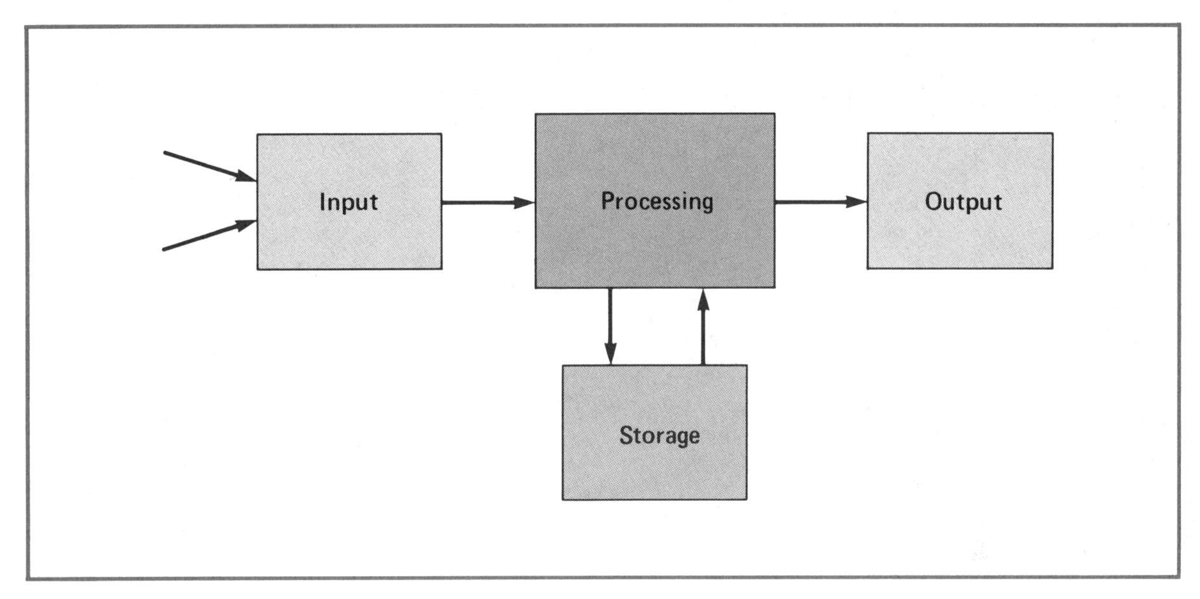

Figure 1-6. A computer system must contain devices for input, processing, output, and storage of data.

includes both data and instructions for operation. These sets of instructions that tell a computer what steps to follow are called programs. The ability to store and follow programs is a feature that separates computers from other processing devices. For example, calculators can perform arithmetic calculations. But calculators do only one computation at a time. Computers, on the other hand, can follow programs through a long series of processing steps.

Think about this function of computer programs. Programs contain sets of instructions that computers follow in sequence to process data and create information. In effect, then, programs are special kinds of algorithms written in languages that computers understand. Thus, in learning the language of mathematics, you are also learning to put together messages for computers.

UNDERSTANDING PROBLEMS IS THE KEY

Computers and calculators are extremely useful tools for processing data and performing mathematical calculations. However, they are only machines. These devices can do only what people tell them to do. You might have the finest calculator money can buy. If you enter incorrect data, however, the calculator cannot produce the correct result.

The rule about accuracy of data is often expressed this way: "Garbage in, garbage out," or, abbreviated, "GIGO." Data processing is useless without reliable data.

Computers process data automatically, changing or combining them into meaningful information.
COURTESY OF NBI, INC.

As a processor of data, you must understand a problem before you can solve it. You must be able to state the problem correctly in English. Then, you must be able to write an algorithm for reaching a solution. The final step is the numerical execution of the problem to reach the solution. In this way, the actual mathematical calculations become a mere exercise. If you enter the correct data, a calculator or computer will perform the calculations for you.

In terms of language, you are translating a problem from the English language to the language of mathematics.

Learning to Be a Problem Solver

This book is about problem solving. Your learning progress will be based mainly upon your ability to devise solutions to realistic problems. In most situations, it should be all right to use a calculator or a computer for ''number crunching.'' After all, you must understand the problem before you can input the proper data or commands to a calculator or computer. The key is in understanding problems and organizing them for solution.

SUMMING UP

- Mathematics is a language. As a language, mathematics has special features for expressing and communicating ideas and images.
- In using a language, you are expressing ideas and images. A listener or reader should be able to understand your messages. When this happens, communication takes place.
- Some ideas can be expressed more easily in the language of mathematics. For example, in English you might say: "Two plus two equal four." In mathematical terms, the statement would be, simply, "2 + 2 = 4."
- An algorithm is a series of steps that are followed in sequence to produce a known result. Algorithms are the sentences and paragraphs of the language of mathematics.
- When you have defined a problem by writing an algorithm, you are ready to apply mathematics to its solution.
- Mathematical tools can be used to execute lengthy and complex mathematical calculations. Calculators and computers greatly increase the speed with which calculations can be performed.
- Computers process data automatically to transform them into useful, meaningful information.
- The basic functions of a computer are input, processing, output, and storage.
- Computer programs are sets of instructions that a computer follows in sequence to process data and create information.
- Understanding problems is the key to finding solutions. You must be able to express the problem in English before writing an algorithm. Once you have an algorithm, the mathematical calculations should produce the solution quickly and easily.

TERMS THAT COUNT

express
communicate
idea
image
comprehend
algorithm
execution
abacus
manual
calculator
mechanical
adding machine
electronic processing
computer
data processing
information
inputs
outputs
programs
storage

TESTING YOUR WORKING KNOWLEDGE

For each question, write an algorithm for solving the problem.

1. Working at Bud's Quick Service on a Saturday, you fill a customer's fuel tank. She hands you a $20 bill. Gas at the full-service pump costs $1.40 a gallon. This sale is for 12 gallons. How much change should the customer receive?

2. You decide to install shelves in the family garage. You want to install eight, each one to be 3 feet long. There are two 9-foot boards and one 6-foot board in the garage. Will those boards be enough to make the eight shelves?

3. You buy four shirts for $9.50 each during a sale at Yoshihara's Fashions. When you get home and unpack the shirts, you check the sales slip. The 6 percent sales tax came to $2.28 and was added to your purchase. You paid a total of $40.68. Was the total correct? If not, how much did you overpay or underpay?

4. Helene wants to open a copy center. She must prepay two months' rent on a store at $735 per month. She must prepay $400 per month for two months rental on a copying machine. She also will need $595 worth of paper and other supplies. Signs and a service counter for the store will cost $835. How much money does Helene need to get started?

5. John and Refugio want to take a trip across the country. They plan to drive a total of 3,000 miles in six days. Their car gets 25 miles per gallon. If gas costs $1.25 per gallon, how much will their fuel cost be?

SKILL DRILLS

A. Find the sum for each problem.

1. 1,435 + 25 + 9,675 + 393 =
2. 3,548 + 134 + 475 + 623 =
3. 4,650 + 1,225 + 875 + 23,893 =
4. 34,847 + 3,425 + 75 + 6,823 =
5. 65 + 29 + 3,875 + 183 =
6. 1,498 + 83,545 + 11,875 + 62,398 =
7. 2,658 + 10,753 + 1,223,325 + 981,875 + 623,534 =
8. 35,324 + 65,876 + 265,826 + 110,753 + 25,638 =

B. Find the difference in each problem.

1. 2,347 − 55 =
2. 12,532 − 5,432 =
3. 82,197 − 2,999 =
4. 93,111 − 32,960 =
5. 106,345 − 99,445 =
6. 762,356 − 132,926 =
7. 1,923,745 − 298,234 =
8. 5,345,812 − 4,919,879 =

C. Find the product in each problem.

1. 7,435 × 33 =
2. 12,235 × 325 =
3. 18,324 × 23 =
4. 199 × 454 =
5. 2,325 × 2,345 =
6. 96,326 × 91,856 =
7. 19,000 × 245 =
8. 2,532 × 10,819 =

D. Find the quotient in each problem.

1. 19,670 ÷ 35 =
2. 59,052 ÷ 532 =
3. 11,388 ÷ 292 =
4. 17,003 ÷ 347 =
5. 98,523 ÷ 123 =
6. 20,736 ÷ 6,912 =
7. 34,132 ÷ 4,876 =
8. 18,549 ÷ 687 =

Communicating With Mathematics

YOUR LEARNING JOB

When you have completed the exercises and assignments for this unit, you should be able to:

- ☐ Describe mathematical methods of presenting information.
- ☐ Read and discuss tables.
- ☐ Explain why scales are used in maps and drawings.
- ☐ Read and explain bar charts and pie charts.
- ☐ Read and discuss scatter graphs and line graphs.
- ☐ Use mathematical forms to present information.

WARM-UP

A. Find the sum in each problem.

1. 45,928 + 17,756 =
2. 3,384,801 + 23,781 =
3. 683,398 + 109,745 + 697,241 + 534,638 =
4. 98,736 + 1,746,825 + 435,121 + 5,908,437 =
5. 3,475,918 + 3,276 + 326 + 341,910 + 9,385,031 =

B. Find the difference in each problem.

1. 645,928 − 33,054 =
2. 301,401 − 299,879 =
3. 97,241 − 34,638 =
4. 98,946,825 − 15,908,437 =
5. 355,276,910 − 88,785,031 =

C. Find the product in each problem.

1. 928 × 76 =
2. 3,801 × 81 =
3. 688 × 745 =
4. 8,125 × 937 =
5. 475 × 3,981 =

D. Find the quotient in each problem.

1. 17,388 ÷ 756 =
2. 25,912 ÷ 328 =
3. 102,627 ÷ 543 =
4. 186,186 ÷ 91 =
5. 479,340 ÷ 5,326 =

Summaries of sales figures on computer printout assist managers in running business operations.
PHOTO BY LISA SCHWABER-BARZILAY

PRESENTING INFORMATION

Mathematics is a special language for presenting and solving problems. As discussed in Unit 1, an algorithm is an efficient tool for stating a problem. An algorithm also provides a step-by-step description of the method for solving a problem.

The basic mathematical functions (addition, subtraction, multiplication, and division) are used to solve problems. When you solve a problem in mathematics, you are developing information.

Mathematics also provides efficient methods for communicating the information developed through mathematical calculations. Techniques are needed for the presentation of numerical information in easy-to-understand formats.

One method of presenting information is through the results of problem solving. Examples are the sums in addition, differences in subtraction, products in multiplication, and quotients in division. Sums, differences, products, and quotients represent solutions of problems. In the language of mathematics, these solutions provide a method of communicating information.

Individual answers, however, can present communication problems. It is often necessary to *summarize* and *compare* results of mathematical calculations. Summaries are totals for sets of numbers. Comparisons involve matching numbers to derive meaning.

As an example of the value of summaries, consider a typical supermarket. In a given week, the market may serve 15,000 customers who buy 220,000 items. The numbers are listed on printed tapes in cash registers. These numbers do communicate information. However, there are so many numbers that people wouldn't have time (or need) to deal with all the figures. This is especially true for supermarket chains, some of which operate hundreds of stores. Instead, a manager at chain headquarters might want to know total daily sales for the stores. These totals are summaries of sales figures.

These store totals, then, would be compared. One comparison would review totals among stores. Other comparisons would be between this week's sales and last week's sales, this month's sales and last month's sales, and so on. Comparison adds meaning to figures.

In addition, special techniques often are used to add meaning to sets of numbers. These methods organize numbers for reference or use graphics (pictures) to portray numeric values. Some of these methods include:

- Tables
- Lines and scales
- Charts
- Graphs.

TABLES

A *table* is a way to show the relationships of different values. A table lists values in two directions: horizontally and vertically. These values may increase or decrease in either direction. Or, the different values along one side of a table may be unrelated.

Withholding Tax Tables

One widely used type of table is the income tax withholding table, an example of which is shown in Figure 2-1. The federal government and many states require employers to *withhold* (keep back or deduct) a percentage of employees' wages for taxes. This tax money must then be sent to the proper government offices. Deductions, or withholding, must be made for income tax and social security. In the table, the horizontal values show the number of withholding allowances claimed. Taxpayers may claim one withholding allowance for themselves, one for a spouse, and one for each child or dependent. A dependent is a person supported by the taxpayer. The left vertical column, specifying wages, represents increasing amounts of income. To locate the amount of income tax to be withheld, you first find your number of withholding allowances on the horizontal line. Then, find your income in the vertical column. By reading down and across, you can then locate where the two values meet. Note how the amount of tax due is determined in the example that follows.

Problem:

Margaret earns $7.50 per hour. In the first half of the month, she worked 84 hours. Margaret claims two withholding allowances.

Solution:

1. Calculate total pay: $7.50 × 84 = $630.
2. Find where income amount is located in the left vertical column of the withholding table: The amount, $630, is located between $620 and $640.
3. Move horizontally across table to the column with two withholding allowances.
4. The number listed in this column represents the amount of tax to be withheld from Margaret's paycheck: $74.

Figure 2-1. An income tax withholding table shows an employer how much to withhold from a worker's paycheck.

SINGLE Persons–**SEMIMONTHLY** Payroll Period

And the wages are–		And the number of withholding allowances claimed is–									
At least	But less than	0	1	2	3	4	5	6	7	8	9
		The amount of income tax to be withheld shall be–									
$520	$540	$71	$63	$55	$48	$41	$34	$28	$22	$15	$10
540	560	75	67	59	51	44	38	31	24	18	12
560	580	79	71	63	55	48	41	34	27	21	15
580	600	83	75	66	58	51	44	37	30	24	18
600	620	87	79	70	62	54	47	40	33	27	21
620	640	92	83	74	66	58	50	43	36	30	23
640	660	97	87	78	69	61	54	47	40	33	26
660	680	101	91	82	73	65	57	50	43	36	29
680	700	106	96	86	77	69	61	53	46	39	32
700	720	110	100	91	81	73	64	57	49	42	35
720	740	115	105	95	85	77	68	60	52	46	39
740	760	120	110	100	90	81	72	64	56	49	42
760	780	125	114	104	94	85	76	67	60	52	45
780	800	130	119	109	99	89	80	71	63	55	48
800	820	135	124	114	104	94	84	75	67	59	51
820	840	141	129	118	108	98	88	79	71	63	55
840	860	146	135	123	113	103	93	83	75	66	58
860	880	151	140	128	117	107	97	87	79	70	62
880	900	156	145	134	122	112	102	92	83	74	66
900	920	161	150	139	128	117	107	97	87	78	69
920	940	167	155	144	133	122	111	101	91	82	73
940	960	172	161	149	138	127	116	106	96	86	77
960	980	177	166	154	143	132	121	110	100	91	81
980	1,000	183	171	160	148	137	126	115	105	95	85
1,000	1,020	189	176	165	154	142	131	120	110	100	90
1,020	1,040	195	182	170	159	148	136	125	114	104	94
1,040	1,060	201	188	175	164	153	141	130	119	109	99

EXERCISE 2-1

In each problem, determine the amount of federal income tax to be withheld from semimonthly payroll checks. Refer to the Table in Figure 2-1 to determine income tax deductions.

1. Terry earns $8.50 an hour and has worked 80 hours. Terry has no dependents and claims one withholding allowance.

2. Geraldine earns $12.25 an hour and has worked 76 hours. Geraldine claims one withholding allowance.

3. Marilyn earns $1.25 for each four pounds of peaches she picks. Marilyn has three withholding allowances. In the first half of the month, she picked 1,800 pounds of peaches.

4. Arturo is a mechanic and has worked 65 hours. He earns $15 an hour and has one withholding allowance.

5. Aida earns $12 an hour as a welder and claims no withholding allowances. So far this month, she has worked 72 hours.

6. Haru earns $8.75 an hour as a secretary and claims no withholding allowances. In the last half of the month, Haru worked 82 hours.

7. Angel is an artist and has just completed 13 portraits for the first half of the month. She earns $70 for each portrait she draws. Angel has four withholding allowances.

8. Carrol earns $6.75 an hour and claims no withholding allowances. Carrol has worked 81 hours.

Mileage Tables

Another type of table is the mileage table, shown in Figure 2-2. This table shows driving distances between cities. Mileage tables generally are used with maps for planning trips. Such tables are helpful for planning stops along the route of a trip. Knowing mileages between points enables you to plan driving times, as well as meal and overnight stops.

Mileage Between Principal Cities

	Albany	Auburn	Binghamton	Buffalo	Cortland	Dansville	Elmira	Erie PA	Geneseo	Geneva	Glens Falls	Ithaca	Jamestown	Kingston
Binghamton	142	88		211	44	135	60	313	158	98	174	54	218	140
Buffalo	301	128	211		169	81	148	96	62	104	333	150	89	351
Jamestown	360	199	218	89	206	118	159	62	137	172	392	184		358
Rochester	227	69	148	75	118	45	109	168	32	46	254	92	145	279
Syracuse	146	27	76	156	34	106	90	240	106	53	161	56	246	198

	Lake Placid	Medina	New York	Niagara Falls	Olean	Oneonta	Oswego	Plattsburgh	Potsdam	Rochester	Syracuse	Utica	Watertown	Wellsville
Binghamton	261	242	213	231	164	62	128	300	222	148	76	92	150	131
Buffalo	395	42	371	20	73	273	148	367	287	75	156	199	217	106
Jamestown	479	105	423	109	54	280	212	466	383	145	246	302	316	93
Rochester	344	46	369	87	108	198	73	303	322		97	134	153	88
Syracuse	251	143	256	163	185	103	40	241	145	97		54	74	155

Figure 2-2. Distances between cities are easy to determine on a mileage table.

When you know the distance between two points, you can calculate driving time between them. Simply divide the distance by your estimated speed. Speed usually is expressed in miles per hour. Suppose that you plan to drive 50 miles per hour between two cities located 200 miles apart. Divide the distance (200) by your anticipated average speed (50). The answer is four hours of driving time.

Travelers can use mathematics to estimate the time that would be required for a trip.

EXERCISE 2-2

Read each question carefully before calculating the answer. Refer to the New York State mileage chart in Figure 2-2 to find distances between cities.

1. You are planning to drive from Buffalo to Ithaca at an average speed of 50 miles per hour. How long will the trip take? Express your answer in hours.

2. On Sunday, Lydia is driving from Elmira to Buffalo. She plans to average 50 miles per hour. She is meeting a friend, Taylor, in Buffalo at 11:30 A.M. What is the latest time that Lydia can leave Elmira and still meet Taylor on time?

3. Kirk is a truck driver. He completed a trip from New York City to Binghamton, to Syracuse, and then back to New York City. How far did Kirk drive?

4. Gloria lives in Plattsburgh and wants to drive to Binghamton. If Gloria averages 50 miles per hour, how long will it take, in hours, for her to reach Binghamton?

5. Charlie is a delivery person. Yesterday, Charlie drove from Buffalo to Jamestown, to Olean, and then back to Buffalo. How many miles did Charlie drive?

6. Tony is driving a loaded truck from Rochester to Utica. Tony plans to average 45 miles per hour. At what time, to the closest hour, will Tony reach Utica if he leaves Rochester at 11 A.M.?

7. Isabel lives in Binghamton and is planning a long drive. Her first stop will be at Syracuse. Her second stop will be at Oswego. Her third stop will be at Buffalo. From Buffalo, she plans to drive home. How many miles will Isabel travel during her drive?

8. Theodore lives in Syracuse. He has an appointment at 10 A.M. in New York City. He plans to leave home at 5 A.M. and drive at 55 miles per hour. Will Theodore make his appointment?

Condition or Truth Tables

You probably see examples of these tables each time you look through a magazine. A *condition table* or *truth table* can be used to present a wide variety of information. One frequent use of these tables is for product information in brochures and advertisements.

For example, an automobile company might use a condition table to list the various types of equipment available on its models. In this type of table, individual items of equipment can be noted as standard or optional. The same table can be used to indicate which items of equipment are available for each model of vehicle.

Think about the possible combinations of equipment that might be ordered for a new car. The following list is just a sample of the types of equipment for which there may be two or more options:

- Body style and model size
- Engine type, design, and size
- Transmission type and model
- Rear axle ratio
- Heating/air conditioning system
- Wheel and tire combinations
- Exterior color or colors
- Interior trim levels and colors
- Radio/cassette player combinations
- Power steering, brakes
- Power windows, seats, locks
- Sunroof, moonroof, T-top designs.

Some of the items in this list may involve dozens of options. Imagine how many words would be required to describe all these options in English. Mathematics provides an easily understood method of presenting this information in a relatively small space. An example of a condition table is shown in Figure 2-3.

Another use for condition or truth tables is a class schedule that you might use at school. Such a schedule might show the times and locations at which each class is offered. The table also

Information Available Through CIDS	Occupational information		Job Bank	Employer	Military	Educational information		Financial aid	Apprenticeships	Approximate number of user sites
	National	State				National	State			
Alabama	●	●	●		●	●	●	●		150
Alaska	●	●			●	●	●	●	●	200
Arizona	●	●	●	●	●	●	●	●	●	200
Arkansas	●	●			●	●	●	●	●	330
Colorado	●	●		●		●	●	●	●	90
Connecticut	●	●	●		●	●	●	●	●	160
Delaware	●	●	●	●	●	●	●	●	●	80
Florida		●	●	●			●	●	●	1,000
Georgia		●				●	●	●	●	130
Hawaii		●	●			●	●		●	100
Idaho		●					●	●	●	100
Illinois	●	●	●	●	●	●	●		●	180
Indiana	●	●			●	●	●	●		180
Iowa	●	●			●		●	●	●	850
Kansas		●		●	●			●	●	130
Maine	●	●	●		●	●	●	●	●	100
Maryland	●	●			●	●	●	●	●	400
Michigan	●	●		●	●		●	●	●	2,000
Minnesota		●					●	●	●	300
Montana	●	●					●	●	●	50
Nebraska	●	●				●	●	●	●	270
New Jersey (all planned)	●	●		●	●	●	●			120
New Mexico	●	●			●	●	●	●	●	20
New York	●	●				●	●	●		300
North Carolina		●	●				●	●		130
North Dakota		●					●	●	●	15
Ohio	●	●			●	●	●	●	●	300
Oklahoma		●				●	●	●		520
Oregon	●	●			●	●	●	●	●	500
South Carolina	●	●	●		●	●	●		●	260
South Dakota	●	●		●	●		●	●	●	350
Vermont	●	●	●		●	●	●	●	●	20
Virginia	●	●			●		●	●	●	860
Washington		●			●		●	●	●	290
Washington, D.C.	●	●	●	●	●	●	●	●		30
Wisconsin	●	●				●	●	●	●	600
Wyoming	●	●				●	●		●	60
Puerto Rico	●	●				●	●	●	●	30

Figure 2-3.
A condition table offers an easy-to-understand summary of information.

might list the name of the teacher for each class period, and any prerequisite classes. A high school class schedule is shown in Figure 2-4.

Tables are used to present many different kinds of information, such as prices, schedules, and statistics. A statistical table is shown in Figure 2-5.

Figure 2-4. A high school class schedule is an example of a condition table.

MASTER SCHEDULE			CHARTER OAK HIGH SCHOOL		SECOND SE
Regular Day	8:00 - 8:55	9:00 - 10:00	10:20 - 11:15	11:20 - 12:15	12:55 - 1
Rally	8:00 - 8:50	8:55 - 9:50	10:10 - 11:00	11:05 - 11:55	12:35 - 1
ART					
Jones	(Math)	(Math)	F-6 Art 3/Art Ds	F-6 Ceramics	(Math)
Spohn	(Math)	Conference	(Math)	(Math)	F-8 Art 1
BUSINESS					
Richards, F.	B-7 "0" Bus.Ap/Comp				
Richards, F.	B-7 Bus. App/Comp.	B-10 Acctg. 1-2	B-10 Acctg. 3-4	B-10 Acctg. 1-2	Conferenc
Kuhlow	(Ind. Ed)	B-7 Bus. App. Comp.	(Ind. Ed)	(Yearbook)	Conferenc
Mussack	B-3 Typing 1-2	B-3 Typing 1-2	B-3 Typing 1-2	B-3 Typing 1-2	B-3 Typir
Redmon	B-8 Typing 1-2	B-8 Shorthand 1-2	B-8 Typing 3-4	Conference	B-8 Typir
Simms	(Math)	(Math)	B-7 Prog/Comp.	B-7 Prog/Comp.	(Math)
ENGLISH					
James	C-1 "0" AP Eng.				
James	C-1 Adv.Comp.	C-1 Adv. Comp.	C-1 Writing I	C-1 Writing I	Conferenc
Collins, M.	C-3 Eng. Fund.	C-3 Eng. Fund.	C-3 Eng. Fund.	C-3 Eng. Fund.	C-3 Basic
Gentz	B-5 Sr. Seminar	B-5 American Lit	B-5 American Lit	B-5 Writing III	Conferenc
Krueger	Conference	B-2 Survey Lit	B-2 Survey Lit	B-2 American Lit	B-2 Eff.(
McCarthy	C-9 Writing I	C-9 Writing I	Conference	C-9 Themes Wl.Lt	C-9 Writi
Sheehy	B-2 Writing III	Conference	C-4 Literature	C-4 Literature	C-1 Conte
Sauvageau	D-2 Amer. Lit	D-2 Writing II	D-2 Writing II	D-2 Writing II	D-2 News
Young	L-3 Writing II	L-3 Eff.Comm.Sk.	C-9 Eff. Comm.Sk.	C-6 Eff.Comm.Sk.	C-6 Adv.(
Baird					L-3 Drama
Farkas					E-3 ESL
Hall	C-7 Literature				
FOREIGN LANGUAGE					
Recio	D-1 Spanish 1-2	D-1 Spanish 3-8	D-1 Spanish 3-4	D-1 Spanish 1-2	D-1 Span
Hall	(English)	C-7 German 1-8	C-7 French 3-8	C-7 French 1-2	Conferen
Woehler	(Science)	B-6 Spanish 1-2	B-6 Spanish 1-2	(Social Studies)	(Science)
HOME ECONOMICS					
Riegel, J.	Conference	(Science)	F-1 Foods	F-1 Foods	F-1 Food
INDUSTRIAL EDUCATION					
Lim	E-2 ROP Cab.Mkg.	E-2 ROP Cab.Mkg.	E-2 Wood 3-8	E-2 Wood 1-2	E-2 Wood
Gilbreath	E-5 Metal 1-8	E-5 Metal 1-8	E-5 Metal 1-8	(Sec. Math)	Conferen
Kuhlow	F-5 Photo 3-8	(Business)	F-5 Photo 1-2	F-5 Yearbook	Conferen
Riegel, R.	E-1 Drafting 1-8	E-1 Drafting 1-8	(Math)	Conference	E-9 ROP
MATH					
Simms	C-4 Adv. Alg.	C-4 Geometry	(Business)	(Business)	C-4 Adv.
Davis, P.	F-4 Sec. Math	F-4 Intro.Alg.	SDR Algebra	(Soc. Studies)	Conferen
Fukagawa	C-8 Algebra	C-8 Algebra	C-8 Algebra	C-8 Algebra	Conferen
Gilbreath	(Ind. Ed)	(Ind. Ed)	(Ind. Ed)	E-5 Sec. Math .	Conferen
Jones	C-2 Math Lab	C-2 Math Lab	(Art)	(Art)	C-2 Math

Table 1. Sources of training needed by workers to qualify for their current jobs, by major occupational g

Occupational group	Sources of training needed for obtaining current job									
		School								
				Post-high school vocational education						
	Any source of training	Any school	High school vocational education	Private	Public	Junior college or technical institute	College, 4 years or more	Formal company[1]	Informal OJT	Armed Forces
Total, age 16 and over										
Number (in thousands)	53,890	28,075	4,692	2,098	1,586	4,965	16,078	9,418	27,004	1,902
Percent of total employment	55	29	5	2	2	5	17	10	28	2
Executive, administrative, and managerial occupations										
Number (in thousands)	7,738	4,674	333	169	134	581	3,638	1,346	4,242	314
Percent of occupational employment	71	43	3	2	1	5	34	12	39	3
Professional specialty occupations										
Number (in thousands)	11,797	10,397	208	367	213	906	8,961	1,184	2,767	281
Percent of occupational employment	93	82	2	3	2	7	70	9	22	2
Technicians and related support occupations										
Number (in thousands)	2,579	1,759	149	168	185	600	744	422	962	152
Percent of occupational employment	85	58	5	5	6	20	24	14	32	5
Sales occupations										
Number (in thousands)	4,867	1,643	185	163	90	356	941	1,315	3,148	90
Percent of occupational employment	43	15	2	1	1	3	8	12	28	1
Administrative support occupations, including clerical										
Number (in thousands)	9,157	5,262	2,659	506	367	1,282	976	1,198	4,945	136
Percent of occupational employment	57	33	16	3	2	8	6	7	31	1
Private household occupations										
Number (in thousands)	81	15	9	2	0	0	4	10	36	0
Percent of occupational employment	8	2	1	0	0	0	0	1	4	0
Service workers, except private household										
Number (in thousands)	4,397	1,604	207	442	195	461	316	1,104	2,233	141
Percent of occupational employment	36	13	2	4	2	4	3	9	18	1
Farming, forestry, and fishing occupations										
Number (in thousands)	862	259	75	15	16	58	128	41	507	7
Percent of occupational employment	28	8	2	0	1	2	4	1	16	0
Precision production, craft, and repair occupations										
Number (in thousands)	7,603	1,817	606	193	280	568	282	1,945	4,710	599
Percent of occupational employment	65	16	5	2	2	5	2	17	40	5
Machine operators, assemblers, and inspectors										
Number (in thousands)	2,742	479	196	45	79	115	69	476	1,957	81
Percent of occupational employment	37	6	3	1	1	2	1	6	26	1
Transportation and material moving occupations										
Number (in thousands)	1,462	97	34	23	10	18	10	311	1,028	80
Percent of occupational employment	36	2	1	1	0	0	0	8	26	2
Handlers, equipment cleaners, helpers, and laborers										
Number (in thousands)	605	69	30	6	16	21	7	68	468	20
Percent of occupational employment	16	2	1	0	0	1	0	2	13	1

[1]A formal company training program such as apprenticeship training or other type of training having an instructor and a planned program.

[2]Informal training from a friend or relative or other experience not related to work.

NOTE: All rows are nonadditive; many workers reported more than one source of training.

Figure 2-5. Large amounts of information are summarized in a statistical table.

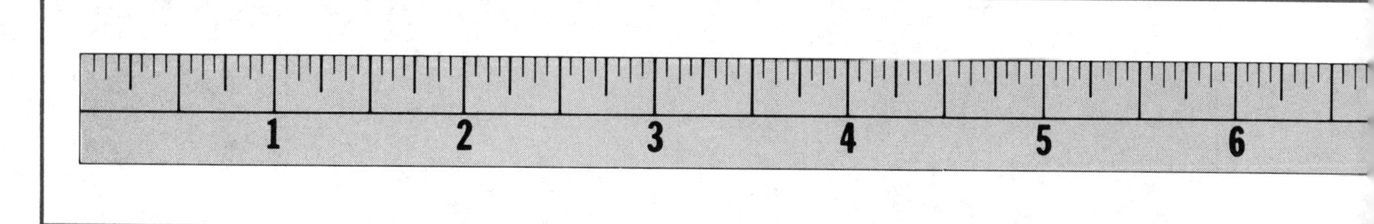

Figure 2-6. One of the most commonly used scales is the ruler.

LINES AND SCALES

Lines and *scales* are other means of presenting information in usable, easily understood formats. A scale is a mathematical tool for measuring graduated values. A ruler with lines that indicate inches and parts of inches is a scale. A ruler is shown in Figure 2-6. Another example of a scale is a thermometer with lines that indicate degrees of temperature. A thermometer is shown in Figure 2-7.

Scales also are used to indicate the relationship between small and large values. A good example is a geographical *map*, which is a small pictorial representation of a much larger area.

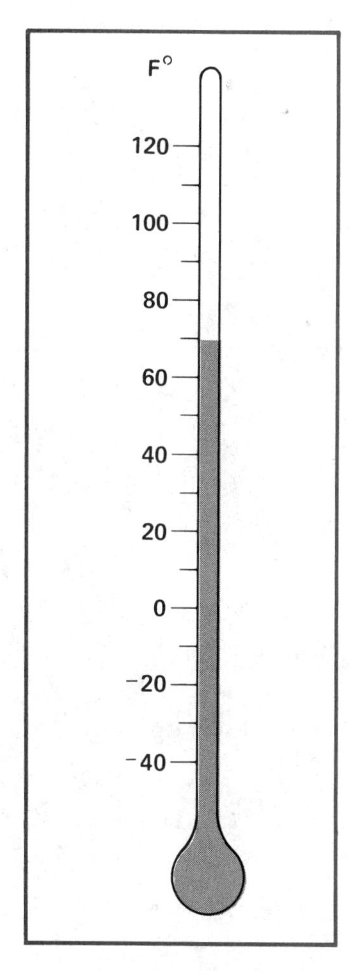

Figure 2-7. Temperature readings are taken from the scale on a thermometer.

Maps

Look at a map of a geographical area. Somewhere on the map, in an open area or near the border, will be a *key*. A map key is a summary of information about the map. Different types of maps have different kinds of information in their keys. However, most maps have a scale indicated in their keys. The scale is the relationship between distances on the map and the actual distances they represent. An example would be 1 inch = 5 miles. This means that, if two cities are 1 inch apart on the map, they actually are 5 miles distant.

Look at the map of the United States, shown in Figure 2-8. Lay a ruler on the map and measure the distance between Chicago and Los Angeles. The map distance is 5 inches. The actual air distance between these cities is approximately 1,750 miles. Knowing these two figures, you can calculate the scale of the map. Simply divide the known distance by the measured distance on the map: 1,750 ÷ 5 = 350. Thus, the scale of this map is: 1 inch = 350 miles.

Maps are a good example of the importance of mathematics in presenting information. If actual distances could not be "scaled down" to usable size, there would be no way to draw maps.

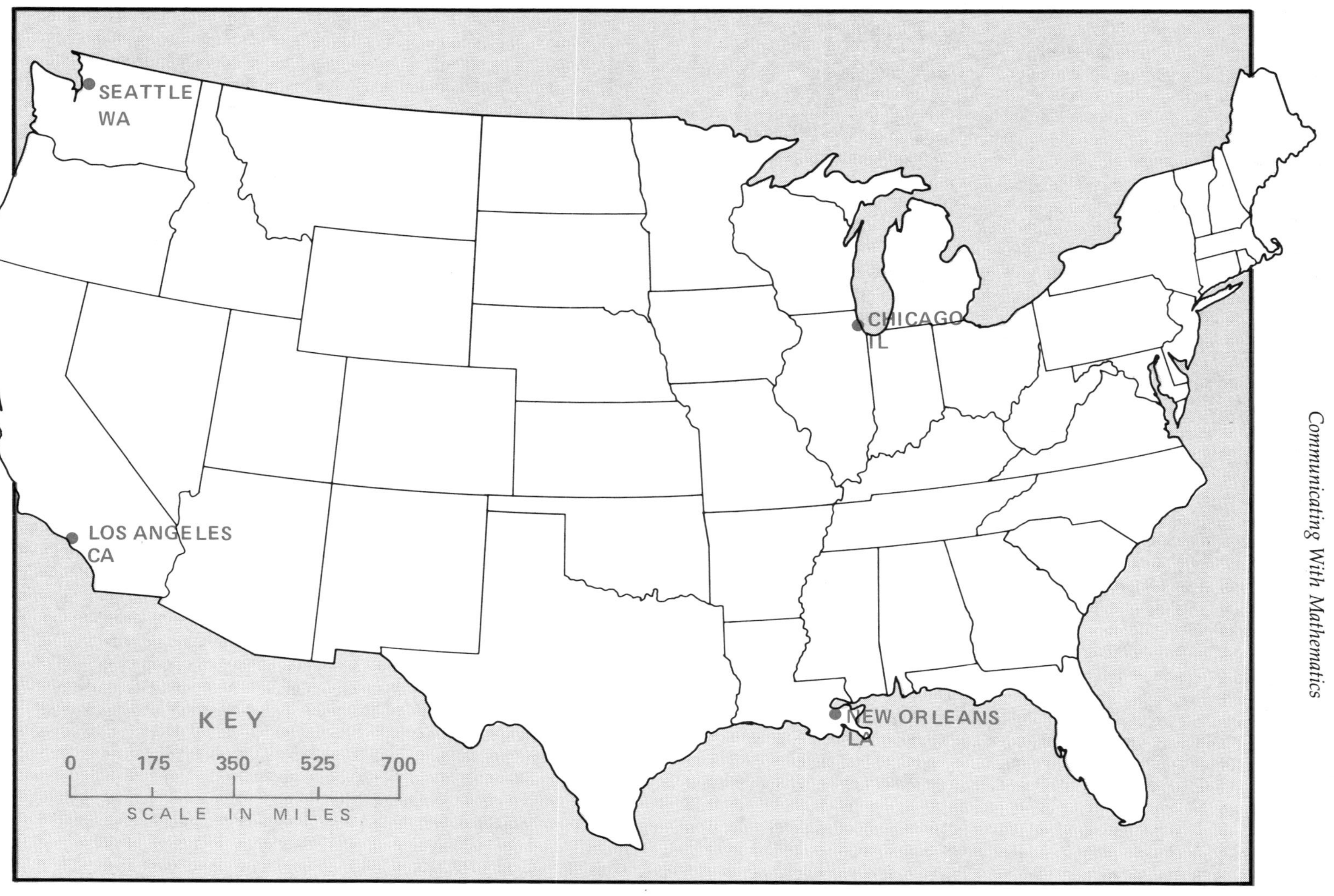

Figure 2-8. Maps are drawn to scale so that large distances can be compressed into small spaces.

Pilots plan trips carefully, using detailed maps to estimate flight times and to calculate fuel requirements. PHOTO BY CLIFF CREAGER

EXERCISE 2-3

Read each question carefully. Then calculate the answer.

1. The scale on a map is 1 inch = 200 miles. What is the distance represented by 2 inches?

2. On a road map, the scale is 1 inch = 50 miles. If the distance between two cities is 200 miles, how many inches separate those cities on the map?

3. The distance between Los Angeles, California, and Albany, New York, is 2,800 miles. On a map, this distance is represented by 8 inches. What is the scale of this map?

4. The scale on a map is 1 inch = 200 miles. What is the distance represented by 4 inches?

5. The scale of a map is 1 inch = 75 miles. What is the distance represented by 4 inches?

6. On Tim's map, the distance between Yellowstone National Park and Albany, New York, measures 6 inches. A mileage

table says that the distance is 2,100 miles. What is the scale of Tim's map?

7. The scale of Alfred's globe is 1 inch = 1,250 miles. If the distance around, or the circumference of, the world is 25,000 miles, what is the circumference of Alfred's globe? Express your answer in inches.

8. On your road map, the distance between two cities measures 3 inches. You know that the actual distance between those cities is 210 miles. What is the scale of your road map? How many miles would 20 inches represent?

Using Maps

Information contained in maps can be used in a variety of ways. For example, suppose you planned to fly your own aircraft from Seattle, Washington, to New Orleans, Louisiana. You would need to determine whether your aircraft could reach New Orleans without stopping to refuel. You also would want to calculate the approximate time required for the flight.

Say that your aircraft had a maximum range of 2,500 miles. You could use the map in Figure 2-8 to determine whether the Seattle-to-New Orleans flight is within the range of your aircraft. Use a ruler to measure the map distance between Seattle and New Orleans (6 inches). Remember, the scale of this map is 1 inch = 350 miles. Simply multiply the scale value by the map distance:

$$6 \times 350 = 2{,}100$$

Thus, the approximate air distance between Seattle and New Orleans is 2,100 miles, well within the 2,500-mile range of your aircraft.

Now, say that your aircraft can cruise at 175 miles per hour. Divide this average speed into the distance to determine approximate flight time in hours:

$$2{,}100 \div 175 = 12$$

Approximate flight time, then, is 12 hours. This is an *estimated* time. An estimate is a calculation based upon known factors. In this case, air distance and normal cruising speed are known factors, or *constants.*

However, actual flight time would be affected by wind speed, weather conditions, and other changing factors. Factors subject to change are called *variables.* You would receive information on weather and other variables just before you took off.

EXERCISE 2-4

Read each question carefully before calculating the answer.

1. A small plane is flying from New York City to Los Angeles. The distance is 2,800 miles. The plane can cruise at 280 miles per hour. The pilot estimates a two-hour layover to refuel and rest. How long will the flight take?

2. Josephine and Ted are driving from Miami, Florida, to Lake George, New York. The scale of their road map shows that 1 inch = 200 miles. On the map, the distance from Miami to Lake George measures 7 inches. How long will their trip take if they average 50 miles per hour?

3. A truck driver has to deliver a load of wood to a construction site at 11 A.M. On the driver's map, the distance measures 3 inches. The scale of the driver's map is 1 inch = 75 miles. If the driver averages 50 miles per hour, what is the latest time he or she can leave and still make the delivery on time?

4. Kim is traveling from Houston to San Antonio by bus. On Kim's map, this distance measures 6 inches and the scale is 1 inch = 30 miles. The bus averages 45 miles per hour for the trip. If Kim leaves Houston at 4:30 A.M., what time will he arrive in San Antonio?

5. A pilot has been hired to pick up a document in Chicago, Illinois, and deliver it to San Francisco, California. Once the document is obtained, the pilot has four hours to make the delivery. The pilot's map has a scale of 1 inch = 310 miles, and the distance between San Francisco and Chicago measures 7 inches. If the aircraft travels at 505 miles per hour, will the document reach San Francisco on time?

6. A ship is returning to home port from its final stop in a three-month cruise. The ship is traveling at 15 miles per hour. The ship's map has a scale of 1 inch = 75 miles. On the map, the distance of this final portion of the cruise measures 15 inches. How many hours will the final leg take?

7. A plane that is traveling to Norfolk, Virginia, will leave Boise, Idaho, at 9 A.M. The pilot's map has a scale of 1 inch = 305 miles. On the map, the distance between Norfolk and Boise measures 8 inches. The plane will average 244 miles per hour. At what time will it arrive in Norfolk? (Add two hours to the trip time to account for the change from Mountain Time to Eastern Time.)

8. Jeremy is traveling from New York City, New York, to Kingston, New York. His map shows that the distance between the cities measures 4 inches. He averages 52 miles per hour and reaches Kingston in two hours. What is the scale of Jeremy's map?

Drawings, Plans, and Blueprints

Another use for lines and scales is in drawings that present information on designs of products or structures. Engineers, architects, and designers use drawings to show sizes, distances, and relationships of parts. Such drawings are copied photographically for use by those who produce the finished products. When developed and printed, the lines and information may appear in white on a blue background. Or, the lines and information may appear in blue on a white background. Such copies are called *blueprints.*

Architectural plans for a house are an example of this mathematical method of presenting detailed information. Architectural blueprints, for example, tell carpenters where to build door and window frames. Blueprints also tell plumbers where to install pipes. Lines are used to represent walls, doors, and other structural features. Depending on the scale used, a 2-inch line might represent a 20-foot wall. These types of information are given in precise measurements. An architectural drawing of a floor plan is shown in Figure 2-9.

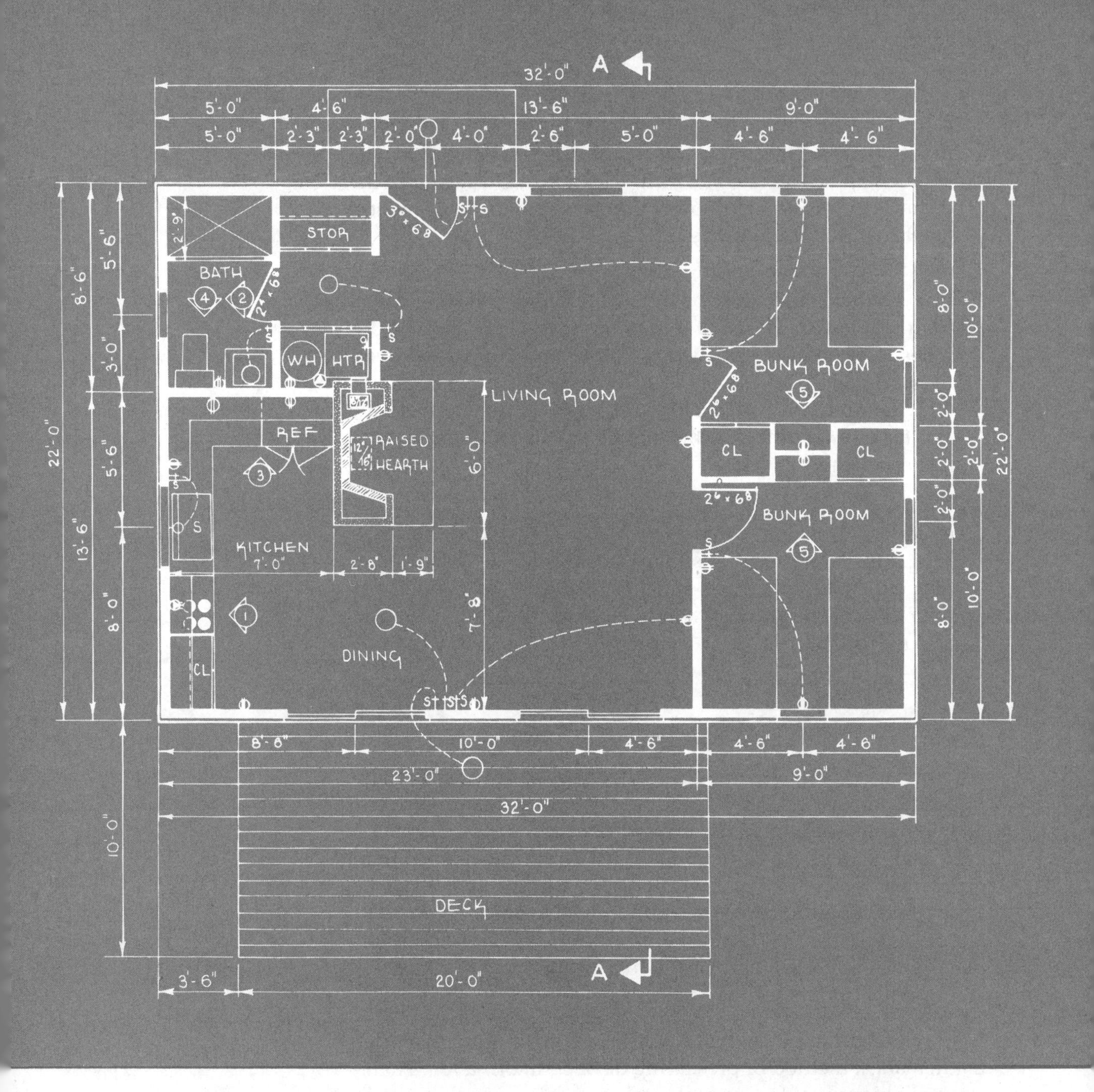

Figure 2-9. Architectural drawings are scaled down to show dimensions of a structure.

Precise information also is presented on engineering drawings and mechanical drawings. These mathematical pictures give exact measurements of length, width, and depth. Precise angles and curves also are shown.

Scales are used for all these types of drawings. For example, an architectural drawing may use a scale such as 1 inch = 1 foot, or 1 inch = 10 feet. Some mechanical drawings are scaled up from the actual size of a part. This occurs when precise measurements are required for manufacturing small parts. Such a scale might read: 12 inches = 1 inch.

Think about an automobile and the thousands of parts that go into it. All those parts must fit together for the automobile to function properly. Imagine the volumes of books that would be required to describe the parts of an automobile in English. Without the special language of mathematics, it would be impossible to design and build such a complicated assembly.

EXERCISE 2-5

Read each question carefully. Then answer the question.

1. You are using a blueprint to cut wood to build a table. The scale is 1 inch = 11 inches. On the blueprint, each leg of the table measures 3 inches long. How long should you cut each leg of the table?

2. You are making a table from a blueprint. The scale of the blueprint is 1 inch = 2 feet. On the blueprint, the dimensions of the table top are 2 inches long by 2 inches wide. What are the actual length and width of the table top?

3. You are looking at a blueprint of a house with a scale of 1 inch = 5 feet. On the blueprint, the dimensions of the house are 6 inches long by 5 inches wide. Calculate the length and width of the house.

4. You are a welder and have been assigned to make an exhaust pipe for a motorcycle. The scale on the blueprint for the vehicle is 1 inch = 6 inches. On the blueprint, the length of the pipe measures 3 inches. How long will the exhaust pipe be?

5. You are a drafter and have been assigned to draw the plans for a part for an airplane engine. The length of the finished part will be 2 inches. You are to draw a plan with a scale of 8 inches = 1 inch. What length will you draw the engine part on the plan?

CHARTS

A *chart* is a picture in which different values are represented. The most familiar types of charts are:

- Bar charts
- Pie charts.

A *bar chart* is a pictorial comparison of values. A *pie chart* illustrates the portions of a whole.

Bar Charts

Bar charts can be used to illustrate many different kinds of information. Bar charts can be either vertical or horizontal. An example of a vertical bar chart is shown in Figure 2-10. A horizontal bar chart is shown in Figure 2-11.

Each bar represents a value. The height or length of a bar is determined by the values of a scale. In Figure 2-10, for example, each line of the scale equals 2 cents. The bar that reaches the fourth line, therefore, has the value 8 cents.

Vertical bar charts usually describe changing values or patterns of change. Horizontal bar charts often describe relative values or quantities of different items. Horizontal bar charts are especially useful for illustrating comparisons, especially when competition of some sort is involved.

EXERCISE 2-6

Read each question carefully. Then draw a bar chart that presents the information accurately.

1. The Tomato Computer Company is a rapidly growing business. In 1976, the company earned $100,000. In 1978, earnings were $140,000; in 1980, $180,000; in 1982, $220,000; and in 1984, $250,000. Draw a vertical bar chart that represents sales from 1976 through 1984.
2. In 1960, the average cost of a gallon of gasoline was 29 cents. In 1968, gasoline sold for 33 cents a gallon. In 1972, the price had increased to 48 cents. By 1978, the price was $1.05. In 1984, gasoline sold for $1.20 a gallon. Draw a vertical bar chart that shows how the price of gasoline has increased.

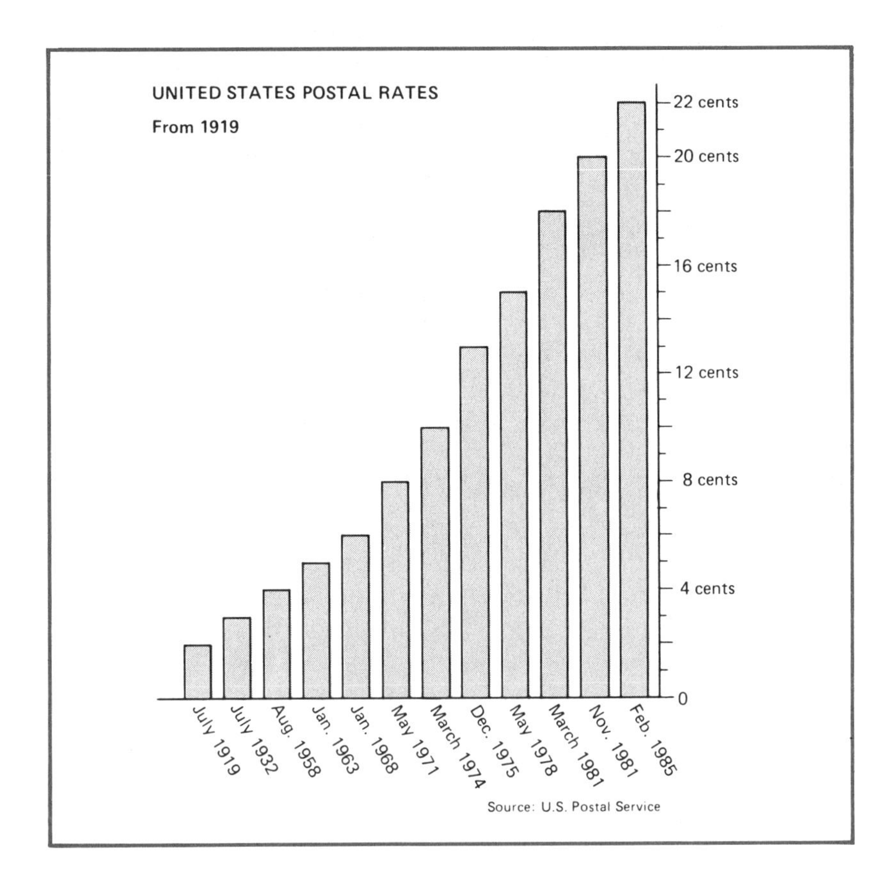

Figure 2-10. Different values are represented in a vertical bar chart.

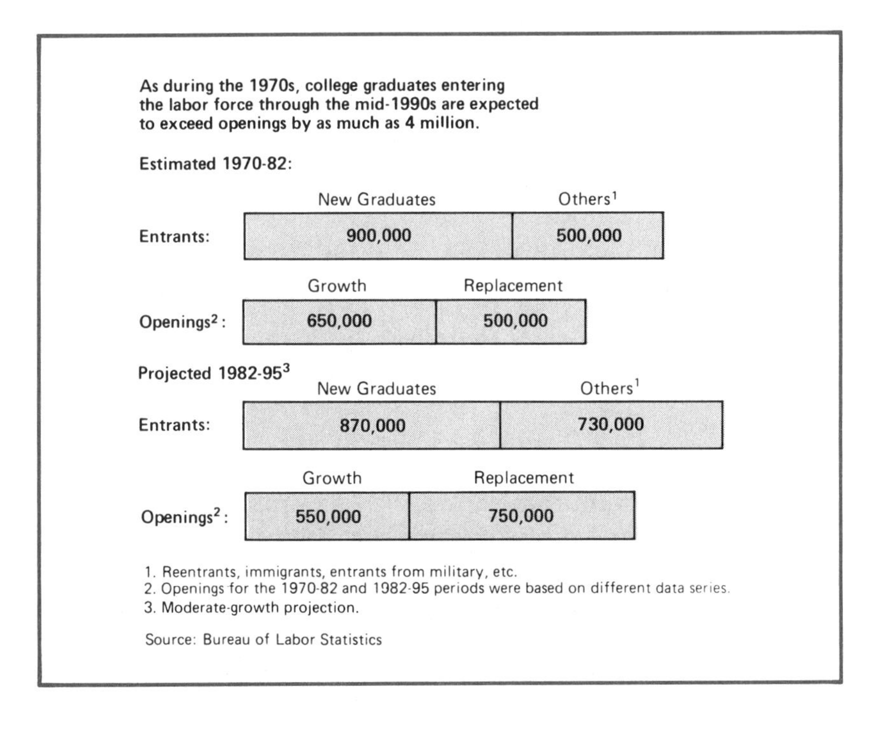

Figure 2-11. A horizontal bar chart provides quick comparisons of values.

3. The world population in 1970 was 3.6 billion. In 1975, it was 4 billion. In 1980, it was 4.4 billion. By 1984, the population had grown to 4.8 billion. Draw a horizontal bar chart that shows how world population has increased.

4. The Guitar Shop employs five craftspeople to make guitars. Last year, Gerard made 75 guitars, Alex made 70, Cindy made 80, Fuji made 75, and Benji made 60. Draw a horizontal bar chart that compares the performances of the employees.

5. The Forest Company sells trees to homeowners. The company has five salespersons. Last year, Salesperson A sold 500 trees, Salesperson B sold 450, and Salesperson C sold 300. In addition, Salesperson D sold 600 trees, and Salesperson E sold 550 trees. Draw a horizontal bar chart that compares the performances of these salespeople.

Pie Charts

A pie chart, or *circle chart,* is used to show the relationships between a whole and its parts. Think about serving a pie for dessert. Four people are sharing the pie. You plan to serve half the pie today and the other half tomorrow.

The first step would be to cut through the center of the pie to divide it into two halves. One of the halves will not be used until the next day. The remaining half must be cut into four pieces. The next step is to cut one of the halves into two quarters. Then, cutting each quarter in half produces four eighths. You now have four equal pieces to serve today. You also have the other half of the pie for the next day. Tomorrow, you can perform the same division to serve four persons again. Figure 2-12 illustrates these relationships. Pie charts are discussed further in Unit 3.

GRAPHS

A *graph* is a diagram that shows the relationship between two sets of values acting upon one another. A graph is produced, or *plotted,* on a *grid.* A grid is a pattern of lines drawn in two directions: horizontal and vertical. One set of values is measured in the horizontal direction. The other set of values is measured in the vertical direction.

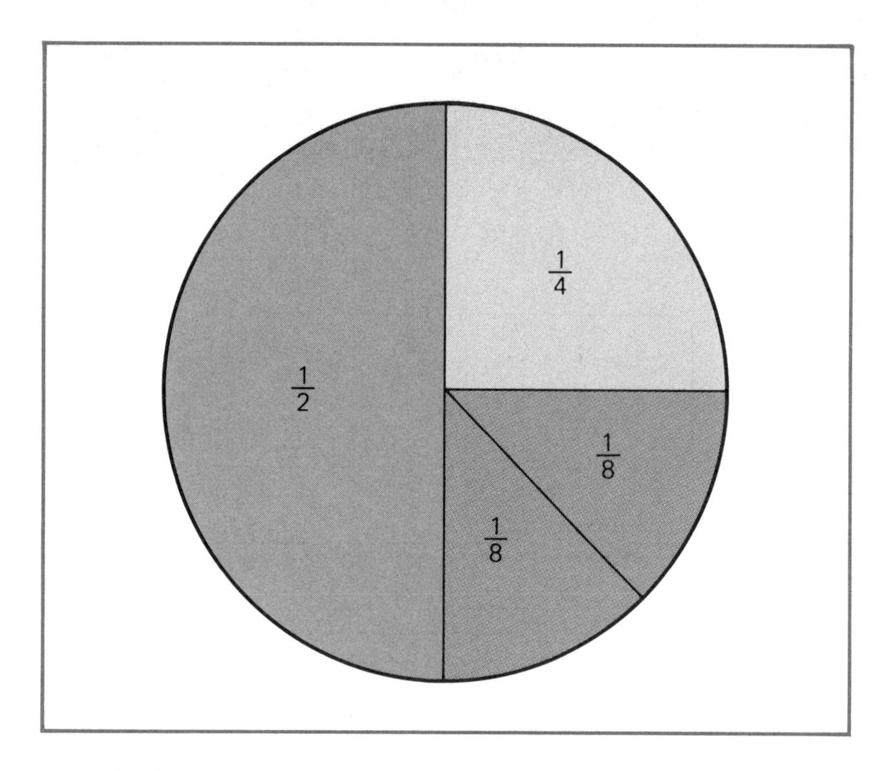

Figure 2-12. Relationships between a whole and its parts are shown in a pie chart.

The lines in either direction are spaced equally apart, the space representing a value. Spaces between lines of a graph can represent distance, time, or any other value being used in a comparison.

Like charts, graphs provide a clear, quickly understood mathematical way of presenting information. Different kinds of graphs can be used to present information in different ways. Common forms of graphs include:

- Scatter graphs
- Line graphs.

Scatter Graphs

A *scatter graph* consists of points plotted on a grid. Scatter graphs can be used to compare values, such as frequencies and distances. Figure 2-13 is an example of a frequency graph. This graph shows the number of students in a class who received varying grades on a test. There are 30 students in the class.

Scatter graphs are useful for organizing data into quickly recognizable information.

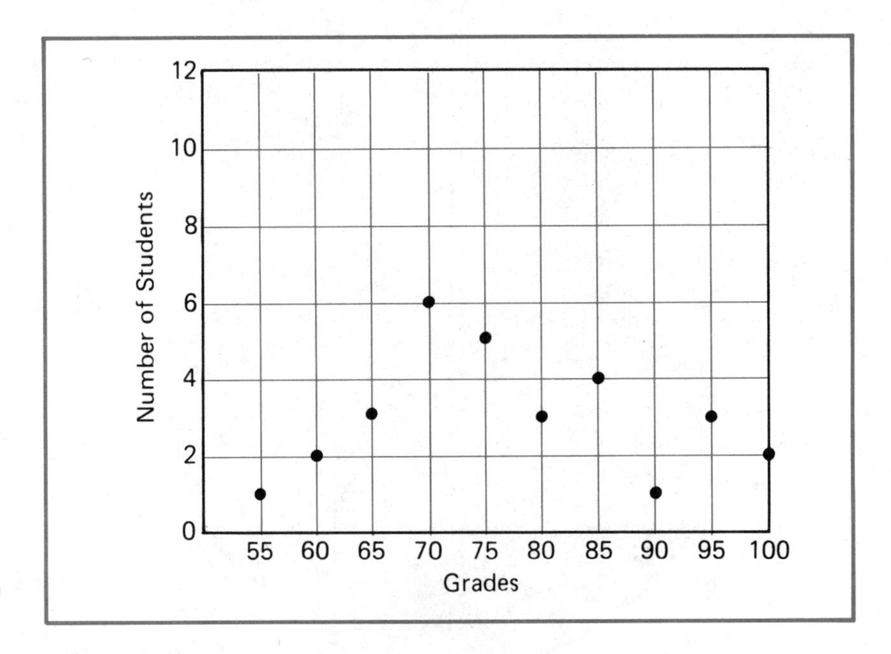

Figure 2-13. A frequency graph is one type of scatter graph.

EXERCISE 2-7

Each question below refers to the scatter graph presented in Figure 2-13. Read each question carefully before answering.

1. How many students scored 70 or below?
2. How many students scored 90 or above?
3. How many students scored 75?
4. What was the test score received most frequently?
5. A score of 70 or above was necessary to pass the test. Did the majority of class members earn a passing grade?

Line Graphs

A *line graph* is plotted in much the same way as a scatter graph. However, on a line graph, the dots or points are connected by lines. The purpose of a line graph is to present information about relationships. A frequent use for line graphs is the presentation of sales figures. A sales graph is shown in Figure 2-14.

Student grading systems are based upon mathematics.

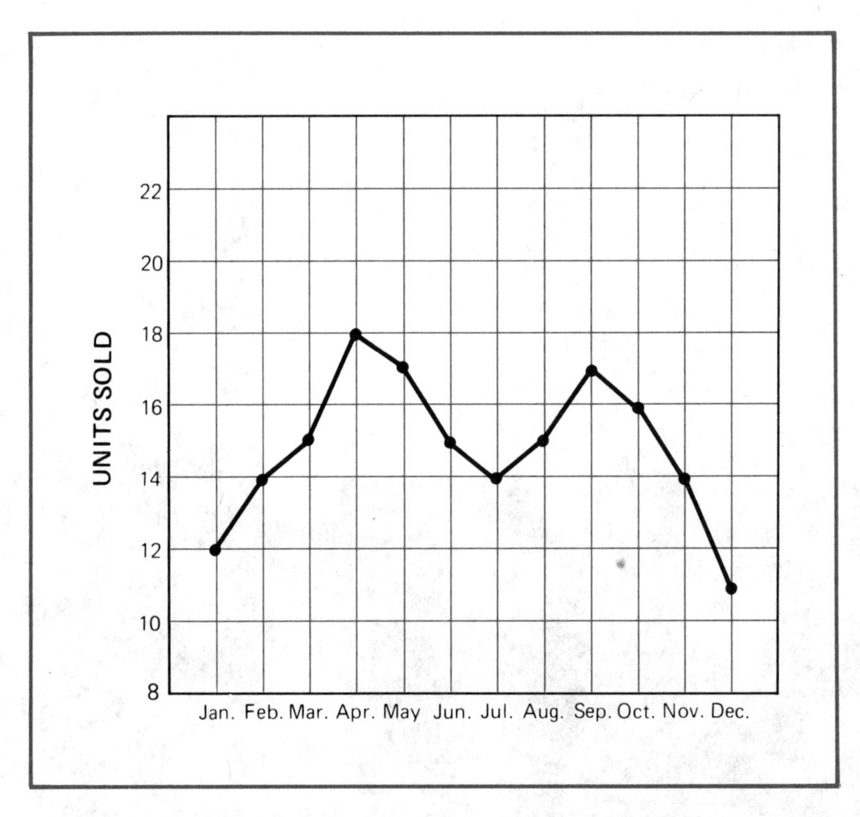

Figure 2-14. Relationships and trends can be shown in a line graph.

The company whose sales figures are shown in Figure 2-14 sells compressors to area businesses. This type of graph helps business executives to plan sales strategies and make other important decisions.

EXERCISE 2-8

Read each question carefully. Present the information given in Questions 2 through 5 in the requested formats.

1. Refer to the line graph in Figure 2-14 and answer these questions:
   - **a.** According to the graph, in which month did the company sell the most compressors?
   - **b.** Looking at the graph, in which month did the company sell the least compressors?
   - **c.** Did compressor sales remain constant for the entire year?

d. If you were president of this company, how many compressors would you have on hand in April of next year?

e. If you were president of this company, what months during the year would you prefer for your salespersons' vacations?

2. Kim is a car salesperson. Kim sold 12 cars in 1980, 15 in 1981, 10 in 1982, 13 in 1983, and 17 in 1984. Draw a vertical bar chart that illustrates Kim's sales history for the years given.

3. Last year, sales for the Bicycle House were as follows: January, 112 bicycles sold; Feburary, 94 sold; March, 86 sold; April, 90 sold; May, 100 sold; June, 125 sold; July, 120 sold; August, 85 sold; September, 75 sold; October, 110 sold; November, 130 sold; and December, 155 sold. Draw a line graph that shows how sales varied monthly during the year. How many bicycles did the company sell last year?

4. Carie is comparing four job opportunities to see which will give her the best weekly income. Job A pays $7.50 an hour for a 30-hour week. Job B pays $6.50 an hour for a 40-hour week. Job C pays $7 an hour for a 35-hour week. Job D pays $8 an hour for a 30-hour week. Draw a horizontal bar chart that compares the weekly incomes of these jobs.

5. The following are test scores for a high school math class. Four students received a score of 70. Five students received a score of 80. Three received a score of 90. Two received a score of 95. Two received a score of 85. And one received a score of 65. Draw a scatter graph that shows class performance for the math test. If 75 was passing, how many students passed the test?

SUMMING UP

- As a language, mathematics offers a number of ways to present information clearly and in easy-to-understand forms. The information communicated by mathematics is generated by the solution of problems.
- The numerical results of problem solving are one method of presenting information. Others include tables, lines and scales, charts, and graphs.
- Tables show the relationships of different values. Among the many uses for tables is the presentation of comparison lists. This type of table is called a condition table.
- Maps use lines and scales to present information in a reduced size. The scale of a map indicates the relationship between the printed information and the area it represents. An example of map scale would be 1 inch = 5 miles.
- Another form of presenting information using lines and scales includes design drawings and plans. Such drawings and plans are guides for workers who make or build things.
- Charts are pictures used to compare different values. A bar chart does this with bars of varying lengths. A pie chart compares values of parts to a whole.
- Graphs are a means of comparing values on a grid. The grid is made up of horizontal and vertical lines. The distance between each line represents a certain value.
- A scatter graph consists of points plotted on a grid. Each point is placed where its horizontal and vertical values intersect.
- A line graph shows comparisons. Points are plotted on a grid and then connected by lines.

TESTING YOUR WORKING KNOWLEDGE

Read each question carefully. Then answer the question. Be sure to present answers in the format requested.

1. Refer to the Withholding Tax Table in Figure 2-1 to determine taxes in the following situations:
   - **a.** Terry earns $7.10 an hour and has worked 75 hours. Terry has one withholding allowance.
   - **b.** Jorge earns $8.45 an hour and has worked 65 hours. Jorge has no withholding allowances.
   - **c.** Mary earns $1.05 for each 12 pounds of peaches she picks and has two withholding allowances. During the first half of the month, she picked 6,300 pounds of peaches.
   - **d.** Carlos earns $15 an hour as a mechanic and has three withholding allowances. Carlos worked 62 hours.

2. The Car Company has five salespeople. Last quarter, Salesperson A sold 25 cars, Salesperson B sold 15, and Salesperson C sold 17. In addition, Salesperson D sold 14 cars and Salesperson E sold 23 cars. Draw a horizontal bar chart that compares these performances.

3. Carrol, Bob, Eduardo, Chen, and Fusako are on the same basketball team. Carrol scored 11 points in the first game, 30 in the second, 15 in the third, and 12 in the fourth. Bob scored 15 points in the first game, 22 in the second, 18 in the third, and 9 in the fourth. Chen scored 9 points in the first game, 24 in the second, 10 in the third, and 19 in the fourth. Fusako scored 13 points in the first game, 14 in the second, 12 in the third, and 17 in the fourth. Eduardo scored 10 points in the first game, 12 in the second, 18 in the third, and 8 in the fourth.
   - **a.** Create a table that shows how many points each player scored in each game. Circle the highest score in each game.
   - **b.** Determine the team total for each game.
   - **c.** Determine how many points each player has scored.

4. National Auto Body earned $145,000 in 1970, $135,000 in 1975, $175,000 in 1980, and $165,000 in 1985. Draw a line graph that

TERMS THAT COUNT

summarize
compare
table
withhold
condition table
truth table
map
key
estimate
constant
variable
blueprint
chart
bar chart
pie chart
circle chart
graph
plot
grid
scatter graph
line graph

shows how the amount of money National Auto Body earned changed from 1970 to 1985.

5. Last week, Elena received a paycheck for $200. She paid $20 in Federal income tax, $10 in State income tax, and $2 for Social Security payments. In addition, she spent $50 on food and $40 on gasoline. Draw a pie chart that shows how the money Elena spent relates to the paycheck she received. Be sure to include the amount of money she hasn't spent yet.

6. The distance between San Francisco, California, and Albany, New York, is 2,800 miles. On a map, this distance measures 8 inches. What is the scale of that map?

7. Todd is riding his bicycle from Plattsburgh, New York, to Lake Placid, New York. On his map, the distance between the cities measures 4 inches. He averaged 24 miles per hour and reached Lake Placid in two hours. What is the scale of Todd's map?

8. Refer to the New York State mileage chart in Figure 2-2 to determine the distances between cities in this question. Kyle is driving from his home in Buffalo to Utica. He wants to stop for one hour to have lunch at noon with his friend in Syracuse. His average driving speed is 52 miles per hour. What is the latest time he can leave Buffalo to meet his friend at noon? From Syracuse, his average driving speed will be 54 miles per hour. How long will his trip to Utica take, including the one-hour stop for lunch?

SKILL DRILLS

A. Find the sum in each problem.

1. 231,578 + 320,366 =
2. 7,214,879 + 19,326 =
3. 713,329 + 439,355 + 665,712 + 326,891 =
4. 518,452 + 9,325,463 + 344,187 + 6,478,102 =
5. 5,325,919 + 4,748 + 430 + 218,456 + 4,721,438 =
6. 41,225 + 25,567 + 345,817 + 2,341,987 + 9,087,712 =
7. 243,546 + 823,419 + 876,123 + 4,213,546 + 39,279 =
8. 325,438 + 985,436 + 361,658 + 592,436 =

B. Find the difference in each problem.

1. 645,928 − 33,253 =
2. 31,231 − 29,879 =
3. 27,241 − 24,389 =
4. 23,436,865 − 15,325,768 =
5. 754,864,389 − 35,423,658 =
6. 213,347 − 212,897 =
7. 354,854 − 176,768 =
8. 56,232,657 − 23,436,878 =

C. Find the product in each problem.

1. 548 × 56 =
2. 1,561 × 21 =
3. 439 × 547 =
4. 8,325 × 232 =
5. 657 × 5,618 =
6. 325 × 54 =
7. 821 × 99 =
8. 376 × 657 =

D. Find the quotient in each problem.

1. 8,100 ÷ 324 =
2. 7,716 ÷ 643 =
3. 59,808 ÷ 712 =
4. 4,277 ÷ 91 =
5. 102,834 ÷ 3,546 =
6. 33,579 ÷ 117 =
7. 381,920 ÷ 352 =
8. 2,795,907 ÷ 543 =

Your Income

YOUR LEARNING JOB

When you have completed the exercises and assignments for this unit, you should be able to:

- ☐ **Use fractions and decimals.**
- ☐ **State a problem (with an algorithm) for verifying gross pay on a paycheck.**
- ☐ **Solve the problem for verifying gross pay on a paycheck.**

WARM-UP

A. Find the sum in each problem.

1. 7,212 + 6,839 + 2,019 + 532 =
2. 3,426 + 3,659 + 4,396 =
3. 12,837 + 72,502 + 34,827 + 45,019 =
4. 938,001 + 38,999 + 5,384,019 =
5. 63,254,512 + 79,768,624 + 3,098 + 34,198 =

B. Find the difference in each problem.

1. 7,123 − 5,125 =
2. 25,627 − 3,019 =
3. 29,154 − 19,789 =
4. 8,234,341 − 3,568,999 =
5. 78,231,546 − 78,142,658 =

C. Find the product in each problem.

1. 2,435 × 8,908 =
2. 25,156 × 3,789 =
3. 29,547 × 9,563 =
4. 598,012 × 4,549 =
5. 234,919 × 8,691 =

D. Find the quotient in each problem.

1. 79,375 ÷ 25 =
2. 682,209 ÷ 27 =
3. 13,068 ÷ 33 =
4. 3,764,708 ÷ 452 =
5. 164,239,944 ÷ 769 =

MONEY MANAGEMENT SKILLS—PAYCHECKS AND PAYROLLS

When you go to work, you agree to sell your time, skills, and efforts to an employer. These resources are yours to sell. Your employer pays for these resources with money. That is the basic exchange of values involved in the world of work. Carrying out this exchange requires use of mathematics.

Most workers are paid a *wage* on a regular basis. A wage is a payment, usually of money, for labor or services performed. Wages may be calculated on an hourly or daily basis, or on a *piecework* basis. Piecework wages are paid according to the value of an item produced or a unit of work performed.

For example, an assembly line worker might be paid 10 cents to install a computer chip on an electronic board. When 10 chips have been installed, the worker has earned $1. This is calculated by multiplying the 10 cents (0.10) by the number of chips installed (10). Thus, $0.10 \times 10 = 1.00$, or one dollar. When 100 chips have been installed, the worker has earned 100 times 10 cents, or $10. This worker's wage depends upon how rapidly he

Salespeople frequently are paid commissions based on their sales volumes. COURTESY OF NCR CORPORATION

or she can install the chips. If the worker installs one chip per minute on the average, that would calculate to 60 chips per hour. At 10 cents a chip, the worker's hourly pay could be determined by the following calculation:

$$\begin{array}{r} 60 \\ \times\ 0.10 \\ \hline 00 \\ 6\ 0 \\ \hline 6.00 \end{array}$$

Thus, the worker earns $6 an hour on the average. However, on a piecework basis, total pay depends entirely upon how many chips the worker installs. Piecework wages often are paid on a daily or weekly basis.

Many workers are paid on a *salary* basis. A salary is a fixed rate of pay for labor or services. Salaries are not necessarily tied to the number of hours worked. Salaried workers usually are paid on a weekly or monthly basis.

Salespersons frequently are paid *commissions,* which are a percentage of their total sales. Many workers in service industries, such as waiters and waitresses, receive *tips* from the customers they serve.

Regardless of the basis on which you are paid, your employer probably will take *deductions* from your *gross pay.* Gross pay is the total income you earn and is also called *earned income.* If you go to work on a job for $4 an hour, your gross pay is that rate multiplied by the number of hours you work. Thus, if you work 40 hours, this number would be multiplied by $4. Your gross pay would be $160. (Check this multiplication. Then figure out what gross pay would be if you worked 35 hours.)

A deduction is money withheld from your gross pay for taxes, insurance premiums, or other purposes. The money that you actually receive, after deductions are taken out, is called *net pay* or *net income.* The pie chart in Figure 3-1 shows that net pay is a part of gross pay. The entire pie, before a slice is removed, is the whole. The pie is gross pay. Deductions represent a part of the whole, or a *fraction.* When that fraction, or part, is taken out as deductions, the part remaining is net pay. Net pay also is a part of the whole, or a fraction. Thus, net pay is a fraction of gross pay.

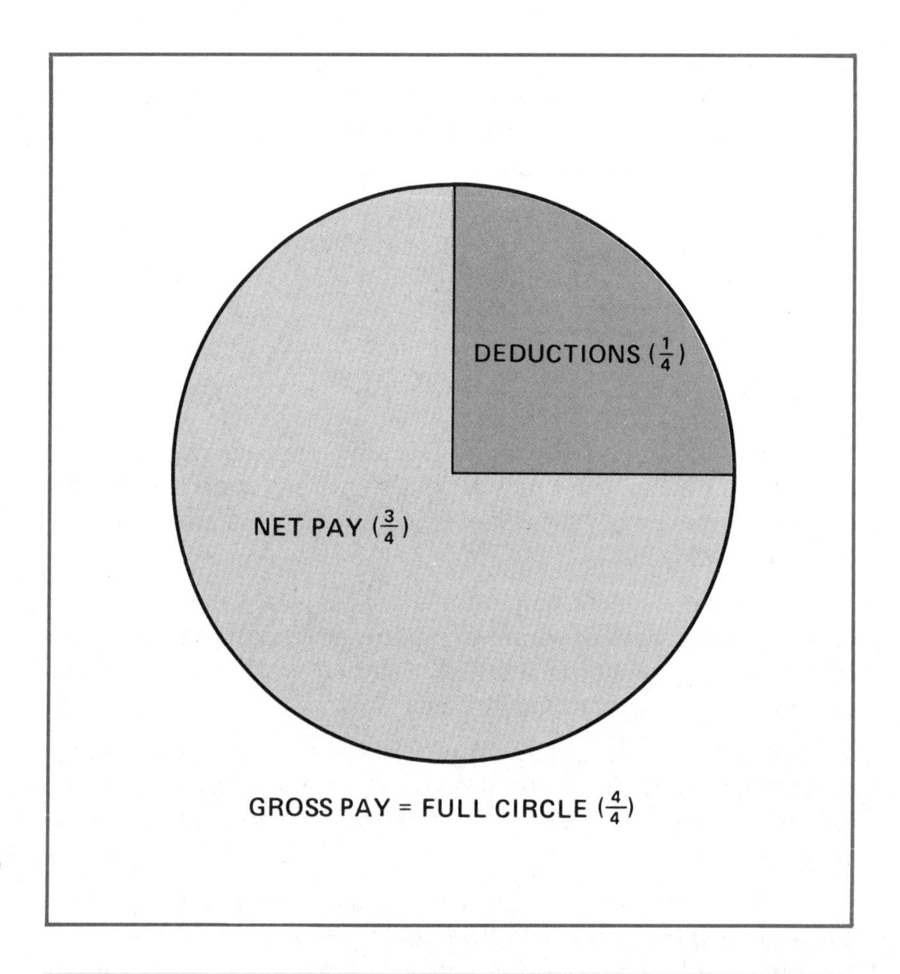

Figure 3-1. Pie chart shows the relationships among gross income (the whole pie), deductions, and net income.

EXERCISE 3-1

Read each problem carefully. Then, answer the questions.

1. Lucy sews dresses in a factory. She is paid $1.50 for each garment she sews. On an average day, Lucy sews 25 dresses.
   a. What is her gross pay for an average day?
   b. What is her gross pay for an average five-day week?
   c. If she works eight hours a day, how much pay does Lucy average per hour?

2. Fred earns $5 an hour as a mechanic's helper.
   a. What is his gross pay for an eight-hour day?
   b. What is his gross pay for a 40-hour week?

3. Manuel works in a hardware store at a wage of $7 an hour. In a typical week, he works 35 hours. Each week, his employer deducts $49 from his total pay for taxes and insurance premiums.
   a. What is Manuel's gross pay for a 35-hour week?
   b. What is Manuel's net pay?

4. Sarah drives her car to make deliveries for a large distribution firm. She gets paid $10 an hour and also receives $0.20 for each mile driven.
   a. What is Sarah's gross pay plus mileage for a 40-hour week when she drives 200 miles?
   b. What is Sarah's gross pay plus mileage for a 27-hour week when she drives 530 miles?

5. Jean picks peaches during her summer vacation. She gets paid $0.10 for each pound of peaches picked. Jean averages 45 pounds per hour.
   a. How much can Jean earn in 43 hours?
   b. How many hours would it take for Jean to earn $990?

6. Ted and Angel are partners in a delivery business. They do shopping for people and then deliver goods to people's homes. They charge clients $10 for each delivery and an additional $0.20 per mile.
   a. What is their total charge to customers if they perform 10 deliveries and travel 200 miles?
   b. What is their total charge to customers if they perform 15 deliveries but travel only 38 miles?

7. Enrique delivers papers for the daily newspaper. He receives $25 for every 1,000 papers delivered.
   a. How many papers does Enrique have to deliver to earn $200?
   b. What is Enrique's gross pay if he delivers 200,000 papers?

CONVERSION SKILLS—FRACTIONS

Think of fractions this way. To describe a fraction, you must first identify the whole of which it is a part. Suppose, for example, that there are 20 students in your class. You represent a fraction of the students enrolled in the class—1 in 20. If your family owns two cars, they might be two parts out of some 50 million cars on the road.

In arithmetic, a fraction has a more specific meaning. A fraction usually is a portion of the smallest whole in the number system. The smallest whole number value is 1. Therefore, a fraction represents a value smaller than the whole number, 1.

Fractions have two parts, usually written one above the other and separated by a line. A fraction also may be written with the two parts on the same line, separated by a slash. The latter form frequently is used within the text of a book.

The *denominator* is the lower, or second, number of a fraction. The denominator is the number of parts into which the whole is divided. The *numerator* is the upper, or first, number. The numerator is the number of parts with which you are dealing. For example, consider the fraction three-eighths, as illustrated in Figure 3-2. This fraction describes three of eight parts into which something has been divided. Written in mathematical form, the fraction would appear this way:

$$\frac{3}{8}$$

A fraction is described by reading the numerator first, then the denominator. Special words are used for the denominator. These words are known as *ordinal numbers.* An ordinal number describes the place of a value in a list. Examples of ordinal numbers include fifth, tenth, twentieth, and so on. To illustrate, the fraction $\frac{2}{5}$ is read as "two-fifths," not as "two-five" or "two-fives." Use the following rule when writing fractions in English: Connect numerator and denominator with a hyphen unless the numerator and/or denominator already contains a hyphen. For examples compare the English expressions of the fractions $\frac{1}{7}$ and $\frac{1}{23}$ in Figure 3-3.

If all the parts of a whole were represented by a fraction, the numerator and denominator would be equal. Thus, if all

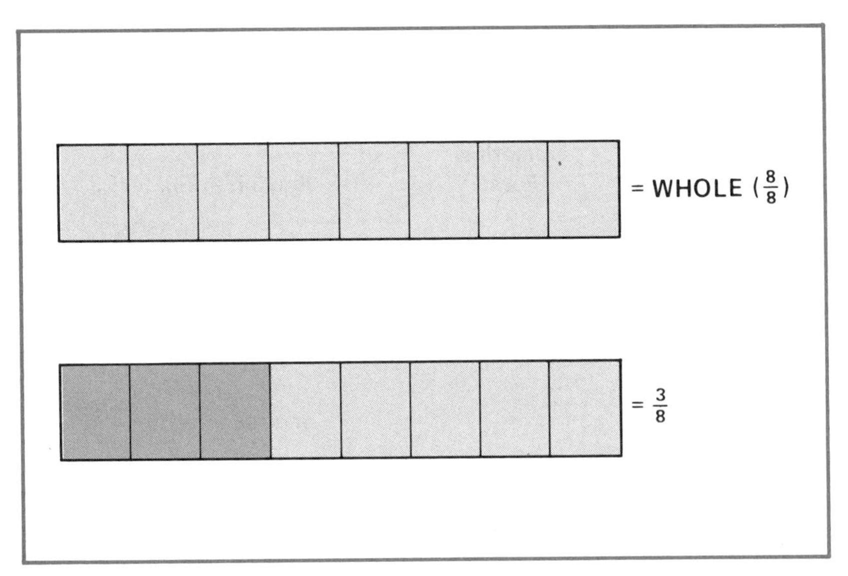

Figure 3-2. A fraction represents part of a whole.

eight parts of the box in Figure 3-2 were represented, the fraction would be written $\frac{8}{8}$. This fraction would represent the whole box, however, not a portion of it. Therefore, the fraction $\frac{8}{8}$ really means 1.

A fraction in which the numerator is smaller than the denominator is called a *proper fraction.* A fraction in which the numerator is equal to, or larger than, the denominator is called an *improper fraction.* Such a fraction is improper because its value represents more than a part of a whole. Improper fractions can be changed to their proper form by removing the whole number or numbers. This is done by dividing the denominator into the numerator, as shown in the following examples:

$$\frac{10}{2} = 10 \div 2 = 5$$

$$\frac{3}{2} = 3 \div 2 = 1 \text{ R } 1 = 1\frac{1}{2}$$

In the first fraction, $\frac{10}{2}$, the denominator can be divided into the numerator an equal number of times. Thus, this improper fraction should be written as a whole number, 5. Figure 3-4 illustrates this. The second fraction above, $\frac{3}{2}$, includes a *remainder* when the denominator, 2, is divided into the numerator, 3. The fraction $\frac{3}{2}$ stands for three halves. Two halves become the whole number, 1, and the third half is the remainder. Thus, the proper way to write $\frac{3}{2}$ is $1\frac{1}{2}$.

EXPRESSING FRACTIONS IN ENGLISH

Numerical Form	English Form	Numerical Form	English Form
$\frac{1}{2}$	one-half	$\frac{1}{15}$	one-fifteenth
$\frac{1}{3}$	one-third	$\frac{1}{16}$	one-sixteenth
$\frac{1}{4}$	one-fourth	$\frac{1}{17}$	one-seventeenth
$\frac{1}{5}$	one-fifth	$\frac{1}{18}$	one-eighteenth
$\frac{1}{6}$	one-sixth	$\frac{1}{19}$	one-nineteenth
$\frac{1}{7}$	one-seventh	$\frac{1}{20}$	one-twentieth
$\frac{1}{8}$	one-eighth	$\frac{1}{21}$	one twenty-first
$\frac{1}{9}$	one-ninth	$\frac{1}{22}$	one twenty-second
$\frac{1}{10}$	one-tenth	$\frac{1}{23}$	one twenty-third
$\frac{1}{11}$	one-eleventh	$\frac{1}{24}$	one twenty-fourth
$\frac{1}{12}$	one-twelfth	$\frac{1}{25}$	one twenty-fifth
$\frac{1}{13}$	one-thirteenth	$\frac{1}{100}$	one-hundredth
$\frac{1}{14}$	one-fourteenth	$\frac{1}{1,000}$	one-thousandth

Figure 3-3. Fractions can be expressed in English.

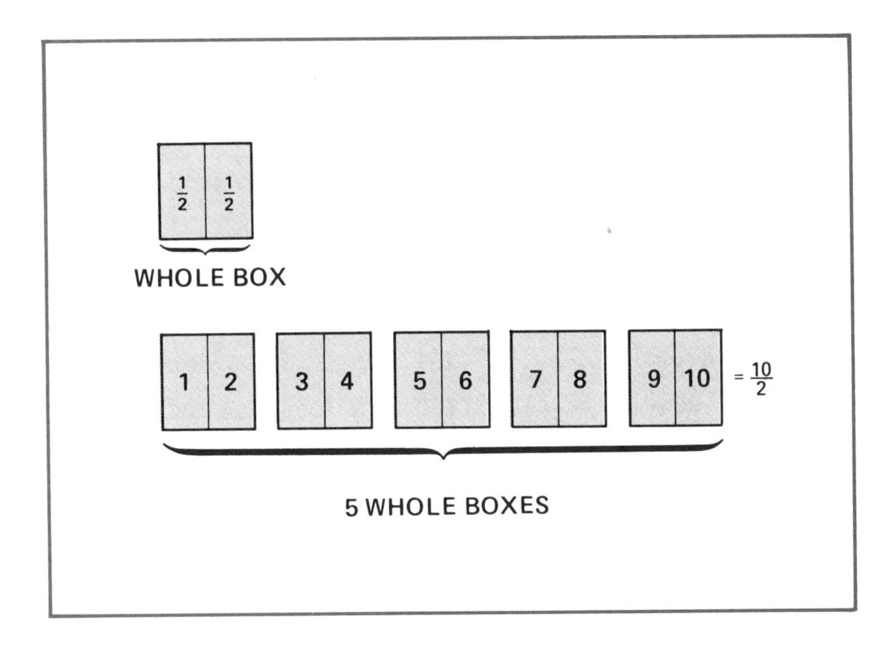

Figure 3-4. Two halves make a whole; ten halves make five wholes.

EXERCISE 3-2

A. Convert these improper fractions to their proper forms.

1.	$\frac{10}{3}$	**7.**	$\frac{9}{2}$	**13.**	$\frac{63}{9}$
2.	$\frac{5}{4}$	**8.**	$\frac{7}{6}$	**14.**	$\frac{41}{8}$
3.	$\frac{4}{3}$	**9.**	$\frac{11}{3}$	**15.**	$\frac{45}{22}$
4.	$\frac{72}{35}$	**10.**	$\frac{89}{9}$	**16.**	$\frac{321}{23}$
5.	$\frac{17}{2}$	**11.**	$\frac{27}{2}$	**17.**	$\frac{199}{101}$
6.	$\frac{16}{3}$	**12.**	$\frac{101}{10}$	**18.**	$\frac{101}{9}$

B. Convert these proper fractions to their improper forms.

1.	$3\frac{2}{3}$	**7.**	$9\frac{1}{2}$	**13.**	$6\frac{2}{9}$
2.	$5\frac{3}{4}$	**8.**	$7\frac{5}{6}$	**14.**	$4\frac{3}{8}$
3.	$4\frac{1}{3}$	**9.**	$11\frac{1}{7}$	**15.**	$4\frac{5}{22}$
4.	$7\frac{2}{35}$	**10.**	$8\frac{5}{6}$	**16.**	$3\frac{1}{234}$
5.	$1\frac{100}{101}$	**11.**	$2\frac{133}{600}$	**17.**	$19\frac{2}{11}$
6.	$16\frac{4}{5}$	**12.**	$10\frac{45}{46}$	**18.**	$1\frac{1}{9}$

C. Write these fractions in numerical terms.

1. three-fourths
2. five thirty-seconds
3. four-fifths
4. twenty-one twenty-thirds
5. two-elevenths
6. seven-ninths
7. eight-fifteenths
8. eleven-twentieths
9. nine-fifths
10. three-sixteenths
11. ten-eighteenths
12. eleven forty-seconds
13. twelve fifty-fourths
14. nine-eighths
15. fifty thirty-thirds
16. twenty-one hundredths
17. one five-hundredth
18. five-hundred thousandths

D. Write these fractions in English terms.

1. $\frac{4}{9}$
2. $\frac{13}{16}$
3. $\frac{9}{10}$
4. $\frac{7}{100}$
5. $\frac{6}{17}$
6. $\frac{7}{32}$
7. $\frac{8}{21}$
8. $\frac{23}{25}$
9. $\frac{3}{1,000}$
10. $\frac{31}{32}$
11. $\frac{16}{19}$
12. $\frac{11}{29}$
13. $\frac{13}{100}$
14. $\frac{99}{100}$
15. $\frac{27}{11}$
16. $\frac{47}{21}$
17. $\frac{21}{300}$
18. $\frac{99}{3}$

Understanding Fractions

An understanding of fractions is necessary for dealing with payrolls. Fractions also are important for many other types of calculations in life and work. Some examples of calculations that involve fractions are measurements, sports statistics, prices of purchases, and sales taxes. (See Appendix A for rules for working with fractions.)

You probably use fractions in a casual way in your everyday life. For example, you may pause during a homework assignment and decide that you are half finished. Think of your homework as the pie shown in Figure 3-1. This means that you have cut the pie into two roughly equal pieces.

You may use fractions in buying or using food products such as butter, cheese, nuts, or cold cuts. For example, butter and margarine are typically sold in one-pound boxes. Each box usually has four separately wrapped bars, or sticks. Thus, the manufacturer has chosen to break a pound of product into four parts. Each bar of butter or margarine is referred to as weighing a quarter-pound. This is because each bar weighs one-fourth, or one-quarter, of a full pound.

EXERCISE 3-3

State each problem and prepare an algorithm for the steps to be followed to solve it. Then solve the problem, expressing the answer in both numerical terms and English.

1. Maria's track coach told her to run three miles each day during training. After running a mile, she stopped to tie a shoe lace that had come loose. Thus, Maria had completed only a fraction of her practice run. What fraction or how much of her daily run had she completed at that point?

2. You worked 40 hours this week at a wage rate of $5 an hour. Your employer deducted $40 for taxes and your insurance premium. Verify your net pay and determine what portion (fraction) of your pay was withheld.

3. Michael and Akemi built a fence for their neighbor, Mr. Ogelthorpe. Michael worked twice as many hours on the fence as Akemi. They decided that Michael should earn twice as much money as Akemi. Mr. Ogelthorpe paid them $99. What fraction of $99 did Michael get? What fraction did Akemi get?

4. Norma, Bertha, Francine, and Linda are on the school track team. Their favorite event is the relay. The race is run over a distance of one mile, or 1,760 yards. Each woman runs the same distance. What fraction of the race does Norma run?

5. Bill borrowed $1,000 from his boss to buy a car. Each week, he works 40 hours. His hourly wage is $7.50. Bill gives his boss $150 a week to repay the money he has borrowed. What fraction of his weekly paycheck does Bill give to his boss?

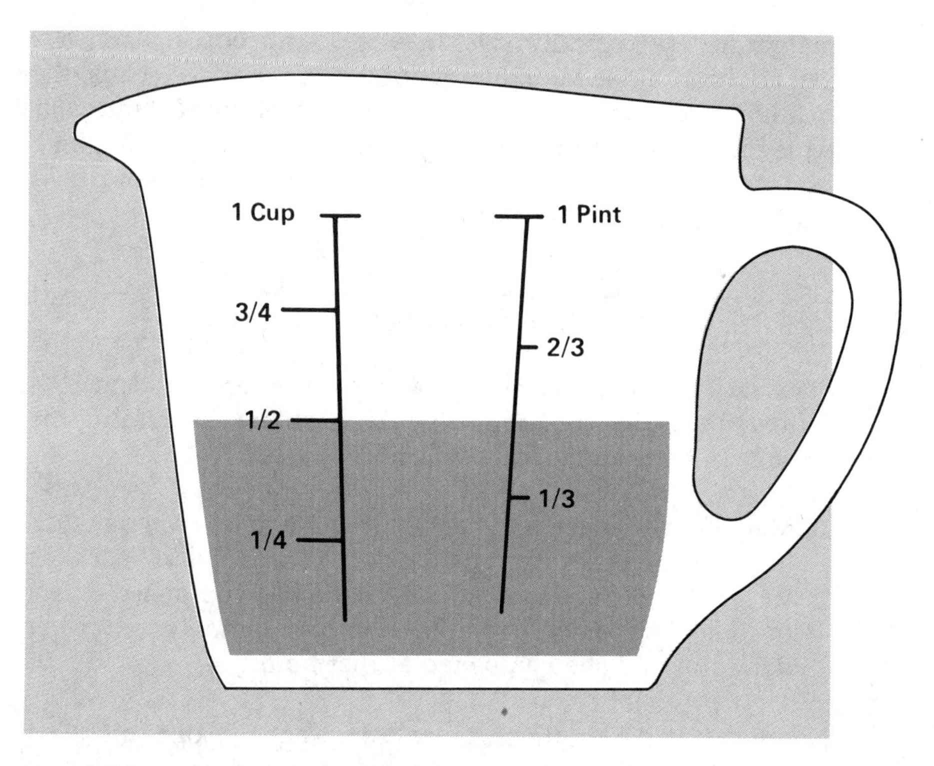

Fractions are parts of a whole.

Using Fractions

Multiplying by fractions with numerators of 1. When a whole number is multiplied by a proper fraction, the result is a number smaller than the original whole number. This is because multiplying by a fraction really is dividing. When you multiply by a fraction, you multiply the whole number (the *multiplicand*) by the numerator of the proper fraction (the *multiplier*). Then you divide by the denominator of the fraction. Because the multiplier is part of a fraction, it is smaller than 1. Thus, the answer, or *product* of the multiplication, will be smaller than the whole number (the multiplicand). As an example, look at the problem below. The whole number, 4, is multiplied by the fraction, $\frac{1}{4}$. As the first step, the whole number is multiplied by the numerator, delivering a product of 4. When 4 is divided by the denominator, 4, the answer, or *quotient*, in the division problem, is 1.

$$4 \times \frac{1}{4} = \frac{4 \times 1}{4} = \frac{4}{4} = 4 \div 4 = 1$$

Note that the whole number in the above problem is multiplied by the numerator. This is because a whole number, any whole number, can be treated as a numerator. Any whole number can be said to have a denominator of 1. Remember that fractions can be handled by dividing the denominator into the numerator. Thus, a denominator of 1 does not change any values. This is illustrated below. A whole number, 5, is converted to a fraction, $\frac{5}{1}$. When the denominator, 1, is divided into five, the quotient is 5. This leaves the value of the whole number unchanged. Using a denominator of 1 thus becomes an easy way to deal with problems involving both fractions and whole numbers.

$$5 \times \frac{1}{1} = \frac{5}{1}$$
$$5 \div 1 = 5$$

EXERCISE 3-4

Perform the following multiplications of whole numbers by fractions. If the answer is an improper fraction (the whole number 1 or greater), convert it to its proper form.

1. $5 \times \frac{1}{5}$
2. $25 \times \frac{1}{10}$
3. $8 \times \frac{1}{15}$
4. $49 \times \frac{1}{7}$
5. $72 \times \frac{1}{8}$
6. $7 \times \frac{1}{5}$
7. $15 \times \frac{1}{3}$
8. $54 \times \frac{1}{9}$

Multiplying by fractions with numerators other than 1. The same method is used when the fraction has a numerator other than 1. Think once more about your class, with 20 students. Your teacher might tell you that an upcoming test will be graded on a *curve.* Grading on a curve means that letter grades will be awarded to designated portions of the class. For example, the top four students might receive a grade of ''A.'' The next four students might earn a ''B,'' and so on. Grading on a curve rates class members on the basis of competition.

Say that your teacher announces that the five letter grades—A, B, C, D, and F—will be divided equally. If A through D are passing grades, how many of the 20 students will pass and how many will fail? The answers are that 16 students will pass and four will fail. These answers were obtained through the following steps:

First, divide the number of students by the number of grades:

$$20 \div 5 = 4$$

Four students will receive each grade. Next, if four of five grades are passing, that means that four of five, or $\frac{4}{5}$, of the students will pass. So, multiply the number of students by the fraction who will pass:

$$20 \times \frac{4}{5} = \frac{20 \times 4}{5} = \frac{80}{5} = 80 \div 5 = 16$$

The same steps are used to determine how many students will fail. One grade out of five will be an F—a failing grade. So, multiply the number of students by the fraction that will fail:

$$20 \times \frac{1}{5} = \frac{20}{5} = 20 \div 5 = 4$$

Reducing Fractions

A fraction should be expressed in its lowest *equivalent* (equal) value, or simplest form. For example, $\frac{2}{4}$, $\frac{4}{8}$, and $\frac{8}{16}$ are all equivalent to $\frac{1}{2}$. To reduce a fraction to its simplest form, divide both the numerator and the denominator by the same number.

$$\frac{2}{4} = \frac{2 \div 2}{4 \div 2} = \frac{1}{2}$$

Sometimes this process has to be repeated several times until the lowest equivalent value is reached.

$$\frac{8}{16} = \frac{8 \div 2}{16 \div 2} = \frac{4}{8} = \frac{4 \div 2}{8 \div 2} = \frac{2}{4} = \frac{2 \div 2}{4 \div 2} = \frac{1}{2}$$

This reduction could be done quicker by dividing the numerator and denominator by 8.

$$\frac{8 \div 8}{16 \div 8} = \frac{1}{2}$$

EXERCISE 3-5

Perform the following multiplications of whole numbers by fractions. If the product is an improper fraction, convert it to its proper form. Express all fractions in their simplest form.

1. $16 \times \frac{3}{8}$

2. $6 \times \frac{2}{3}$

3. $18 \times \frac{2}{3}$

4. $24 \times \frac{5}{6}$

5. $3 \times \frac{2}{3}$

6. $5 \times \frac{2}{5}$

7. $2 \times \frac{17}{18}$

8. $27 \times \frac{2}{9}$

9. $120 \times \frac{3}{4}$

10. $8 \times \frac{9}{24}$

11. $64 \times \frac{3}{4}$

12. $12 \times \frac{5}{6}$

13. $15 \times \frac{2}{5}$

14. $26 \times \frac{7}{13}$

15. $12 \times \frac{12}{13}$

16. $18 \times \frac{5}{6}$

17. $54 \times \frac{4}{9}$

18. $6 \times \frac{7}{12}$

CONVERSION SKILLS—DECIMALS

Fractions come in other forms that represent parts of whole numbers. One of these forms is called a *decimal.* A decimal is a fraction in which the parts of the whole have been divided into 10, 100, or 1,000 pieces. In the decimal form, the denominator is not written as a number beneath a line. Instead, a point, called a *decimal point,* is used to separate the whole numbers from the fractional parts. (See Appendix A for rules for working with decimals.)

To illustrate a decimal, start with a whole number, 98,765. The decimal point is placed at the right of the smallest digit in a whole number. Reading from right to left, the 5 stands for the number of units, or single objects. This is a total value of 5. The 6 stands for six groups of 10 objects, or 60. The 7 stands for seven groups of 100 objects, 700. The 8 stands for eight groups of 1,000 objects, 8,000. The 9 stands for nine groups of 10,000 objects, or 90,000. When this number is read, the values are given from left to right. The number, in English, is ninety-eight thousand seven hundred sixty-five. The place values of the digits 98,765 are shown in Figure 3-5.

Decimal, or fractional, values are expressed to the right of a decimal point. The digits to the right of a decimal also have different place values. Consider the number 98,765.4321 as an illustration. The first number to the right of the decimal, 4, represents four-tenths of a whole unit. The second number, 3,

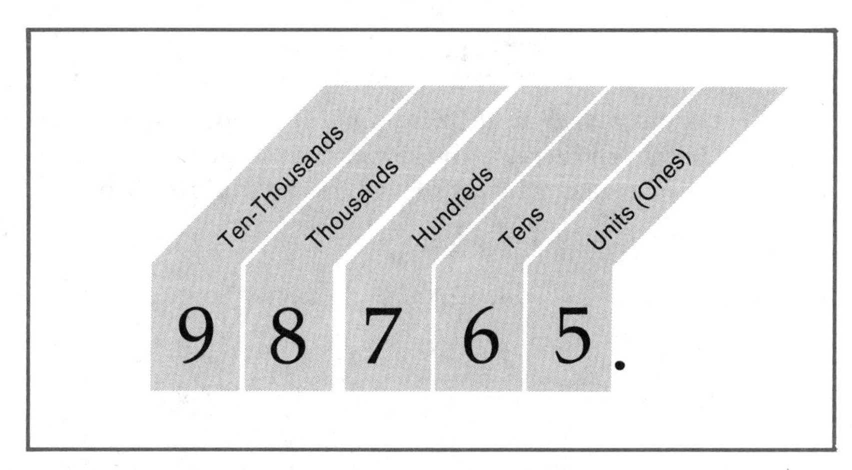

Figure 3-5. Each digit in a number has a different value, depending upon its location.

represents three-hundredths of a unit. The third number, 2, represents two-thousandths of a unit. And the fourth number, 1, represents one ten-thousandth of a unit. Therefore, 98,765.4321 means 90,000 plus 8,000 plus 700 plus 60 plus 5 plus the fractions $\frac{4}{10}$, $\frac{3}{100}$, $\frac{2}{1,000}$, and $\frac{1}{10,000}$. The decimal value is read as a fraction according to the total number of places shown. In this example, the decimal 0.4321 is read as forty-three hundred twenty-one ten-thousandths. Note that when there is no whole number, a zero is placed to the left of the decimal, in the unit position. These positions and values are shown in Figure 3-6.

When you read decimals and whole numbers, "and" indicates the position of the decimal point. Thus, three and two-tenths can be shown as 3.2.

Number values rise to the left of a decimal point and decrease to the right. To illustrate the declining values to the right of the

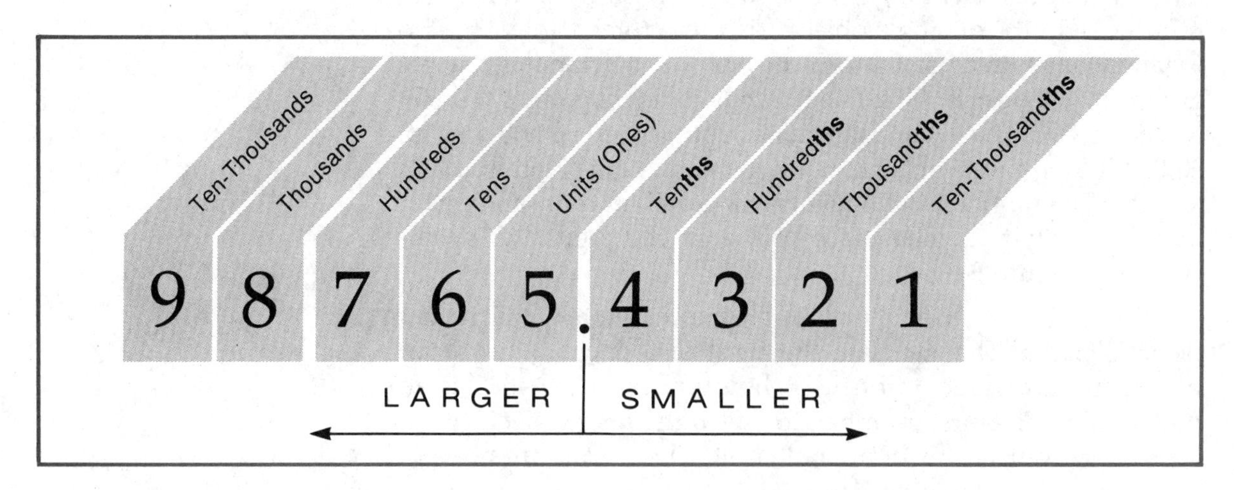

Figure 3-6. Values of digits rise as they move left from a decimal point. Values get smaller as they move farther to the right of a decimal point.

decimal, consider the individual values of the fractional numbers. This comparison can be shown by converting the decimal values to fractions:

$$0.100 = \frac{100}{1000}$$
$$0.003 = \frac{3}{1000}$$

Standing alone, the numbers as fractions are $\frac{100}{1000}$ and $\frac{3}{1000}$. So, the 3 at the right of .123 has a far smaller value than the 1.

EXERCISE 3-6

A. Write each decimal in English.

1. 5.25
2. 17.34
3. 1.89
4. 0.321
5. 3.876
6. 6.37
7. 0.547
8. 56.56
9. 10.9
10. 0.107
11. 23.999
12. 25.1
13. 0.001
14. 1.003
15. 13.13
16. 12.321
17. 15.9
18. 54.01

B. Write each decimal in numerical form.

1. five-hundredths
2. six and seven-tenths
3. two-hundred fifty-five thousandths
4. four and nine-tenths
5. twenty-seven hundredths
6. sixteen-thousandths
7. one and thirty-four hundredths
8. seventy-five thousandths
9. six hundred and four-thousandths

Decimals are used in figuring averages. An example is the batting average in baseball.
COURTESY OF CALIFORNIA ANGELS

Converting Fractions to Decimals

Fractions can be converted to decimals through division. Using simple long division, divide the denominator into the numerator. Place the decimal point of your quotient directly above the decimal point in the dividend. For proper fractions, this division produces a quotient smaller than a whole number. A whole-number quotient is impossible because the divisor is larger than the dividend. Also, remember that your purpose is to convert from one form of fraction to another. Therefore, the quotient of your division will have a value smaller than 1.

The decimal that results from this division is called a *fraction equivalent*. An equivalent is an equal value. An equivalent value can be interchanged with the original that it replaces. This is true for all mathematical problems you solve. Equal values can be exchanged at any time. Thus, in working problems, fractions and decimals, as equivalents, can be exchanged freely.

Decimal equivalents are used regularly in sports. A popular example is the baseball batting average. To illustrate, suppose a player comes to bat five times in a game and gets two hits.

In word problems involving fractions, your whole, or total, is used as the denominator. Thus, the total at bats, 5, is the denominator in this situation. The number of hits, 2, is part of this whole and is used as the numerator. The fraction, then, is $\frac{2}{5}$. To derive a decimal equivalent, divide 5 into 2. As shown below, the result is 0.4. (In baseball, a player's average is carried out to three decimal places. Thus, a 2-for-5 performance is written .400.)

$$\frac{2}{5} = \begin{array}{r} 0.4 \\ 5\overline{)2.0} \\ \underline{0\ \text{x}} \\ 2\ 0 \\ \underline{2\ 0} \end{array} = 0.4$$

Other examples of fractions converted into decimals are shown below.

$$\frac{3}{4} = \begin{array}{r} 0.75 \\ 4\overline{)3.00} \\ \underline{2\ 8\text{x}} \\ 20 \\ \underline{20} \end{array} = 0.75 \qquad \frac{5}{8} = \begin{array}{r} 0.625 \\ 8\overline{)5.000} \\ \underline{4\ 8\text{xx}} \\ 20 \\ \underline{16} \\ 40 \\ \underline{40} \end{array} = 0.625 \qquad \frac{3}{5} = \begin{array}{r} 0.6 \\ 5\overline{)3.0} \\ \underline{3\ 0} \end{array} = 0.6$$

Fractions to decimals in payrolls. In working with payrolls, fractions are converted to decimal equivalents. This is done because money amounts are written in decimals. For example, if you earn four and one-half dollars an hour, that figure is written $4.50. To simplify payroll calculations, portions of hours worked also are expressed as decimals. If you worked thirty-six and one-half ($36\frac{1}{2}$) hours, your time would be expressed as 36.5 hours. In this way, your pay rate ($4.50) can be multiplied by your hours (36.5) to compute your gross pay ($164.25). Check this multiplication and then figure out what your gross pay would be if you worked 40 hours.

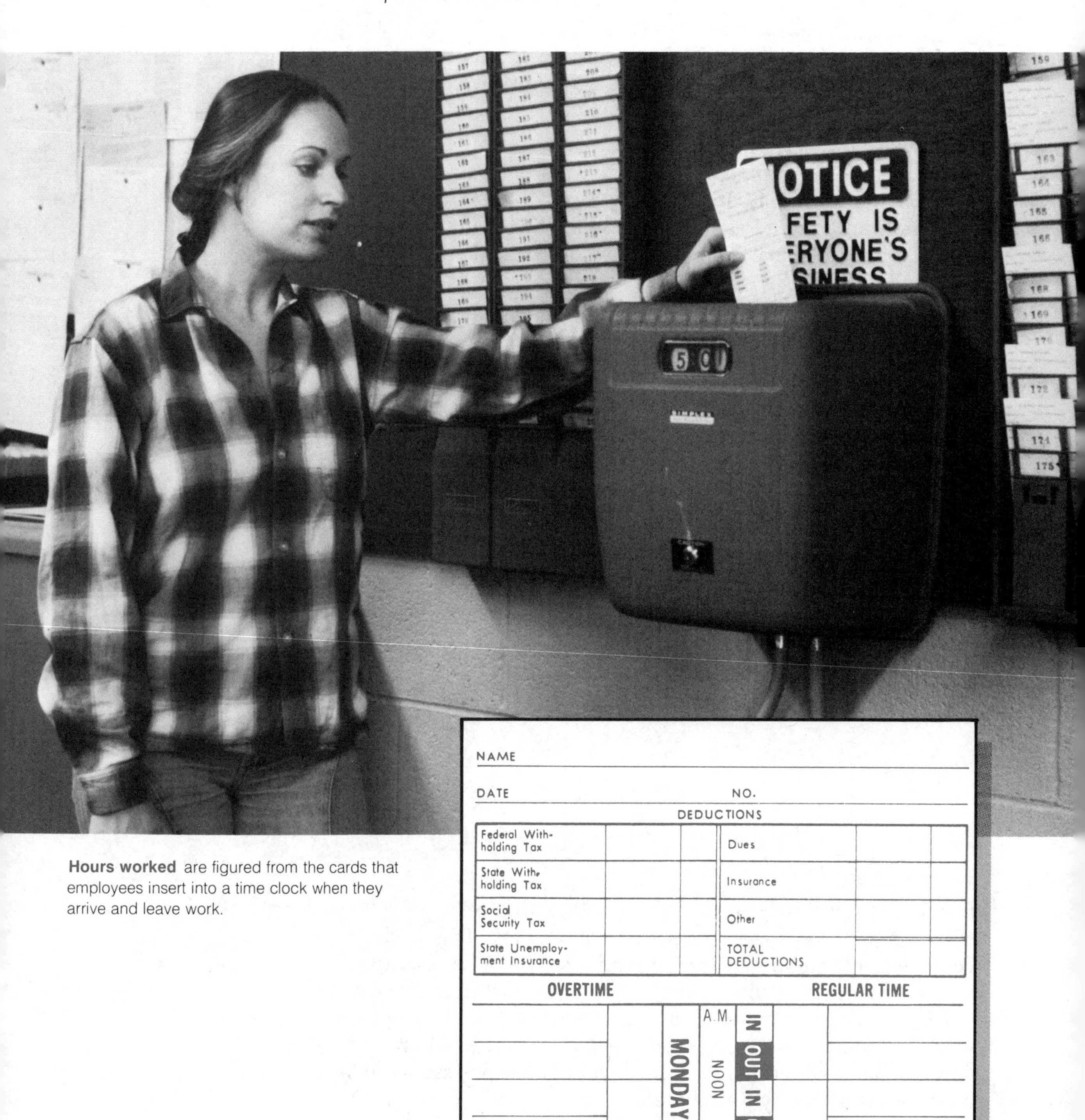

NAME

DATE NO.

DEDUCTIONS

Federal Withholding Tax			Dues		
State Withholding Tax			Insurance		
Social Security Tax			Other		
State Unemployment Insurance			TOTAL DEDUCTIONS		

OVERTIME REGULAR TIME

MONDAY A.M. IN OUT NOON IN OUT P.M.

TUESDAY A.M. IN OUT NOON IN

Hours worked are figured from the cards that employees insert into a time clock when they arrive and leave work.

EXERCISE 3-7

Convert each fraction to a decimal and compute gross pay in each of the following situations.

1. Maryann works six and one-quarter hours on Saturday at $5 an hour. What are her earnings on Saturday?

2. Your company operates on a work week of $37\frac{1}{2}$ hours. Your hourly pay rate is $6.50. What is your gross pay for the week?

3. Jose works an average of $6\frac{3}{4}$ hours a day for a five-day work week at $5.80 an hour. What is Jose's weekly gross pay?

4. Janine gets paid $3.50 for each $3\frac{1}{2}$ pounds of peaches she picks. The first day on the job, Janine picks 35 pounds of peaches. How much has she earned?

5. Fujio is a truck driver. He earns one and one-half dollars for each mile driven. On Fridays, Fujio drives 60 miles. What is his gross pay for Fridays?

6. Martha is a writer. She earns two and one-half cents for each word written. Each day, she writes a one-thousand word article for the local newspaper. How much does she earn for each article?

7. Mrs. Feldstein owns a dump. She charges people two and one-half dollars per truckload to dump trash. Today, three hundred fifty trucks went to Mrs. Feldstein's dump. How much did she earn today?

8. Kathy baby-sits for her neighbor's little girl from 7:00 to 10:30 two nights a week while her neighbor is at night school. Kathy gets $3.50 an hour. How much does Kathy earn every week?

SUMMING UP

- Going to work involves an exchange of values. An employee receives wages as payment for labor or services performed.
- Wages usually are paid on an hourly basis. Piecework wages are paid on the basis of an item produced or a unit of work performed. Some workers are paid a salary, which is a fixed rate of pay for labor or services.
- Gross pay is earned income, which may be expressed by multiplying an hourly wage rate by the number of hours worked.
- Net pay is the money a worker receives after deductions are taken from gross pay. Net pay is a fraction, or part, of gross pay.
- A fraction is a part of a whole. In mathematics, a fraction usually is part of the smallest whole number, 1.
- The denominator of a fraction is the number of parts into which a whole is divided.
- The numerator of a fraction is the number of parts with which you are dealing.
- A fraction in which the numerator is smaller than the denominator is called a proper fraction. An improper fraction has a numerator larger than the denominator.
- When a whole number is multiplied by a fraction, the result is smaller than the original whole number.
- Decimals are fractions in which parts of the whole have been divided into multiples of 10.
- A decimal point is used to separate whole numbers from fractional parts. Whole numbers are expressed to the left of a decimal point, increasing in value from right to left. Fractional, or decimal, values are expressed to the right of a decimal point, decreasing in value from left to right.
- Fractions are converted to decimals through division. The quotient, called a fractional equivalent, will have a value less than 1.

TESTING YOUR WORKING KNOWLEDGE

Read each question carefully. Then solve the problem.

1. Linda earns $6 an hour at the amusement park. Her wages are calculated to the quarter-hour. During one week, she works the following hours: Monday, 8; Tuesday, $6\frac{1}{2}$; Wednesday, $7\frac{1}{4}$; Thursday, $7\frac{1}{2}$; Friday, 8.

a. How many hours did Linda work, expressed in decimal form?

b. What was her gross pay for the week?

c. How many hours did Linda work during the week, expressed in fractional form?

2. Convert each number as directed.

a. To numerical form: seven-twentieths; five-ninths; three-fourths; one-quarter; twenty-seven thirty-seconds.

b. To English form:

$\frac{2}{9}$ $\frac{21}{25}$ $\frac{1}{18}$ $\frac{79}{100}$ $\frac{7}{16}$ $\frac{9}{6}$ $\frac{11}{12}$ $\frac{200}{245}$

c. To decimals:

$\frac{4}{16}$ $\frac{4}{5}$ $\frac{2}{4}$ $\frac{10}{80}$ $\frac{12}{24}$ $\frac{3}{5}$ $\frac{3}{4}$ $\frac{6}{8}$ $\frac{12}{16}$ $\frac{8}{32}$

3. At the end of a regular work week, Tony's employer deducts $52 from his gross pay for taxes and insurance. Tony earns $6.50 an hour for a 40-hour work week. Based upon this information:

a. What was Tony's gross pay?

b. What was his net pay?

c. What fraction of his gross pay was deducted?

d. Express that fraction as a decimal.

e. What fraction of Tony's gross pay is represented by his net pay?

f. Express that fraction as a decimal.

4. Amelia, a programmer, has worked for the Tomato Computer Company for four weeks. Her paycheck the first week was $500. Her paycheck the second week was $550. Her paycheck

TERMS THAT COUNT

wage
piecework
salary
commission
tip
deduction
gross pay
earned income
net pay
net income
fraction
denominator
numerator
ordinal number
proper fraction
improper fraction
remainder
multiplicand
multiplier
product
quotient
curve
equivalent
decimal
decimal point
fraction equivalent
decimal equivalent

the third week was $400. Her paycheck the fourth week was $550.

a. How much did Amelia earn in four weeks?

b. What fraction of her four-week gross pay did Amelia earn in the first week? Express the answer in decimal form.

c. What fraction of her four-week gross pay did Amelia earn in the third week? Express the answer both as a proper fraction and in decimal form.

5. Benjamin washes windows of large businesses for Cecilia's Window Service. He gets paid $3 per window and $0.25 for every mile he drives. Today, Benjamin washed 20 windows and drove 240 miles.

a. What was Benjamin's gross pay plus mileage?

b. What fraction of his total gross pay plus mileage resulted from washing windows?

c. Express that fraction as a decimal.

d. What fraction of his gross pay resulted from driving?

e. Express that fraction in English terms and decimal form.

f. Did Benjamin receive more money for mileage driven or for washing windows?

PRESENTING INFORMATION

Read each question carefully and solve the problem as directed.

1. Jake and Thelma just got back from their first vacation to the tropics. They spent a total of \$3,000. Plane fare was \$1,500. They also spent \$500 on hotels and meals. Thelma spent \$600 on gifts for her family and friends, while Jake spent \$400 on fishing trips. Draw a pie chart that shows how Jake and Thelma's vacation expenses relate to one another.

2. Billy earns \$1,500 a month as an airplane mechanic. He spends one-third of his monthly salary on rent. He spends one-fifth of his monthly salary on food. He spends one-fifteenth of his monthly salary on gasoline for his truck. Billy puts the rest in his savings account.
   - **a.** How much money does Billy spend on rent each month?
   - **b.** How much money does Billy put away in savings each month?
   - **c.** What fraction of his gross pay does Billy put away in savings? Express your answer in decimal form.
   - **d.** Draw a pie chart that shows how Billy spends his money.

SKILL DRILLS

Perform the following multiplications of whole numbers by fractions. Express each answer as a proper fraction and in its simplest form.

1. $6 \times \frac{3}{8}$

2. $9 \times \frac{1}{3}$

3. $8 \times \frac{2}{33}$

4. $24 \times \frac{5}{6}$

5. $35 \times \frac{2}{13}$

6. $52 \times \frac{2}{15}$

7. $21 \times \frac{7}{18}$

8. $274 \times \frac{2}{9}$

9. $12 \times \frac{4}{54}$

10. $81 \times \frac{9}{4}$

11. $4 \times \frac{3}{4}$

12. $2 \times \frac{15}{6}$

13. $15 \times \frac{1}{3}$

14. $2 \times \frac{15}{16}$

15. $124 \times \frac{6}{13}$

16. $8 \times \frac{50}{75}$

17. $5 \times \frac{119}{123}$

18. $6 \times \frac{71}{29}$

4

Your Paycheck

YOUR LEARNING JOB

When you have completed the exercises and assignments for this unit, you should be able to:

- ☐ **Convert decimals to percentages.**
- ☐ **Use the earned-hours method of computing gross pay involving overtime.**
- ☐ **State a problem (with an algorithm) for determining gross pay that involves commissions or shift differentials.**
- ☐ **Solve the problem for determining gross pay involving commissions or shift differentials.**

WARM-UP

A. Find the sum in each problem.

1. 23.54 + 45.09 + 23.11 + 109.73 =
2. 114.087 + 43.807 + 1,398.99 + 386.834 =
3. 574.98 + 3,876.02 + 9,230.78 + 983.836 =
4. 436.87 + 345.817 + 3,109.75 + 9,769.43 + 5,748.33 =
5. 34,736.019 + 45,639.57 + 23,567.39 + 72,397.88 =

B. Find the difference in each problem.

1. 3,234.23 − 2,879.87 =
2. 6,810.31 − 3,726.42 =
3. 34,783.05 − 17,994.01 =
4. 73,385.019 − 29,287.58 =
5. 4,271.352 − 3,436.467 =

C. Find the product in each problem. Carry the decimals to five places.

1. 287.29 × 2,879.87 =
2. 465.72 × 26.19 =
3. 490.26 × 14.3 =
4. 35.678 × 7.655 =
5. 71.362 × 0.465 =

D. Find the quotient in each problem. Carry the decimals to five places.

1. 75.168 ÷ 8.7 =
2. 810.85 ÷ 6.42 =
3. 3,783.546 ÷ 17.1 =
4. 3,718.26 ÷ 27.3 =
5. 2,715.12 ÷ 6.48 =

FIGURING YOUR GROSS PAY

If you become a wage earner, your pay probably will be based on an hourly rate. It is easy to calculate *gross pay,* or total pay before *deductions,* if you work a 40-hour week. Say you are paid $4 an hour. Your gross pay for a 40-hour week would be calculated as shown below:

$$40 \times 4 = 160$$

Now, say that you earn time-and-a-half for any hours over 40 in a given week. The time you work over 40 hours in one week is called *overtime.* Thus, your overtime wage rate is $1\frac{1}{2}$ times your regular wage rate. If your hourly wage is $4, time-and-a-half would be $4 multiplied by 1.5, or $6 an hour. So, one way to figure your gross pay would be to multiply 40 hours by 4 to calculate your pay for the first 40 hours worked. Then, find gross overtime pay by multiplying overtime hours worked by the time-and-a-half pay rate. Then, add the two gross pay figures to obtain gross pay for the week.

This method of calculation is a natural one for an individual, especially if you are performing the arithmetic using pencil and paper. However, if you use a calculator to perform these steps, you will notice a problem. You are working with different pay rates. This forces you to use the device's memory to complete your calculations. This is because you have two sets of hours to multiply by different rates. Thus, the result of the first multiplication must be saved and added to the result of the second multiplication. This saving of totals is done in memory. An algorithm for this computation method, along with actual computation, is shown in Figure 4-1.

Earned Hours: Measuring the Value of Your Work

There is a more efficient method of computing gross pay. Instead of calculating with two different pay rates, you can use the *earned-hours* method. Earned hours describes the total value of the hours you work. The concept is simple: Instead of using two pay rates, you add to the number of hours worked to allow for overtime pay. That is, you increase the number of paid hours to reflect the amount of overtime earned. This increased number of hours is the earned-hours figure.

	Descriptive	Numerical
1.	Multiply regular hours by regular hourly wage.	40 × $4
2.	Save the result, regular gross pay.	$160
3.	Multiply regular hourly rate by 1.5.	$4 × 1.5
4.	Save the result, overtime hourly wage.	$6
5.	Subtract regular hours from total hours.	52 − 40
6.	Save the difference, overtime hours.	12
7.	Multiply overtime hours by overtime hourly wage.	12 × $6
8.	Save the result, overtime gross pay.	$72
9.	Add regular gross pay and overtime gross pay for total gross pay.	$160 + $72 = $232

Figure 4-1. Computing regular and overtime pay separately is an inefficient method of determining gross pay.

	Descriptive	Numerical
1.	Subtract regular hours from total hours.	52 − 40
2.	Save the result, overtime hours.	12
3.	Multiply overtime hours by 1.5 to get overtime earned hours.	12 × 1.5 = 18
4.	Add regular hours and overtime earned hours to get total earned hours.	40 + 18 = 58
5.	Multiply total earned hours by regular hourly wage for total gross pay.	58 × $4 = $232

Figure 4-2. The earned-hours method of computing gross pay is a logical way to combine regular and overtime hours in a payroll.

The earned-hours method is the best way to compute total gross pay. The algorithm in Figure 4-2 shows the advantages of the earned-hours method over calculating with overtime pay rates.

In Figure 4-1, the regular hours (40) are multiplied by the regular hourly wage ($4), giving $160. Then, the hourly rate is mutiplied by 1.5 to determine the overtime hourly wage. Next, the regular hours are subtracted from total hours to determine the number of overtime hours. That number (12) is then multiplied by the overtime wage rate ($6) to produce overtime gross pay. Finally, overtime gross pay is added to regular gross pay to determine total gross pay.

Figure 4-2 illustrates the earned-hours method of computing total gross pay. First, regular hours are subtracted from total hours to determine the number of overtime hours. Next, overtime hours are multiplied by 1.5 to give them the value of time-and-a-half. This figure is added to regular hours to determine total earned hours. Finally, the earned-hours total is multiplied by the regular wage rate to determine total gross pay.

Compare the two algorithms. Notice that the second algorithm, using earned hours, is shorter and simpler than the first algorithm. There is another important difference. The sequence of calculations in the first algorithm cannot be performed on most hand-held calculators. Two separate memory banks would be needed to record the different values being computed.

Overtime pay usually is calculated by the earned-hours method.

If the earned-hours method simplifies calculations for a single paycheck, think how necessary it is for a large payroll.

EXERCISE 4-1

For each problem, write an algorithm and then solve the problem.

1. Figure your gross pay for a 45-hour week if wages for the first 40 hours are $4.50 an hour. Your overtime pay is figured at time-and-a-half.

2. Don earns $6 an hour. He is asked to work during a weekend emergency, for which his employer pays double time. He works six hours on Saturday. If he also worked his usual 40 hours during the week, what was his gross pay for the week?

3. Barbara worked 46 hours last week. Her pay rate was $4.35 per hour for the first 40 hours. She receives time-and-a-half for hours worked over 40. What was Barbara's gross income last week?

4. Juan earns $8.50 per hour working for a company that operates on a 35-hour work week. His crew is called in to work eight hours on a holiday. Juan and his fellow workers are paid double time for the holiday work. What is Juan's gross pay for the 35-hour normal work week plus the eight hours of overtime?

5. Your normal work week is 40 hours, at a rate of $9.50 an hour. However, a parts shortage has delayed production in your plant, so your crew works overtime to meet delivery deadlines. You work 10 hours a day, Monday through Friday, and another 10 hours on Saturday. Your employer pays time-and-a-half for all hours over 40. What is your gross pay for the week?

USING FACTORS

You have seen how an employer would calculate your gross pay. Now, take another look at the earned-hours method and why it works.

The key to the solution is the *factor* used in the calculations. A factor is a number or a value that is used to multiply other numbers. For example, if your employer pays you time-and-a-half for overtime, the factor in your calculations is 1.5. If this factor remains the same, it doesn't matter which value—pay rate or hours—is multiplied first. The solution will be the same.

To illustrate, suppose you wanted to figure your overtime gross pay in a given situation. You might be faced with a choice of working eight hours overtime on a Saturday or taking a weekend trip. Your regular gross pay is not a factor in your decision. Your normal hourly wage is $4, and overtime on Saturday pays time-and-a-half. There are three different methods you can use to determine overtime gross pay. The important factor in each solution is 1.5.

In the first example, use the factor of 1.5 to calculate the overtime hourly wage rate:

$$4 \times 1.5 = 6$$

Then, multiply $6 by the number of overtime hours:

$$6 \times 8 = 48$$

In the second example, use the earned-hours method. Multiply overtime hours by 1.5 to determine earned hours:

$$8 \times 1.5 = 12$$

Then, multiply your regular pay rate by earned hours:

$$4 \times 12 = 48$$

Finally, use the 1.5 factor a third way. Multiply your regular pay rate by the number of overtime hours:

$$4 \times 8 = 32$$

Now, multiply that figure by the overtime factor:

$$32 \times 1.5 = 48$$

EXERCISE 4-2

For each problem, write an algorithm and then solve the problem.

1. Nancy works 2.5 hours of overtime on Monday and 1.5 hours of overtime on Thursday. Her regular wage is $7.50 per hour for a 40-hour week. If she is paid 1.5 times her normal wage for overtime, what is Nancy's gross pay for the week?

2. You receive a regular day's pay for a mid-week holiday. To maintain production schedules, the management of your company schedules a special shift on Saturday. You work 6.5 hours of overtime that day. Overtime pay is 1.5 times your normal hourly rate of $7.00. Your normal work week is 40 hours. Compute your gross pay for the week.

3. Albert, an assistant art director, works a 37.5-hour week at Pete's Periodicals. To meet a magazine deadline, Albert and his crew work overtime one week. They work 2.5 hours overtime on Wednesday, and 2.25 hours overtime each on Thursday and Friday. Pete's Periodicals pays time-and-a-half, or 1.5 times regular pay, for overtime. Albert earns $12.50 an hour for regular time. What is his gross pay for the week?

4. The city crew works overtime for two days, installing sidewalks, curbs, and gutters. Long workdays are needed when concrete is poured at the job site. Mary, a concrete finisher, earns $9.50 an hour for the first eight hours each day in a five-day work week. She earns 1.5 times that amount for overtime. What is Mary's gross pay that week if she works 12.75 hours Thursday and 13.25 hours on Friday?

5. Herman, a clerk-typist for the county probation department, works late one evening to prepare reports for the court. His regular salary for a 40-hour week is $7.80 an hour. He works an extra 3.5 hours, at time-and-a-half pay, typing the reports. What is his gross pay for the week?

PERCENTAGES

A *percentage* is a decimal value expressed as a part of 100. A denominator of 100 is used for all percentage conversions from fractions to decimals. The numerator, then, represents the units of 100.

The denominator of 100 was chosen because this value proved most convenient. The goal was a method of fraction-to-decimal conversion that would be easy to use and have meaning. The choices were 10, 100, or 1,000.

If an object is divided into 10 parts, there are not enough parts to split up many ways. Dividing by 1,000 gives too many parts to manage easily. A denominator of 100 seems just right. Finally, the United States system of money is based on 100 cents to the dollar.

Conversions: Decimals to Percentages

A decimal, such as 0.25, which indicates part of a whole, can be changed easily to a percentage. This is done by moving the decimal point two places to the right (multiplying by 100) and adding a symbol. The symbol is the percent sign, %, which stands for a denominator of 100. The slash in the percent sign represents the number one. The zeros represent the two zeros in the number, 100. So, the percent symbol, %, means that the preceding number has been multiplied by 100. This eliminates the denominator of 100. The decimal above, 0.25, becomes 25%, or 25 percent. Other examples of converting decimals to percentages are shown below:

$$0.1 = \frac{1}{10} = 10\%$$

$$0.01 = \frac{1}{100} = 1\%$$

$$0.001 = \frac{1}{1000} = 0.1\%$$

$$0.39 = \frac{39}{100} = 39\%$$

$$0.572 = \frac{572}{1000} = 57.2\%$$

Conversions: Fractions to Decimals and Percentages

Percentages and decimals are converted from fractions in the same basic way. To develop a percentage value, simply divide to create a decimal of at least two places. Then move the decimal point two places to the right. Finally, add a percent sign. This conversion process is illustrated below:

$$\frac{4}{5} = 5\overline{)4.00} = 0.80 = 80\%$$

$$\begin{array}{r} 0.80 \\ 5\overline{)4.00} \\ \underline{4\ 0x} \\ 00 \\ 00 \end{array}$$

$$\frac{3}{4} = 4\overline{)3.00} = 0.75 = 75\%$$

$$\begin{array}{r} 0.75 \\ 4\overline{)3.00} \\ \underline{2\ 8x} \\ 20 \\ \underline{20} \end{array}$$

Conversions: Percentages to Decimals

Percentages can be converted to decimals by moving the decimal point two places left and dropping the percent sign. Moving the decimal point two places to the left is dividing by 100 (or multiplying by $\frac{1}{100}$). Examples of converting percentages to decimals are shown below:

$$64\% = \frac{64}{100} = 100\overline{)64.00} = 0.64$$

$$\begin{array}{r} 0.64 \\ 100\overline{)64.00} \\ \underline{60\ 0x} \\ 4\ 00 \\ \underline{4\ 00} \end{array}$$

$$26\% = \frac{26}{100} = 100\overline{)26.00} = 0.26$$

$$\begin{array}{r} 0.26 \\ 100\overline{)26.00} \\ \underline{20\ 0x} \\ 6\ 00 \\ \underline{6\ 00} \end{array}$$

Percentages and Money Values

In dealing with percentages, think about money. A dollar has 100 parts, called pennies. A penny is 1% of a dollar. Much of the math you use after you leave school will involve money. Your paycheck is probably the most important example.

Think about *monetary* (money) values. How much of a dollar is a quarter? A quarter of something is $\frac{1}{4}$. That fraction also can be written as $\frac{25}{100}$, or 0.25, or 25%. Therefore, a quarter is 25% of a dollar, or 25 cents. This is the same as saying that there are four quarters in a dollar. Multiply 25 cents by 4 and the product is 100 cents, or $1.00.

In dealing with money, it often is necessary to determine the percentage of an amount that includes a decimal. For example, you might have to calculate 10% of $8.40. Ten percent is really 0.1 or $\frac{1}{10}$. So, the problem can be expressed as finding one-tenth of $8.40. The solution is to multiply 8.40 by 0.1 to obtain 0.840, or 84 cents. This process is illustrated below:

$$\$8.40 \times 10\% = 8.40 \times \frac{1}{10} = \begin{array}{r} 8.40 \\ \underline{\times .10} \\ .8400 \end{array} = 0.84 = \$0.84$$

EXERCISE 4-3

A. Convert each fraction to a decimal equivalent.

1. $\frac{2}{4}$
2. $\frac{9}{10}$
3. $\frac{1}{5}$
4. $\frac{9}{100}$
5. $\frac{1}{4}$
6. $\frac{5}{8}$
7. $\frac{21}{30}$
8. $\frac{3}{5}$
9. $\frac{47}{50}$

B. Convert each percentage to a decimal equivalent.

1. 87%
2. $5\frac{3}{4}$%
3. $88\frac{3}{5}$%
4. $2\frac{1}{2}$%
5. $1\frac{3}{4}$%
6. $93\frac{1}{2}$%
7. 1%
8. 36%
9. $20\frac{4}{5}$%

C. Compute each percentage of monetary value.

1. 15% of $89
2. 6% of $2
3. 4.5% of $26
4. 12.5% of $550
5. 6% of $21,111
6. 5% of $12.80
7. 6.5% of $84
8. 8% of $112.50

PERCENTAGES AND PAYCHECKS

Percentage calculations are used to determine the gross pay of many workers. Examples of payroll factors in which percentages are used include:

- Commissions
- Bonuses
- Shift differentials
- Cost-of-living raises.

Commissions

A *commission* is a fee paid to an employee or agent for completing a business transaction or performing a service. The most common use of commissions is for determining the gross pay of salespersons. A commission usually is a percentage of the value of a sale. For example, a salesperson in a clothing store might receive a commission of 10%. This means that the salesperson is paid 10% of the retail price of each item he or she sells.

A commission really represents a percentage of the income of a business. Traditionally, commissions have benefitted both salespersons and their employers. As a salesperson's sales increase, so does that individual's income. Increased sales also benefit the business.

Some sales jobs pay a modest salary plus a small commission on sales. Other salespersons work without a salary. This is called working on *straight commission.* On a straight commission basis, your gross pay would equal the value of your sales multiplied by your commission percentage. Suppose your commission was 10%. If your sales totaled $5,000 in a week, your gross income would be $500. This simple calculation is shown below:

$$5{,}000 \times 10\% = 5{,}000 \times 0.10 = 500$$

If you worked on a salary-plus-commission basis, your gross pay calculation would involve an added step. Your salary would be added to the total of your commissions for the pay period.

Pay for some salespersons is based upon a complicated formula. However, percentages are the basis for calculating most commissions.

To find the gross pay when just commission and rate are known, divide the commission by the percentage rate. Thus, $500 divided by 10% equals $5,000.

EXERCISE 4-4

For each problem, write an algorithm and then solve the problem.

1. Susie earns a salary of $150 a week at the Buy More Department Store. She also is paid a commission of 5% on her sales. During a recent week, her sales totals were: Monday, $750; Tuesday, $1,275; Wednesday, $890; Thursday, $1,435; and Friday, $1,180.

   a. What was Susie's total commission income for the week?

   b. What was Susie's gross income for the week?

2. Richard is a salesperson at Honest Hank's Used Cars. He works on a straight commission basis, earning 10% of the price of each car he sells. Richard earns $250 in commissions one week. What were his gross sales?

3. Job openings are available at Olaf's Home Center and at the Fix-Up Emporium near your home. Olaf's pays its salespersons a salary of $5 an hour for a 40-hour week, plus 5% commission on all sales. The Fix-Up Emporium pays its salespersons straight commission of 15% on all sales. One of your friends is a salesperson at Olaf's. Another friend works at the Emporium. Both say that their gross sales average about $2,000 a week.

   a. What is the average weekly gross pay of your friend at Olaf's?

   b. What is the average weekly gross pay of your friend at the Emporium?

   c. Based on your pay projections, where would you prefer to work? Explain.

Bonuses

A *bonus* is money or something else of value given an employee in addition to usual pay. Many companies give bonuses to employees at the end of the year as a way of sharing profits. The higher the company's profit, the greater the bonuses. Some employers also give bonuses to reward job performance.

Bonuses may be based on a complicated formula. The formula may include several factors: salary level, length of service, and type of work performed are typical. Whatever the formula, however, bonuses will reflect a percentage of the company's income, or profit.

For example, suppose the Citywide Computer Service Company recorded year-end profits 10% higher than those for the previous year. The company's management decides to distribute bonuses to key employees. The company's sales representatives, Diane and Jeff, are awarded bonuses based upon their increased sales totals. Diane and Jeff both work on a basis of salary plus commissions.

The bonus formula includes two weeks' extra salary plus an additional 1% commission on gross sales for the year. Diane has been with the company a year longer than Jeff, so her salary is higher. Diane receives a salary of $1,000 a month. Jeff's salary is $900 monthly. However, Jeff signed two major new accounts during the year, so he received more commission income than Diane. Her gross sales were $65,000; Jeff's were $78,000. Each earns 15% commission on gross sales.

Diane's bonus was the sum of two weeks' salary and 1% commission on gross sales. At $1,000 a month, her annual salary is $12,000. Her weekly salary is $\frac{1}{52}$ of $12,000, or $230.77. (Note that the answer to three decimal places is 230.769. However, in working with money, amounts of 0.005 or over become an additional cent, or 0.01.) Two weeks' salary is $230.77 multiplied by 2, or $461.54. The commission portion of her bonus would be $65,000 multiplied by 1%, or 0.01, which equals $650. Adding $461.54 and $650 gives a total bonus of $1,111.54.

Jeff's bonus is computed in the same manner. His $900 monthly salary is multiplied by 12 to obtain an annual salary of $10,800. This figure is divided by 52 to obtain a weekly salary of $207.69. (Note that multiplying by the fraction, $\frac{1}{52}$, is the same as dividing by the whole number, 52.) Multiplying $207.69 by 2 gives two weeks' salary, or $415.38. One percent of Jeff's gross sales of $78,000 is $780. Thus, Jeff's bonus equals $415.38 plus $780, or $1,195.38. These computations are shown in Figure 4-3.

Diane	Jeff
$1,000 × 12 = $12,000	$900 × 12 = $10,800
$12,000 ÷ 52 = $230.77	$10,800 ÷ 52 = $207.69
$230.77 × 2 = $461.54	$207.69 × 2 = $415.38
$65,000 × 0.01 = $650	$78,000 × 0.01 = $780
$461.54 + $650 = $1,111.54	$415.38 + $780 = $1,195.38

Figure 4-3. A formula for salespersons' bonuses may be based upon salary level and sales performance.

EXERCISE 4-5

For each problem, write an algorithm and then solve the problem.

1. Anita had a monthly salary of $1,000 and was paid 15% commission on gross sales of $65,000. Tony had a monthly salary of $900 and was paid 15% commission on gross sales of $78,000.
   a. What was Anita's gross income for the year?
   b. What was Anita's average monthly gross income?
   c. What was Tony's gross income for the year?
   d. What was Tony's average monthly gross income?

2. Nationwide Industries pays its employees an end-of-the-year bonus based upon salary and length of service. First-year employees receive 10% of their monthly gross earnings. For each year of service thereafter, 1.5% of monthly gross pay is added to the 10%. On that basis, compute the bonuses paid to the following employees:
   a. Miguel has worked for the company for 10 years and earns a monthly gross salary of $2,200.
   b. Sally earns $1,850 monthly. She has been employed at Nationwide for five years.
   c. Bruce Watenabe is the assistant plant manager. A 23-year employee, he earns $3,200 a month.

Some industries operate more than one shift. Workers usually earn higher pay on swing or graveyard shifts. COURTESY OF INLAND STEEL COMPANY

Shift Differentials

When you look for a job, you may have several factors to consider. One of your decisions may involve working days or nights. Many organizations operate more than a single eight-hour *shift.* A shift is a regular, scheduled period of work.

You may be asked to work nights. Most companies offer extra pay for employees who work a night shift. This extra pay usually is called a *shift differential.* Most employers base shift differentials on percentages of regular day-shift pay.

Many organizations operate 24 hours a day. Around-the-clock operations usually have three shifts: In addition to the day shift, there is a *swing shift* and a *graveyard shift.* The swing shift usually occupies the evening hours. The graveyard shift operates from late night until morning.

Shift differentials of 10% or more can make a significant difference in your income. Being able to work with percentages will help you evaluate such decisions.

EXERCISE 4-6

For each problem, write an algorithm and then solve the problem.

1. Joe earns $7.50 an hour on the assembly line at Peach Computer Works. Demand for the company's products is high, and a second shift is being added. Joe is offered a position on the new night shift at 10% additional pay. If he continues working a 40-hour week:
   a. Compute his weekly gross pay on the night shift.
   b. How much more money would Joe earn by moving to the night shift?

2. Maxine and Rachel are neighbors who both work as dispatchers for the city police department. Maxine has worked for the department for six years. Her base salary is $8.50 an hour. Rachel has been a dispatcher for three years. Her base salary is $7 an hour. However, Rachel works the graveyard shift and receives a 20% pay differential.
   a. What is Maxine's gross salary for a 40-hour week?
   b. What is Rachel's gross salary for a 40-hour week?
   c. Rachel takes a one-week vacation, and Maxine fills in for her on the graveyard shift. What is Maxine's gross salary for that week?

Cost-of-Living Increases

Many employees receive *cost-of-living* wage adjustments from year to year. Such adjustments recognize increases in living costs due to rising prices. The federal government publishes a *consumer price index (CPI)* each year. The consumer price index compares the prices of some 300 goods and services with the costs of those items in 1967. The index figures for each year are percentages of those 1967 prices. For example, the index hit 150% in 1974. This meant that average prices were 1.5 times as high as 1967 prices. An increase in the general level of prices is called *inflation*.

Labor unions frequently negotiate cost-of-living increases in their contracts with employers. When the consumer price index is rising slowly, a typical cost-of-living raise might be 2% or 3%. In times of rapid inflation, cost-of-living adjustments may be higher.

Cost-of-living increases are especially important in types of jobs where *merit increases* are uncommon. A merit increase is a pay raise based upon increasing job skills, improved job performance, or other competitive factors. Being assured that an employer provides cost-of-living increases may be an important factor in your job search.

EXERCISE 4-7

For each problem, write an algorithm and then solve the problem.

1. Your hourly wage at Jones & Smith Products is $7.20. Members of your union receive annual cost-of-living adjustments tied to the consumer price index. At the end of the current contract year, your employer grants a 3% cost-of-living wage increase.
   - **a.** What is your new hourly wage rate?
   - **b.** Based on 40 hours, what is your new weekly gross pay?
   - **c.** On a weekly basis, how much of a pay raise did you receive?

2. Larry earns a salary of $650 a week. His employer announces a cost-of-living adjustment of 4%.
   - **a.** How much more will Larry earn in a week?
   - **b.** What is Larry's new weekly salary?
   - **c.** What is Larry's new annual gross pay?

3. Ethel's employer has agreed to give all workers a 2.5% cost-of-living wage adjustment. Prior to the adjustment, her normal work week was 40 hours, at an hourly wage of $5.20.
   - **a.** What is Ethel's new hourly wage rate?
   - **b.** Based on normal working hours, how much will Ethel earn in the first year following the adjustment?

Decisions Involving Percentages

When you go to work, you may have to consider commissions, bonuses, and shift differentials in selecting a job. For example, consider the following situation:

You have applied for positions at two companies. Company A pays $5.50 an hour and time-and-a-half for overtime. Workers at Company A average about two hours of overtime a week. Company B pays $5.75 an hour, and you have the choice of working the swing shift, for which the company pays an 8% differential. You decide that you wouldn't mind working the swing shift at Company B, if you can make more money. The problem is to determine which job offers the higher gross income.

First, figure out how much you are likely to earn at Company A. Two hours of overtime at time-and-a-half are equal to three earned hours. Adding 3 to 40 gives 43 earned hours per week on the average. Then, multiply 43 by the base pay rate of $5.50 to obtain $236.50. These computations appear below:

$$2 \times 1.5 = 3$$
$$40 + 3 = 43$$
$$43 \times 5.50 = 236.50$$

Now, figure out your probable pay at Company B. First, multiply the base wage rate, $5.75, by the 8% shift differential factor. Then, add the product, 46 cents, to the base wage to obtain $6.21 per hour. Finally, multiply the night shift hourly rate, $6.21, by 40 hours to obtain $248.40 gross pay. These computations appear below:

$$5.75 \times 0.08 = 0.46$$
$$5.75 + 0.46 = 6.21$$
$$6.21 \times 40 = 248.40$$

Thus, in a typical week, you would earn $11.90 more at Company B. Your decision then becomes a matter of personal preference. Are you willing to work a swing shift for a pay differential of about $12 a week?

EXERCISE 4-8

Percentages may figure in the calculations you do to decide among job opportunities.

For each problem, write an algorithm and then solve the problem.

1. Karla earns a salary of \$160 a week at the garden shop, plus 10% commission on her sales. If her sales total \$1,700 in one week, what is her gross pay?

2. Jose earns \$6.50 an hour working for Smith & Jones. His friend, Ron, works for Brown & Green at \$6.25 an hour. However, Ron works the swing shift, which pays a 5% differential. Who has a higher gross pay if both individuals work a 40-hour week?

3. Your company just announced a 4% cost-of-living wage increase for the coming year. Your current hourly wage is \$5.35, and your normal work week is $37\frac{1}{2}$ hours. How much additional income will the increase bring you over a full year?

SUMMING UP

- In payroll preparation, fractions are converted to decimal equivalents. For example, a quarter of an hour would be converted to 0.25.
- Earned hours is an important concept in payroll computations involving overtime. Earned hours means the total value of hours worked.
- A factor is a number or value used to multiply or divide other numbers.
- A percentage is a decimal value expressed as a part of 100.
- A simple process is used to convert decimals to percentages: Multiply the decimal value by 100 and add the percent symbol, %.
- Percentages are necessary for several types of payroll computations. Included are computations used to figure commissions and shift differentials. Bonuses and cost-of-living adjustments also may be calculated on the basis of percentages.

TESTING YOUR WORKING KNOWLEDGE

Write an algorithm for each problem. Then solve the problem.

1. Martha, a salesperson at Tamara's Fashions, earns $4.50 an hour for a 40-hour week. She works six hours overtime during a two-day "moonlight sale," for which she receives time-and-a-half. What was her gross pay for the week in which the sale was held?

2. Akira works overtime on three occasions during a week. He works an extra half-hour on Monday. On Wednesday, his overtime amounts to three-quarters of an hour. Akira also works an extra quarter-hour on Thursday. He receives time-and-a-half pay for hours over 40 a week. Using decimal equivalents, compute his total earned hours.

3. Your base pay is $5.50 an hour. In recent negotiations, your employer and your union agreed to a 3% annual cost-of-living adjustment. This increase is to become effective on January 1 of each of the next three years. What will your hourly wage be at the end of that three years?

TERMS THAT COUNT

gross pay
deduction
overtime
earned hours
factor
percentage
monetary
commission
straight commission
bonus
shift
shift differential
swing shift
graveyard shift
cost-of-living
consumer price index (CPI)
inflation
merit increase

PRESENTING INFORMATION

For each question, draw a bar chart, pie chart, or line graph that accurately presents the correct answer.

1. Judith has started working for Acme Incorporated, a company manufacturing toys. Her starting salary was $2,000 a month. In addition, Acme pays its employees an end-of-the-year bonus based upon yearly salary and length of service. First-year employees receive 5% of their yearly gross earnings. For each year of service thereafter, 1.5% of yearly gross is added to the 5%.

   a. Draw a bar graph showing how Judith's yearly gross income increases over a five-year period.

   b. Show the same data on a line graph.

2. Marcus works as a salesperson for Carrot Computer Company. Marcus earns $6.75 an hour and also receives 10% commission on sales. In addition, Marcus gets time-and-a-half for any hours over 40 in a given week.

   a. In his first week on the job, Marcus worked 40 hours and sold $5,000 worth of computer equipment. Draw a pie chart showing what percentage of his gross pay came from hourly wages and what percentage was from commissions.

   b. In his second week, Marcus worked 50 hours but sold only $1,000 worth of computer equipment. Draw a pie chart showing what percentage of his gross pay was from hourly wages, what percentage was from overtime wages, and what percentage was from commissions.

SKILL DRILLS

A. Compute each percentage.

1. 7% of 70
2. 6% of 100
3. 8% of 24
4. 24% of 256
5. 50% of 12
6. 21% of 100
7. 25% of 60
8. 75% of 40

B. Convert each fraction to a percentage equivalent.

1. $\frac{3}{4}$
2. $\frac{9}{10}$
3. $\frac{7}{8}$
4. $\frac{15}{30}$
5. $\frac{125}{1,000}$
6. $\frac{5}{8}$
7. $\frac{3}{50}$
8. $\frac{3}{12}$

C. Convert each percentage to a decimal. Then, convert the decimal equivalent to a proper fraction.

1. 17%
2. 63%
3. 81%
4. 24%
5. 25%
6. 9%
7. 75%
8. 66.6%

P A R T

Mathematics and Your Life

IN THIS PART

Mathematics plays a big role in your everyday life. As you develop into adulthood, you will use mathematics increasingly in your daily routine. Everyday activities, such as shopping, cooking, credit purchases, and leisure, all involve mathematics. The valuable skills presented will add to the quality of your life.

Understanding how interest works, and how savings accounts grow through compounding interest, will help you in planning for your future security.

In this part, you also are introduced to life on the selling side of the retail counter. Things happen fast when you gain responsibility for operating a cash register. Mathematics skills enable you to handle this responsibility successfully.

Finally, you will see how mathematics is related to your health. Exercise and diet are two ways in which you can use mathematics to maintain a healthy body.

Mathematics In Your Everyday Life

YOUR LEARNING JOB

When you have completed the exercises and assignments for this unit, you should be able to:

- ☐ **Use ratios, proportions, and equations to make conversions.**
- ☐ **Solve problems by using proportions and equations.**
- ☐ **Write algorithms for finding the unknowns in equations.**
- ☐ **Solve an equation.**
- ☐ **Simplify numbers by rounding off.**
- ☐ **Use measurements to estimate the size, quantity, or value of objects in everyday life.**

WARM-UP

A. Find the sum in each problem.

1. 12.34 + 88.94 + 4.5 + 23.10 + 6.5 + 24.3 + 2.543 =
2. 23.327 + 32.43 + 39.563 + 435.43 + 66.342 + 678.44 =
3. 546.293 + 442.54 + 367.745 + 549.007 + 34.441 =
4. 74,653.634 + 2,653.653 + 56,266.039 + 12.029 + 9,324.226 =
5. 734.523 + 23.592 + 95,635.033 + 5,315.636 + 43,643.038 =

B. Find the difference in each problem.

1. 523.029 − 214.987 =
2. 425.965 − 425.876 =
3. 3,216.003 − 326.049 =
4. 22,765.009 − 22,012.324 =
5. 75,930.613 − 69,941.777 =

C. Compute each percentage.

1. 5% of 100
2. 2% of 10
3. 25% of 24
4. 2% of 12
5. 5% of 40

D. Find the product in each problem. Carry the decimals to four places.

1. $2 \times \frac{2}{23} =$
2. $5 \times \frac{2}{15} =$
3. $9 \times \frac{3}{27} =$
4. $7 \times \frac{5}{36} =$
5. $8 \times \frac{3}{31} =$

Ratios can be used to compare product volumes. For example, the ratio between the gallon milk container on the left and the quart container is 4:1.

MAKING COMPARISONS

Mathematical comparisons are part of everyday life. Think about going to a grocery store. You probably compare quantities and values before making purchases. You seek the best value for your money.

In the language of mathematics, comparisons often are called *ratios*. A ratio is a mathematical way of comparing the size or amount of two objects.

For example, one gallon of milk is equal to four quarts. In the language of mathematics, this English statement can be expressed as a ratio. Ratios can be written as fractions or by using a colon. Thus, the ratio of quarts to gallons can be expressed in English as 4 to 1, and mathematically as $\frac{4}{1}$ or 4:1.

Money can be compared in the same manner. A $5 bill is equal to five $1 bills. The ratio of five-dollar bills to one-dollar bills can be expressed as the ratio $\frac{5}{1}$ or 5:1.

Normally, the second number in a ratio is the denominator in the fraction that expresses the same relationship.

EXERCISE 5-1

Express the following comparisons as both ratios in English, and in mathematical forms as fractions *and* ratios.

1. $100 compared with $20.
2. Two gallons (eight quarts) of ice cream with one quart of ice cream.
3. A 440-yard run with a 100-yard dash.
4. A five-pound bag of potatoes with a two-pound bag.
5. A $20 bill with a $50 bill.

CONVERSIONS THROUGH PROPORTIONS AND EQUATIONS

Ratios, when expressed in the language of mathematics, include only numerical values. Ratios do not contain the names of objects. Therefore, ratios of unlike items can be the same in mathematics. Consider another ratio involving money. One $20 bill is equal to four $5 bills. Written as a fraction, this comparison would be $\frac{20}{5}$, or $\frac{4}{1}$. Thus, the ratio of a $20 bill to a $5 bill is 4:1.

Now, recall that the ratio of a gallon of milk to a quart also is 4:1. Thus, the ratio involving money and the ratio involving milk cartons are the same, 4:1. This sameness, or equality, of ratios is called *proportion*. Mathematically, this proportion can be expressed as follows:

$$\frac{\text{gallon}}{\text{quart}} = \frac{\$20}{\$5} = \frac{4}{1} = 4:1$$

Remember, when two items are identical, or equal, the language of mathematics uses an equal sign (=) to express this equality. A statement in mathematics that contains an equal sign is called an *equation*. An equation simply says that two things are equal.

Solving Problems With Proportions

The language of mathematics exists largely to solve problems. The idea of proportion can be useful in solving many of the problems you may encounter in your everyday life.

Say that a gallon of milk costs $2. However, you wish to purchase only a quart. What price would you want to pay for a quart of milk to obtain value equal to the cost of the gallon?

The solution to the problem is this: The ratio of a gallon to a quart is 4:1. You want to pay an amount for a quart so that the prices are in the same ratio of 4:1. Written as a proportion, this equation might be expressed as follows:

$$\frac{4}{1} = \frac{2}{?}$$

That equation states that a gallon is four times as much as a quart. Dividing the $2 by four gives 50 cents. Therefore, the value of the question mark in the equation is 50 cents. Stated mathematically, the equation is:

$$\frac{4}{1} = \frac{2}{0.50}$$

If the store charges more than 50 cents for a quart of milk, you know that the gallon is the better value. If the store charges less than 50 cents for a quart, the quart is the better buy.

EXERCISE 5-2

In each question, state each given proportion in mathematical terms. Then use the proportions to solve the problem.

1. John stopped at an outdoor food stand to buy fresh eggs. The stand offered three different packages. Package number one contains four eggs and costs 44 cents. Package number two contains a dozen eggs and costs $1.26. Package number three contains 36 eggs and costs $3.56. Which package of eggs is the best buy?

2. Sophia is shopping for material to make dresses for her daughters. In one store, she found four different kinds of white material. Each felt and looked exactly the same. Material number one costs $30 for 5 yards. Material number two costs $21 for 3 yards. Material number three costs $35 for 7 yards.

Material number four costs $26 for 4 yards. Which material is the best buy?

3. Charlie was out looking for the best place in town to buy gasoline. He went to Joe's Gas Stop and bought 10 gallons for $11.30. Then he went to Gina's Self Serve and bought 5 gallons for $6.25. After Gina's, Charlie went to the MiniGas and bought 4 gallons for $4.96. Finally, Charlie drove to Stop And Gas and bought 2 gallons for $2.80. Which gas station sold gas at the best rate?

4. Fredette stood in the produce section of a grocery store trying to decide which bag of potatoes was the better buy. One bag contained 12 pounds and cost $3. A second bag contained 15 pounds and cost $4. Which bag of potatoes did Fredette buy?

5. A local store is having a sale on rope. Two brands are being offered. Brand A is selling at $36 for 120 feet. Brand B is selling at $36 for 40 yards. Which brand, if either, is a better buy?

ALGORITHMS FOR FINDING THE UNKNOWN IN AN EQUATION

Proportions are very helpful in making everyday consumer decisions. Remember, a proportion represents the sameness of two ratios. A proportion has four parts, or quantities. If you know three of the parts, the fourth part, the *unknown*, can be determined. In the language of math, this is known as *solving the proportion* or *solving the equation.*

As an example, suppose the price of gasoline is $1.25 per gallon. That is a ratio of 1.25:1. If you purchase 6 gallons, how much would you expect to pay?

The price, $1.25 per gallon, stays the same as you pump gasoline into the tank. Therefore, the ratio of price to volume remains the same. The proportion, therefore, can be expressed:

$$1.25:1 = \$:6$$

or

$$\frac{1.25}{1} = \frac{\$}{6}$$

In this equation, the $ is the unknown—the total price of the 6 gallons. *Cross multiply* the two fractions to determine this price. Cross multiplying consists of two steps: Multiply the numerator of the fraction to the left of the equal sign by the denominator of the fraction to the right of the equal sign. Then, multiply the denominator of the fraction on the left by the numerator of the fraction on the right.

$$1.25 \times 6 = 1 \times \$$$
$$7.50 = \$$$

In this case, you multiply 1.25 by 6 to obtain 7.50. Multiplying the dollar sign by 1 gives $. Therefore, the value of $ is 7.50, and you would expect to pay $7.50 for 6 gallons of gas. The equation is written this way:

$$\frac{1.25}{1} = \frac{7.50}{6}$$

Unknown as a Numerator

In the example above, the numerator was unknown. The following equation also has a numerator as the unknown:

$$\frac{7}{2} = \frac{N}{5}$$

An algorithm can be used to solve an equation with an unknown numerator. This is shown and solved in Figure 5-1. Understanding the reason for Step 2 is the key to using this algorithm. When the known side is converted to a decimal, the decimal actually is a numerator with a denominator of 1. Remember, equations containing two fractions are solved by cross multiplying the fractions. In this case, the decimal is the numerator on the left side

Figure 5-1. Algorithm shows a three-step method for finding the value of an unknown numerator in an equation.

Algorithm	Numerical
1. Convert the known side to a decimal.	$\frac{7}{2} = 3.5$
2. Multiply the decimal by the known number on the right.	$3.5 \times 5 = 17.5$
3. Result is the value for N.	$N = 17.5$

of the equation. Multiplying the decimal by the denominator on the right side, 5, produces the value 17.5. The other step in cross multiplication involves multiplying the other numerator, N, by the other denominator, 1. Therefore, 1 × N, or N, equals 3.5 × 5, or 17.5.

EXERCISE 5-3

A. Write the algorithm for solving an equation in which the numerator is the unknown.

B. Solve each equation for the unknown variable.

1. $\frac{35}{100} = \frac{N}{1000}$
2. $\frac{9}{18} = \frac{N}{36}$
3. $\frac{3}{5} = \frac{N}{33}$
4. $\frac{2}{7} = \frac{N}{6}$
5. $\frac{5}{9} = \frac{N}{13}$
6. $\frac{12}{13} = \frac{N}{14}$
7. $\frac{13}{15} = \frac{N}{100}$
8. $\frac{2}{9} = \frac{N}{3}$
9. $\frac{1}{33} = \frac{N}{54}$
10. $\frac{8}{300} = \frac{N}{100}$
11. $\frac{0.5}{25} = \frac{N}{2}$
12. $\frac{1.35}{3.12} = \frac{N}{4.23}$
13. $\frac{15.9}{11} = \frac{N}{56.7}$
14. $\frac{N}{3} = \frac{.006}{1.34}$
15. $\frac{N}{3,133} = \frac{14}{1,333}$
16. $\frac{N}{10} = \frac{8.34}{9}$
17. $\frac{N}{49} = \frac{3}{19}$
18. $\frac{N}{6.1} = \frac{43}{1.34}$

Unknown as a Denominator

In some equations, the unknown may be in the denominator. An equation with an unknown denominator is shown below:

$$\frac{7}{2} = \frac{5}{N}$$

An algorithm can be used to solve an equation with an unknown denominator, as presented in Figure 5-2.

Algorithm	Numerical
1. Convert the known side to a decimal.	$\frac{7}{2} = 3.5$
2. Divide the known number on the right by the decimal.	$5 \div 3.5 = 1.43$
3. Result is the value for N.	N = 1.43

Figure 5-2. Solving an equation with an unknown value as a denominator also requires only three steps.

Cross multiplying the fractions in this equation produces a new mathematical statement. The new statement is $3.5 \times N = 5$. Using long division, the equation also can be stated this way:

$$N = \frac{5}{3.5}$$

```
        1.428
3.5.)5.0.000   = 1.42 or 1.43
     3 5 xxx
     1 5 0
     1 4 0
       1 00
         70
        300
        280
         20
```

Thus, dividing the known number on the right, 5, by the decimal, 3.5, produces the value of N, 1.43.

EXERCISE 5-4

A. Write the algorithm for solving an equation in which the denominator is the unknown.

B. Solve each equation for the unknown variable.

1. $\frac{3}{N} = \frac{300}{1000}$
2. $\frac{9}{N} = \frac{13}{36}$
3. $\frac{3}{5} = \frac{113}{N}$
4. $\frac{3}{7} = \frac{4}{N}$
5. $\frac{5}{9} = \frac{313}{N}$
6. $\frac{2}{13} = \frac{14}{N}$
7. $\frac{3}{15} = \frac{100}{N}$
8. $\frac{12}{9} = \frac{34}{N}$
9. $\frac{11}{33} = \frac{9}{N}$
10. $\frac{8}{300} = \frac{23}{N}$
11. $\frac{0.5}{25} = \frac{2}{N}$
12. $\frac{1.56}{3.52} = \frac{2.3}{N}$
13. $\frac{15.9}{41} = \frac{4.5}{N}$
14. $\frac{3}{N} = \frac{0.6}{1.4}$
15. $\frac{333}{N} = \frac{1.4}{2}$
16. $\frac{10}{N} = \frac{8.4}{9.5}$
17. $\frac{49}{N} = \frac{3}{1.999}$
18. $\frac{6.1}{N} = \frac{3}{4}$

Solving Problems With Equations

In some purchasing situations, you may have a certain amount of money available to spend. In this case, you might want to determine how much of a product you can buy.

For example, say that 2 pounds of mixed nuts cost $3.20. If you have $5, how many pounds of mixed nuts can you buy?

In solving this problem, you are interested in the ratio between money and pounds of nuts. This ratio is 3.20:2, or $\frac{3.20}{2}$. Knowing this, you can write an equation to express the proportion. The unknown in this equation will be the number of pounds of nuts you can buy for $5. Let N equal the answer, and the equation is written this way:

$$\frac{3.20}{2} = \frac{5.00}{N}$$

The 3.20:2 ratio can be simplified and expressed as 1.60:1, or $\frac{1.60}{1}$, or simply 1.60. Writing a fraction in decimal form often makes computations easier. In simplified form, the equation becomes:

$$1.60 = \frac{5.00}{N}$$

The solution is computed by dividing 5.00 by 1.60 to obtain 3.13.

To see why this is done, think of a simpler equation, such as $3 = \frac{6}{2}$. If 2 goes into 6 three times, it also is true that 3 goes into 6 two times. That is, if $\frac{6}{2} = 3$, then $\frac{6}{3} = 2$.

Therefore, in the case of the mixed nuts, if $\frac{5.00}{N} = 1.60$, you also know that $\frac{5.00}{1.60} = N$. Using long division or a calculator, you can calculate that N = 3.13, as shown below:

$$\frac{5.00}{1.60} = N$$

$$5.00 \div 1.60 = N$$

$$5.00 \div 1.60 = 3.125 \text{ or } 3.13$$

EXERCISE 5-5

In each problem, write an equation to express the proportion given. Then solve the equation for the unknown. Do not carry answers beyond two decimal places.

1. Carl's automobile has a 15-gallon fuel tank. His car has a range of 330 miles on 15 gallons of gas. If Carl has 4 gallons of fuel, how far will he be able to drive?

2. Amanda loved jelly beans. Her mother gave her $5 to go to the candy store. Amanda skipped all the way. She gave the man behind the counter her money and asked for a bag of red jelly beans. The man told her that red ones cost $1.30 for 2 pounds. How many pounds of jelly beans did she receive?

3. Saul took his son's soccer team to Lillian's Ice Cream Store. Lillian's was having a special sale: four double-decker cones for $3. There were 20 members on the team. How much money did Saul spend?

4. Terry needed to buy milk. She had $3.54 to spend. The local grocery store charged $1.18 for 2 quarts. How many quarts of milk was Terry able to buy?

5. Jason and Manuel plan on sailing around the world. They will travel 25,000 miles. If they travel an average of 350 miles per week, how many days will their trip take?

6. Eva is an airline pilot. On the average, her plane travels at a speed of 1,845 miles every three hours. The distance from New York City to Los Angeles is 2,700 miles. How many hours does it take Eva to fly from New York City to Los Angeles?

7. Efrain went looking for work. He found a factory hiring people to assemble computers. Pay was on a piecework basis: For every four parts put together, Efrain would be paid $5. How many parts would Efrain have to assemble to make $545?

8. Janet can ride her bicycle 5 miles in 20 minutes. How long would it take her to ride 60 miles?

Rounding simplifies calculations. On a gasoline pump, the total cost is rounded to the nearest cent.
PHOTO BY CLIFF CREAGER

ROUNDING

Many everyday mathematical calculations do not require extremely precise results. This is true especially for calculations involving money. You need calculate only to two decimal places to figure cents. The process of simplifying answers by eliminating unnecessary decimals is called *rounding*.

The general rule for rounding numbers is simple. Just look at the first, or smallest, digit that you don't need. For example, you may want to round the number 123.456 to represent an amount of money. Only two decimal places are needed to express cents. You don't need the 6.

The rule for rounding is this: If the digit to be eliminated is 5 or larger, increase the last digit you use by 1. If the digit you don't need is 4 or smaller, the last digit remains the same. In the example 123.456, the 6 is greater than 5. Therefore, rounding produces $123.46.

Before practicing in Exercise 5-6, study the following examples of rounding numbers:

$$
\begin{aligned}
1.536 &= 1.54 \\
2.043 &= 2.04 \\
6.3548 &= 6.35 \\
56.7839 &= 56.78 \\
0.0045 &= 0.005
\end{aligned}
$$

EXERCISE 5-6

A. Round each number to two decimal places.

1. 0.0435
2. 1,234.456
3. 3.567
4. 0.006
5. 5.119
6. 2.313
7. 390.9909
8. 132.127
9. 11.3361
10. 0.999
11. 0.888
12. 1.563
13. 15.415
14. 3.42589
15. 8.496
16. 4.999

B. Round each number to one decimal place.

1. 0.0635
2. 1,234.456
3. 3.567
4. 5.119
5. 2.313
6. 390.9909
7. 132.127
8. 11.3361
9. 0.999
10. 0.888
11. 1.563
12. 15.415
13. 3.614
14. 3.42589
15. 8.496
16. 6.134

C. Write an algorithm for rounding numbers.

ESTIMATING: THE USE OF MEASUREMENTS

Many everyday situations call for another mathematical skill. That skill, called *estimating,* involves judging the approximate size, weight, value, number, or extent of something. Estimating usually involves calculations of measurements using proportions or equations. Calculations performed in estimating generally use numbers that have been rounded.

Estimating is a common practice in the world of work. For example, a building contractor must estimate the amount of materials needed for a project. A printer estimates the cost of a job for a customer.

Estimating skills are important when ordering materials for a project.

In your everyday life, you might estimate the amount of food and beverages needed for a party. You can estimate the amount of fuel required for a trip. Expanding a recipe to feed more people calls for estimating skills. The list is almost endless.

In mathematics, estimating usually involves a *constant,* or a known value that does not change. Think about planning a weekend trip. The constant involved in the trip plan is the average fuel consumption, or miles per gallon, you can expect from your automobile. Dividing this constant, average fuel consumption, into the number of miles to be traveled, gives your estimated fuel requirement.

Faces in a Crowd

Have you ever attended a sports or other spectator event and wondered about the size of the audience? There is a simple procedure for estimating crowd size.

Say that you attend a football game in a large stadium. Most spectator arenas are divided into sections. You can estimate the seating capacity of the arena through a two-step process. First, determine the number of seats in a section. Second, multiply that figure by the number of sections. This process can be described in an algorithm, as is done in Figure 5-3.

1. Count the number of seats in a row within a section.
2. Count the number of rows in the section.
3. Multiply the number of seats in a row by the number of rows to compute the approximate number of seats in the section.
4. Count the number of sections in the arena.
5. Multiply the product in Step 3 by the number of sections to estimate the seating capacity of the arena.

Figure 5-3. Estimating the crowd in an arena is easy using this algorithm.

The first step in Figure 5-3 is important. It is best to choose a row in the middle of a section. Typically, rows become wider toward the top, or rear, of a section in an arena. The rows near the middle of the section should represent the average row size.

Section sizes vary in many arenas, so it may be necessary to count seats in more than one section. However, remember that you are only estimating. Your goal is not to determine the precise seating capacity.

Once you have estimated the seating capacity, you can estimate the size of the crowd by judging how many of the seats are occupied.

Estimating and Rounding

Estimating is a process of determining approximate values. Therefore, it usually is not necessary to use precise numbers in your calculations. For example, you might want to estimate the gross pay of a worker for 39 hours at an hourly rate of $7.10.

First, round off the numbers to simplify your calculations. Change the numbers to 40 hours and $7 an hour. These numbers can be multiplied easily:

$$40 \times 7 = 280$$

Your estimate, then, is that the worker will earn gross pay of $280. Now, use a calculator or pencil and paper to multiply the exact figures. Your answer should be $276.90, which is very close to your estimate of $280. The purpose of estimating is to obtain a rough, or approximate, figure.

Estimating can be used to check the accuracy of multiplication. Calculating with rounded-off factors lets you establish a reasonable value for comparison with your exact figures. If the estimate and the exact multiplication product are close, you know the multiplication probably is correct.

Estimating as an Everyday Activity

Estimating is useful for many everyday activities. For example, say you have 20 mathematics problems to solve for homework tonight. You also may want to watch a special television program. Your task is to estimate how much time you will require to complete your homework before the show begins.

One method would be to work one of the problems and time your effort. Say you spend 5 minutes working one problem. Twenty problems should require about 100 minutes, or an hour and 40 minutes. To be safe, you might round off that figure to two hours.

Measuring Your Mileage

Estimating is particularly valuable for planning trips. Say that you plan to drive from Albany, New York, to Pittsburgh, Pennsylvania. You want to estimate how much fuel you need for the round trip. You also want to know how much the fuel will cost.

You need to determine only two factors to estimate your fuel requirement. One is fuel consumption: How many miles will your vehicle travel on a gallon of fuel? The other is distance: How many miles will you travel?

The road atlas lists the distance between Albany and Pittsburgh as 453 miles. Thus, your round trip would be 906 miles. Your car usually gets about 26 miles per gallon under highway driving conditions. Before you do any calculating, remember that in estimating you can round off.

Assuming that you will do some driving while in Pittsburgh, you can round off your total mileage at 950. You also can round off your anticipated fuel mileage at 25 miles per gallon. From handling money, you know 25 is one-quarter of 100. Therefore, you can round your consumption into 100. That is, you can travel 100 miles on 4 gallons. You can then multiply 4 by 9 to determine that you can drive 900 miles on 36 gallons. The remaining 50 miles will take 2 gallons. The trip, then, will require 38 gallons. Estimating puts small, easily figured values together, like building with blocks.

Estimating fuel costs is an important part of planning a trip.

To estimate your fuel cost, one more step is necessary: Multiply the amount of fuel by the price per gallon. Again, you can round off and break down the values for estimating. Say fuel costs $1.19 per gallon. Round to $1.20. Now multiply by 10, an easy division of 40 and a rounded value of 38. The result is $12. Multiply $12 by 4 for the cost of 40 gallons: $48. Then subtract the cost of 2 gallons, $2.40, from $48: $45.60. You can even subtract 40 cents from this total to adjust for rounding the price. That is, by rounding, you added one cent per gallon to the price of gasoline. Now, for a more exact estimate, you can make a correction: $45.20.

EXERCISE 5-7

For each question, write an algorithm for solving the problem. Then solve it.

1. Mr. Thomais is planning to fly from New York City, New York, to Los Angeles, California. The flight plan includes a 1 hour and 10 minute layover in Chicago, Illinois, and another 54 minute layover in Denver, Colorado. The distance from New York City to Chicago is 725 miles. The distance from Chicago to Denver is 812 miles. The distance from Denver to Los Angeles is 1,014 miles. If Mr. Thomais' plane averages 508 miles per hour, approximately how long will it take him to reach his destination?

2. Francine decided to bake brownies for her brother's birthday party. Fifty-nine people had been invited. Her recipe called for 1.1 cups of butter, 1.26 cups of sugar, 3.49 cups of flour, 2 eggs, 0.5 cups of melted chocolate, and 1.5 cups of walnuts. This recipe makes enough brownies for 15 people. Estimate the quantity of each recipe ingredient Francine will need to make enough brownies for all 59 people.

3. Chi and Ching-yu want to place new carpet in their home. They decide to purchase and install new carpet themselves. Their home has five rooms. The living room is 290 square feet. The den is 196 square feet. The master bedroom is 232 square feet. Each of the other two rooms is 144 square feet. The carpet they want costs $8.48 per square yard. There are 9 square feet in 1 square yard. Estimate how much it will cost to purchase the carpeting.

4. Howard and Pauline are planning to paint walls in four rooms. The rooms have 532, 628, 704, and 496 square feet of wall space. Instructions on the paint can say that each can will cover 600 square feet of wall space. How many cans of paint will Howard and Pauline need? If each can sells for $16.87, approximately how much will their painting project cost?

5. Jack and Victoria are driving from New York City, New York, to Atlanta, Georgia. The distance is 891 miles. Their car gets 19 miles per gallon of gasoline. Gasoline is $1.28 a gallon. They plan to drive at 55 miles per hour. Estimate how much Jack and Victoria will have to spend on gas. Then, estimate how long it will take them to complete the drive.

6. Mrs. Burns is planning to make two shirts each for her three sons. Her pattern calls for 1.54 yards of material to make one shirt. She is considering three materials. Material A costs $10.97 for 3 yards. Material B costs $5.99 for 1.51 yards. Material C costs $9.05 for 2.1 yards of material. Estimate how much it will cost Mrs. Burns to purchase enough of each type of material to make shirts for her sons. Then, choose the type of material you would advise her to purchase.

SUMMING UP

- Comparisons frequently are called ratios in the language of mathematics.
- Ratios involving unlike objects can be equal. This sameness of ratios is called proportion. In mathematics, proportions are written with an equal sign between the ratios.
- A mathematical statement that contains an equal sign is called an equation. An equation simply states that two things are equal.
- Equations frequently contain an unknown. If three parts of an equation are known, the unknown can be determined by solving the equation. An equation is solved by stating the ratios as fractions and cross multiplying the numerators and denominators.
- Algorithms can be written for solving the unknown in an equation. The steps differ, depending upon whether the unknown is a numerator or a denominator.
- Rounding off numbers is a useful method of simplifying mathematical calculations. If the first digit that is not needed in a decimal number is 5 or greater, the last digit used is increased by 1. If the unneeded digit is 4 or less, the last digit remains the same.
- Estimating is a mathematical method of determining approximate values. Numbers usually are rounded to simplify the task of estimating.

TESTING YOUR WORKING KNOWLEDGE

For each question, write an algorithm for solving the problem. Then, solve the problem.

1. Mr. Jackson has to travel from Hollywood, California, to Seattle, Washington, to perform a benefit concert. The distance is 1,146 miles. Mr. Jackson is driving his Rolls Royce, which averages 16 miles to a gallon of gas on the highway. How many gallons will Mr. Jackson use?

2. The population of the United States increases by one person every six seconds. Estimate the population growth for each 24-hour day.

3. Waldo is a marathoner. He wants to run 26 miles, 385 yards in 2 hours, 10 minutes. In planning for the race, he needs to estimate an average running time for each mile. Using rounding methods, estimate an average time per mile for the race.

4. A car that averages 18 miles per gallon is to be used for a trip of 1,600 miles. How many gallons of gasoline will be needed? At $1.11 per gallon, estimate the overall fuel cost.

5. Horacio is preparing an estimate for electrical wiring of a new house. He uses a system based on an average cost for each electrical outlet. Horacio counts outlets and multiplies by $21.19 to cover the cost of parts and labor. When he develops this total, he multiplies by 79% and adds the result to the previous total. This allows for business costs and profits. Using rounding methods, figure estimates for jobs with the following numbers of outlets:

   a. 110
   b. 96
   c. 72
   d. 88

TERMS THAT COUNT

ratio
proportion
equation
unknown
solving the proportion
solving the equation
cross multiply
rounding
estimating
constant

PRESENTING INFORMATION

1. The Foot Factory makes shoes. It costs the shoe factory $4 to make one pair of shoes. For each pair made, 40% of the cost is for materials, 30% is for labor, and 30% is for advertising.
   a. Draw a pie chart showing how materials, labor, and advertising costs for one pair of shoes relate to total costs.
   b. Draw a bar graph showing how much it costs the Foot Factory to make 100, 500, 700, and 1,000 pairs of shoes.

2. The Foot Factory sells each pair of shoes for $6. This means the company earns $2 profit on each pair sold.
   a. Draw a pie chart relating profit and costs for labor, materials, and advertising to the selling price.
   b. Draw a bar graph showing how profits grow when The Foot Factory makes and sells 100, 500, 700, and 1,000 pairs of shoes.

SKILL DRILLS

A. Solve to three decimal places each equation for the unknown variable. Round answers.

1. $\frac{5}{100} = \frac{N}{11}$
2. $\frac{1}{8} = \frac{N}{3}$
3. $\frac{3}{5} = \frac{N}{9}$
4. $\frac{12}{7} = \frac{N}{4}$
5. $\frac{5}{19} = \frac{N}{13}$
6. $\frac{2}{3} = \frac{N}{14}$
7. $\frac{1}{5} = \frac{N}{12}$
8. $\frac{2}{29} = \frac{N}{3}$
9. $\frac{11}{33} = \frac{N}{3}$
10. $\frac{8}{13} = \frac{N}{14}$
11. $\frac{0.5}{25} = \frac{N}{9}$
12. $\frac{3.15}{5.32} = \frac{N}{4.03}$
13. $\frac{5.3}{11.7} = \frac{N}{5.7}$
14. $\frac{N}{3.3} = \frac{.336}{1.34}$
15. $\frac{N}{3.32} = \frac{14}{1.333}$
16. $\frac{N}{19} = \frac{8.34}{21}$
17. $\frac{N}{75} = \frac{2}{3}$
18. $\frac{N}{6.1} = \frac{4.3}{134}$

B. Solve to three decimal places each equation for the unknown variable. Round answers.

1. $\frac{35}{N} = \frac{5}{100}$
2. $\frac{9}{N} = \frac{1}{36}$
3. $\frac{3}{5} = \frac{1.1}{N}$
4. $\frac{2}{7} = \frac{6}{N}$
5. $\frac{15}{19} = \frac{13}{N}$
6. $\frac{12}{13} = \frac{14}{N}$
7. $\frac{13}{15} = \frac{100}{N}$
8. $\frac{2}{9} = \frac{3}{N}$
9. $\frac{11}{133} = \frac{54}{N}$
10. $\frac{3}{0.634} = \frac{1.34}{N}$
11. $\frac{0.5}{2.53} = \frac{2.7}{N}$
12. $\frac{1.35}{43.12} = \frac{4.32}{N}$
13. $\frac{5.9}{12.11} = \frac{56.47}{N}$
14. $\frac{3}{N} = \frac{.006}{1.34}$
15. $\frac{3{,}133}{N} = \frac{14}{1{,}333}$
16. $\frac{170}{N} = \frac{8.34}{9.765}$
17. $\frac{49}{N} = \frac{3}{19}$
18. $\frac{61}{N} = \frac{43.54}{0.34}$

Mathematics for Leisure

YOUR LEARNING JOB

When you have completed the exercises and assignments for this unit, you should be able to:

- ☐ Describe the role of mathematics in sports and games.
- ☐ Read and understand tables used to present information on leisure activities.
- ☐ Use tables to present information on team standings, individual performances, and schedules.
- ☐ Describe percentage-based grading systems used in sports.
- ☐ Use measurements in grading performances for sports events based on time and distance.
- ☐ Explain how handicapping is used to equalize competition.

WARM-UP

A. Find the sum in each of the following problems.

1. 2,319 + 8,918 + 428 + 2,310 + 645 + 243 + 4,824 =
2. 23,429 + 1,297 + 34,927 + 82,726 + 81,829 + 54,274 =
3. 546,293 + 43,453 + 815,934 + 546,834 + 3,452,845 =
4. 43,273,814 + 4,238,734 + 83,284,375 + 54,823 + 4,542,612 =
5. 384,593 + 43,564 + 23,342,974 + 4,543,958 + 83,982,744 =

B. Find the difference in each of the following problems.

1. 345,464 − 298,679 =
2. 645,375 − 425,876 =
3. 4,637,821 − 567,768 =
4. 45,376,274 − 21,867,689 =
5. 56,325,756 − 55,456,887 =

C. Find the product in each of the following problems.

1. 4,573 × 87 =
2. 6,823 × 215 =
3. 7,045 × 3,734 =
4. 6,666 × 7,777 =
5. 3,745 × 17,328 =

D. Find the quotient in each of the following problems.

1. 359,784 ÷ 456 =
2. 285,769 ÷ 313 =
3. 1,084,336 ÷ 976 =
4. 132,006 ÷ 1,347 =
5. 3,020,376 ÷ 3,176 =

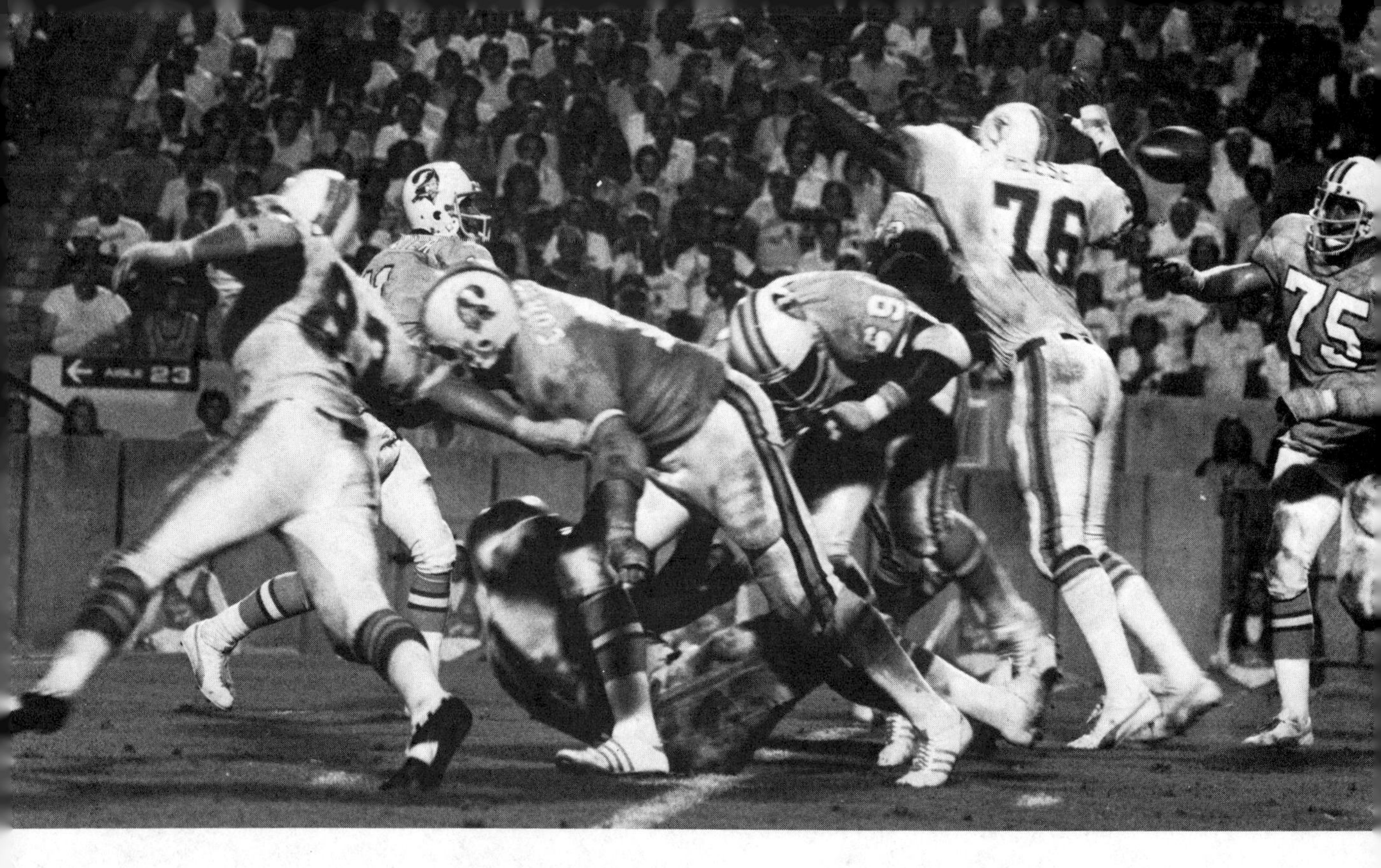

Mathematics plays a major role in most sports and games. Both team and individual performances are measured mathematically.
COURTESY OF FLORIDA DEPARTMENT OF COMMERCE/DIVISION OF TOURISM

MATHEMATICS = GAMES PEOPLE PLAY . . . AND WATCH

Think about your favorite leisure time activities. Do you participate in organized sports? Are you a sports fan? Do you bowl or play golf? Do you enjoy games involving playing cards or dice?

Whatever your favorite pastimes are, there is a good chance they involve mathematics. Pastimes are activities you enjoy during leisure hours. Think about a typical conversation involving any sport. The first question usually is: ''Who won?'' The very next question almost always is: ''What was the score?''

Mathematics plays a major role in all competitive sports and in most games. In sports, mathematics is used to present information, such as scores, and for grading performances by individual athletes and teams. Measurements are important in many sports, such as the distance a discus is thrown. In some sports, conversions are important because measurements may be given in both the metric and English systems. Examples are given later in this unit.

In most sports, *statistics* are of major interest to participants, coaches, and fans. Statistics are collections of numerical data.

KEEPING SCORE: PRESENTING INFORMATION

Pick up the sports section of your local newspaper, and one fact will become apparent. The most important aspect of sports news is scores. From archery to yachting, most headlines contain scores. Most sports sections contain different examples of presenting information in mathematical form.

For instance, team standings for various sports are presented as tables of numbers. A typical table of team standings is shown in Figure 6-1. These tables usually contain three or more columns of information. The first column on the left lists wins, often indicated by the heading "W." The next column, with an "L" as a heading, lists losses. A third column usually gives the performance percentage of each team. The performance percentage is calculated by dividing the number of wins by the number of games played. Some team standings tables also contain a fourth column, which usually has the heading "GB." The GB stands for games behind the leading team.

The games behind figure is calculated by adding the difference in wins to the difference in losses and dividing by 2. If Team A beats Team B today, Team A is said to have picked up a full game on Team B. However, if Team A wins today and Team B does not play, Team A picks up half a game on Team B.

Ties That Count

Some sports, notably hockey, assign values to tie games. In professional hockey, a win is worth 2 points in the standings, and a loss is worth zero. A tie is worth 1 point in the standings. Thus, a team with fewer victories still can stay close in the standings by tying rather than losing games. In hockey, the standings are

Team	W	L	Pct.	GB
Boston	14	1	.933	
Philadelphia	12	4	.750	$2\frac{1}{2}$
Washington	12	7	.632	4
New Jersey	8	8	.500	$6\frac{1}{2}$
New York	8	12	.400	$8\frac{1}{2}$

Figure 6-1. A team standings table typically lists wins, losses, winning percentages, and games behind the leader.

based upon total points, rather than on percentages. An example of a hockey standings table is shown in Figure 6-2.

Can You Find the Leader?

In baseball, football, basketball, and most other team sports, championships are decided on the basis of won-lost percentage.

Figure 6-2. Instead of percentages, hockey standings are based upon 2 points for each victory and 1 point for each tie game. Points for wins and ties are combined to grade each team's performance.

NATIONAL HOCKEY LEAGUE

Campbell Conference

Smythe Division

	W	L	T	Pt.
Edmonton	24	7	3	51
Winnipeg	19	12	4	42
Calgary	18	14	3	39
Los Angeles	15	13	7	37
Vancouver	8	24	4	20

Norris Division

	W	L	T	Pt.
Chicago	16	16	3	35
St. Louis	15	13	5	35
Detroit	12	18	4	28
Minnesota	11	18	6	28
Toronto	6	24	5	17

Wales Conference

Patrick Division

	W	L	T	Pt.
Washington	19	10	7	45
Philadelphia	19	10	5	43
N.Y. Islanders	19	13	1	39
Pittsburgh	13	16	3	29
N.Y. Rangers	11	18	5	27
New Jersey	11	18	4	26

Adams Division

	W	L	T	Pt.
Montreal	19	9	6	44
Quebec	15	15	5	35
Buffalo	13	12	9	35
Boston	14	15	5	33
Hartford	13	15	4	30

Wednesday's Results

Los Angeles 3, Van. 3 (OT)
Hartford 5, New Jersey 3
Buffalo 6, Toronto 0
Pittsburgh 6, N.Y. Islanders 5
Detroit 5, N.Y. Rangers 2
Washington 6, Philadelphia 0
St. Louis 4, Chicago 3
Minnesota 4, Winnipeg 0
Edmondon 6, Calgary 5
Only games scheduled

Tonight's Schedule

Boston at Los Angeles
Montreal at Quebec
Washington at N.Y. Islanders
Toronto at New Jersey
Only games scheduled

Early in a season, this can lead to some confusion in the standings. For example, it is possible for the true leader of a league to be games behind another team, early in a schedule. This unusual situation is illustrated in Figure 6-3.

Other Tables

Many kinds of sports information are presented in *tabular* form. Tabular means in the form of a table. In addition to tables of standings, a sports section may contain schedules, team rosters, and a variety of individual and team performance statistics. Most large daily papers devote one or more pages entirely to the presentation of such tabular information.

Team	W	L	Pct.	GB
Rockets	2	1	.667	$\frac{1}{2}$
Jets	5	3	.625	
Raiders	3	2	.600	$\frac{1}{2}$
Giants	2	3	.400	$1\frac{1}{2}$
Lobos	1	2	.333	$1\frac{1}{2}$
Panthers	2	4	.333	2

Figure 6-3. In most team sports, winning percentage is the value that determines a league championship. Early in a season, it is possible for the leader in percentage to be behind in games won and lost.

EXERCISE 6-1

Read each question carefully before calculating the answer. Round off calculations to three decimal places.

1. The Giants have won 5 games and lost 2. The Rebels have won 6 games and lost 3. Which team has the better won-lost percentage?

2. The Yellow Jackets have won 8 games and lost 3. The Sparrows have won 9 games and lost 4. The Cardinals have won 8 games and lost 4. What is each team's won-lost percentage?

3. The following won-lost-tied records are for four top teams in a hockey league. The Eagles have won 8 games, lost 3, and tied 2. The Blue Jays have won 8 games, lost 4, and tied 1. The

Ducks have won 7 games, lost 1, and tied 5. The Flyers have won 9 games and lost 4. Determine team standings. Then, construct a table that shows the won-lost-tied records for each team.

4. Two baseball teams are competing for the league championship. Team 1 has won 16 games and lost 2. Team 2 has won 13 games and lost 3. What is each team's won-lost percentage? By how many games is the second-place team behind the first-place team?

5. There are five hockey teams in the Middletown League. Team 1 has won 8 games, lost 4, and tied 3. Team 2 has won 7 games, lost 3, and tied 5. Team 3 has won 9 games, lost 4, and tied 2. Team 4 has won 7 games, lost 6, and tied 2. Team 5 has won 4 games and lost 11. Construct a table that shows the team standings, games won, games lost, and games tied for each team.

6. The Tri-State Football League reported the following won-lost records after the first month of play: The Lions had won 3 games and lost 1; the Condors had won 2 games and lost 1; the Elks had won 1 game and lost 1; and the Bears had won 3 games and lost 2.

a. Which team has the best won-lost percentage?

b. By how many games are the Elks behind the first place team?

c. How many games behind the first-place team is the second-place team?

d. Construct a table that shows team standings, games won, games lost, and games behind leader for each team.

7. The Falcons have won 9 games and lost 2. The first-place team has won 12 games and lost 1. By how many games do the Falcons trail the first-place team?

8. Two baseball teams are battling for first place. Team 1 has a won-lost percentage of .750 and has played all of its league games. Team 2 has won 7 games and lost 3 and has one league game left. If Team 2 wins the last season game, will they win the championship?

GRADING SYSTEMS IN SPORTS

Competition is the heart of most sports and games. Thus, these activities require methods of grading performances. Once again, mathematics plays the major role.

Sometimes, competition is so close that mathematical calculations are required to separate a winner from a loser or runner-up. A famous example occurred in the 1949 baseball season. Two players went into the final game of the season in a virtual tie for the American League batting championship. The players, both now in the Hall of Fame, were Ted Williams and George Kell. When the final day of the season ended, Kell had won the batting title by the narrowest margin in history: .3429 to .3427. Both Williams and Kell were credited with batting averages of .343 for the season.

Baseball Statistics

Baseball is the most statistics oriented of all team sports. Statistics are kept for almost anything that can happen on a baseball diamond. Major statistical areas used to grade individual players' contributions include:

- Batting average
- Runs scored
- Runs batted in (RBI)
- Pitcher's won-lost percentage
- Earned run average (ERA)
- Fielding percentage.

Runs scored and RBI are simple statistics from a mathematical standpoint. You add each run or RBI to the player's previous total. A pitcher's won-lost percentage is calculated in the same manner as team won-lost percentages. Divide the number of wins by the total number of games won and lost. Calculations for the other grading systems are presented below.

Batting average. A *batting average* is a percentage of the number of hits divided by the number of at-bats. For example, say that a player gets 2 hits in 4 at-bats in a game. The player's average for that game is:

$$2 \div 4 = \frac{2}{4} = .500$$

Batting averages are carried out to three decimal places because of the closeness of percentages. The example cited above of the 1949 season underlines this closeness.

EXERCISE 6-2

Read each question about batting averages carefully before calculating the answer. Round off calculations to three decimal places.

1. Cathy made 2 hits in 5 at-bats. What is her batting average?

2. Fernando made 1 hit in 3 at-bats. What is his batting average?

3. During her first season, Claire made 11 hits in 36 at-bats. What is her batting average?

4. Tony has played in three games this season. In the first game, he made 1 hit in 4 at-bats. In the second game, he made 3 hits in 3 at-bats. In the third game, he made no hits in 3 at-bats. What is Tony's batting average?

5. Silvia has played in four games. In the first game, she made 3 hits in 3 at-bats. In the second game, she had no hits in 4 at-bats. In the third game, she made 2 hits in 3 at-bats. In the last game, she had 1 hit in 5 at-bats. What is Silvia's batting average?

6. Raleigh has played in four games. In the first game, he made 2 hits in 3 at-bats. In the second game, he went 1 for 4 at-bats. In the third game, he made 1 hit in 3 at-bats. In the fourth game, he had no hits in 4 at-bats. What is Raleigh's batting average for Game 1? Game 2? Game 3? Game 4? What is Raleigh's batting average for all four games?

7. Raquel has 7 hits in 29 at-bats. Leroy has 5 hits in 17 at-bats. Hortencia made 6 hits in 19 at-bats. Leona has 3 hits in 11 at-bats. Construct a table that lists the batters according to batting average, number of at-bats, and number of hits.

8. Write an algorithm for calculating a batting average.

Earned run average. An *earned run average (ERA)* is a calculation of the frequency with which a pitcher allows opposing teams to score runs. An earned run is a run scored by the offensive team without benefit of an error by the defensive team. An earned run average consists of the number of earned runs for each nine innings pitched. A pitcher's earned run average requires two calculations. First, divide the total number of earned runs by the number of innings pitched. Then multiply by nine. For example, say a pitcher gave up 120 runs in 300 innings. The pitcher's ERA would be 3.60. The solution is based on these calculations:

$$120 \div 300 = 0.40$$
$$0.40 \times 9 = 3.60$$

EXERCISE 6-3

Read each question carefully before calculating the answer. Round off calculations to two decimal places.

1. Andre has pitched in 180 innings and has given up 60 runs. What is Andre's ERA?

2. Rodney pitched in 190 innings and gave up 40 runs. Joey pitched in 210 innings and gave up 80 runs. Yoko pitched in 140 innings and gave up 27 runs. Which pitcher has the lowest ERA? Which has the highest?

3. Carrol has pitched in five games. In the first game, she gave up 1 run in 5 innings. In the second game, she gave up 3 runs in 9 innings. In the third game, she gave up 2 runs in 7 innings. In the fourth game, she pitched 9 innings without a run being scored. In the fifth game, she pitched 5 innings and gave up 2 runs. What is Carrol's ERA?

4. Hayato has pitched in three games. In the first game, he gave up 2 runs in 9 innings. In the second game, he gave up 1 run in 4 innings. In the third game, he pitched 8 innings without a run being scored. What is Hayato's ERA?

5. Carlos pitched in 210 innings and gave up 47 runs. Virginia pitched in 245 innings and gave up 59 runs. Amelia pitched in 330 innings and gave up 83 runs. Which pitcher has the lowest ERA?

6. Julie, Mary, and Oki are pitchers. Each has given up 27 runs. However, Julie has pitched 163 innings, Mary has pitched 137 innings, and Oki has pitched 97 innings. Determine each pitcher's ERA.

7. David has pitched in four games. In Game 1, he allowed 2 runs in 9 innings. In Game 2, he gave up 1 run in 5 innings. In Game 3, he pitched 8 innings without giving up a single run. In Game 4, he allowed 3 runs in 8 innings. What is David's ERA for each game? What is David's ERA for all four games? Draw a line graph that shows how David's ERA changed during the four games described.

8. Write an algorithm for calculating a pitcher's earned run average.

Fielding percentage. A player's *fielding percentage* is calculated in a manner similar to won-lost percentage. Fielding opportunities on batted or thrown baseballs are called "chances." The number of successful fielding plays is divided by the number of chances to determine fielding percentage. Unsuccessful fielding attempts are called "errors."

Top major league players may commit only a few errors in an entire season. For example, an outfielder might have 320 chances in a 162-game season. Say that the outfielder makes 9 errors. This means that he or she handled 311 chances successfully. That would be a fielding percentage of .972, based upon the following calculation:

$$311 \div 320 = .972$$

EXERCISE 6-4

Read each question about fielding percentages carefully before calculating the answer. Round off calculations to three decimal places.

1. Fred made 15 errors in 175 chances. What is Fred's fielding percentage?

2. Karen made 11 errors in 132 chances. What is her fielding percentage?

3. During five baseball seasons, Don has made 67 errors in 532 attempts. What is Don's fielding percentage?

4. Michi has played in four games. In Game 1, Michi made 1 error in 9 attempts. In Game 2, she made 2 errors in 9 attempts. In Game 3, she made 1 error in 7 attempts. In Game 4, she made 1 error in 6 attempts. Determine Michi's fielding percentage for each game. Then, draw a horizontal bar graph that compares her fielding percentage for each game. In which game did she have the best percentage?

5. Conrad has a fielding percentage of .850. He had 200 chances at fielding the ball. How many balls did Conrad field successfully? How many errors did he make?

6. Bonnie made 9 errors in 85 attempts. Joyce made 15 errors in 123 attempts. Leonore made 7 errors in 73 attempts. Sandra made 5 errors in 63 attempts. Which player has the highest fielding percentage? Which player has the lowest fielding percentage?

7. Two players are competing to see who can earn the best fielding percentage. Player 1 made 1 error in 6 attempts in the first game and 2 errors in 9 attempts in the second game. In the last three games, no errors were made in 30 chances. Player 2 made 3 errors in 7 attempts in the first game and no errors in 27 chances in the second, third, and fourth games. In the last game, 1 error was made in 8 attempts. Which player will win the competition?

8. Write an algorithm for determining fielding percentage.

Basketball

The most common statistics in basketball are based on per-game averages. These include points scored, rebounds, assists, and steals. Averages are calculated by dividing the total number of each type of offensive or defensive action by the number of games played. A player who averages 22 points a game would have scored 220 points in 10 games.

Other common basketball statistics are field goal percentage and free throw percentage. Each is calculated by dividing the total number of attempts into the number of successful attempts. If a player hit 400 field goals on 800 attempts, that player would have a .500 field goal percentage.

EXERCISE 6-5

Read each question about basketball statistics carefully before calculating the answer. Round off calculations to three decimal places.

1. In today's game, Rudy hit 9 field goals in 15 attempts. Calculate Rudy's field goal percentage.

2. Judy made 67 field goals during the season. She attempted 133. What is Judy's field goal percentage?

3. Timothy has played in four games. In Game 1, he made 11 out of 24 field goals. In Game 2, he attempted 17 and made 6. In Game 3, he hit 5 of 15 field goals. In Game 4, he attempted 11 field goals and made 3. What is Timothy's field goal percentage for each game? In which game did he have the best percentage?

4. Caroline has played in four games. In Game 1, she scored 22 points. In Game 2, she made 14 points. In Game 3, she hit 5 baskets for 10 points. In Game 4, she scored 13 points. What is Caroline's points per game average?

5. Lewis made 25 of the 43 free throws he attempted. Calculate his free throw percentage.

6. Jerry made 12 of 21 free throws he tried. Kim hit 7 of 13 free throws. Thomas made 5 out of 9 free throws. Ben attempted

15 free throws and made 6. Which player has the best free throw percentage?

7. Wendy has played in four games. In Game 1, she made 4 rebounds. In Game 2, she grabbed 1. In Game 3, she didn't make a single rebound. In Game 4, however, she had 3. Calculate Wendy's rebounds per game average.

8. Two basketball players are competing for an award for best rebounds per game average. So far, Player 1 has made 44 rebounds and has played in 19 games. Player 2 has made 37 rebounds and has played in 15 games. In the last game of the season, Player 1 made 3 rebounds and Player 2 made 1 rebound. Which player won the award?

Football

Mathematics is used to present a wide variety of statistical information about football. Individual football statistics include *pass completion percentages* for quarterbacks and yards gained per carry (and per game) for running backs. Each of these is simple to calculate.

For example, suppose a quarterback attempted 200 passes in a season and completed 100. Simply divide the number of completions by the number of attempts to calculate the pass completion percentage:

$$100 \div 200 = \frac{100}{200} = .50 = 50\%$$

Thus, this quarterback has a 50% pass completion percentage. Yards per carry is equally simple. Divide the total number of yards gained by the number of times the running back carried the ball.

For example, say that a back gains 1,600 yards in a season. If he carried the ball 400 times, the yards per carry average would be calculated this way:

$$1600 \div 400 = \frac{1600}{400} = \frac{4}{1} = 4$$

Thus, the back averaged 4 yards per carry. To determine his average gain per game, divide the total-yards figure by the number of games in which he played. If this back played in the National Football League, he probably appeared in 16 games. The following calculation produces the yards per game average:

$$1600 \div 16 = \frac{1600}{16} = \frac{100}{1} = 100$$

Other football statistics can be calculated in a similar manner. Examples are a quarterback's average yards gained per pass completion and average yards passing per game.

EXERCISE 6-6

Read each question about football statistics carefully before calculating the answer. Round off calculations to two decimal places.

1. Eli plays quarterback on his high school team. He has attempted 47 passes and completed 21. What is his pass completion percentage?

2. Toby is a running back and has carried the ball 35 times and gained 133 yards.
   a. What is Toby's yards per carry average?
   b. Toby has gained all 133 yards in two games. What is Toby's yards per game average?

3. Otis has completed 67 of 111 passes to gain a total of 603 yards. What is Otis' pass completion percentage? Calculate Otis' average yards gained per pass completion.

4. Ron has caught 36 passes for a total of 252 yards. What is his average yards gained per catch?

5. Clay carried the ball 45 times and gained 176 yards. Brian carried the ball 34 times and gained 143 yards. Joshua carried the ball 41 times and gained 179 yards. Which running back has the best yards per carry average?

6. Rory has played quarterback in four games. In Game 1, he completed 6 of 11 passes. In Game 2, he completed 7 of 9 passes. In Game 3, he hit on 5 of 5 passes. In Game 4, he made 3 of 6 passes. What is Rory's pass completion percentage?

7. Ramon has played in four games. In Game 1, he carried the ball 20 times and gained 82 yards. In Game 2, he gained 74 yards in 17 carries. In Game 3, he ran the ball 11 times for a total of 52 yards. He carried the ball 21 times and gained 124 yards in Game 4. Draw a line graph that shows how Ramon's yards per carry average differed in each game. From your graph, determine if Ramon's average increased or decreased?

8. During his first season, Taylor gained 1,250 yards in 289 carries. In the second season, he carried the ball 256 times for 1,198 yards. During his third season, Taylor gained 2,100 yards and carried the ball 375 times. What is Taylor's yards per carry average for each season? What is Taylor's overall yards per carry average?

MEASUREMENTS AND CONVERSIONS FOR LEISURE

Measurement is an important activity in many sports and leisure activities. Some sports have grading or scoring systems based upon time and/or distance. Examples include:

- Track and field
- Swimming
- Horse racing
- Auto racing
- Downhill ski racing and ski jumping.

Think about how many mathematical functions might be used in scoring a track meet. Running events over short, medium, and long distances are graded in terms of time. A sprint race may be decided by a thousandth of a second. In the mid-1980s, two hours stands as a ''magic barrier'' in the men's marathon.

Field events are graded in terms of distance, either horizontal or vertical. Some participants compete to see how far (long jump,

Conversions are common in some sports, especially those in which performances are graded in terms of time and distance.
COURTESY OF UNIVERSITY OF SOUTHERN CALIFORNIA

triple jump) or how high (high jump, pole vault) they can leap. Others compete to see how far they can throw an object (shot put, discus, javelin, hammer throw).

International track and field events are conducted under the *metric system* of measurement. Examples are the 100-meter dash, 1500-meter run, and the 4 × 100 relay. In the United States, *English measurements* were used for many years, until the national movement toward metrics gained momentum. In the 1980s, American track and field tended to be split between the two measuring systems.

Running events are measured primarily in metric terms, with some exceptions. The mile run, for example, remains a popular American event. The *conversion*, or change, to metrics has aided American track athletes. Americans formerly were at a disadvantage in international events where metric events were run. American sprinters were accustomed to running the 100-yard dash. For international competition, they had to change their training for the 100-meter dash, a distance approximately 10% greater than 100 yards.

The same was true for other races. A mile run, for example, covers 1,760 yards. Converting to metric terms, the mile is approximately 1609 meters. The international distance, frequently called the "metric mile," is only 1500 meters. Again, runners can be placed at a disadvantage when they have to adjust to different distances.

Unlike running events, field events in the United States continue to be measured most often in English terms: feet and inches. Sports reports continue to refer to 18-foot pole vaults, 70-foot shot puts, and 28-foot long jumps as outstanding performances. In coverage of foreign events, graded in metrics, conversions are made to English measurements for American sports fans.

EXERCISE 6-7

Read each question carefully before performing the required conversion. Round off calculations to two decimal places. Refer to the Conversion Chart in Appendix B for English-to-metric ratios.

1. The marathon distance is 26 miles, 385 yards. How many meters is this?
2. How many meters do sprinters run in the 100-yard dash?
3. Jackie has to make a 22-foot jump to win the long jump. If she makes a 7-meter jump, will she win the event?
4. The 1500-meter run often is called the metric mile. How many yards are in 1500 meters? Is this distance more or less than a mile?
5. A popular bicycling event is the 4-Kilometer Pursuit. How many yards do contestants travel to complete this event? How many miles?
6. In football, Steve O'Neal set the record for longest punt with a 98-yard kick. Convert this record from yards to meters.
7. Measured in feet, how much longer is the 100-Meter Dash than the 100-Yard Dash?
8. In the 1952 Olympics, Cy Young won the Gold Medal in the Men's Javelin Throw with a 243-foot toss. How many meters did Cy throw the javelin?

Other Sports Conversions

Swimming is another sport in which time and distance are the standards of performance. American swimming officials switched their sport to metric distances earlier than track and field officials. Growing international competition was one reason, but there was another. Only one English-to-metric conversion is involved in swimming: yards to meters. Thus, it was easy for competitors and spectators alike to accept the conversion. They didn't have to deal with feet and inches, as is necessary in track and field.

Auto racing also has experienced some movement toward metrics in the 1980s, especially in road racing. Some races have been shortened by converting from miles to kilometers. A 500-mile road race, for example, might be changed to 500 kilometers. The conversion shortens the race to approximately 310.7 United States miles.

EXERCISE 6-8

Read each question carefully before performing the required conversion. Round off calculations to two decimal places. Refer to the Conversion Chart in Appendix B for English-to-metric ratios.

1. A popular diving event is the 10-Meter Platform Dive. From how many feet above the water do contestants dive?

2. In the 400-Meter Medley Relay, four swimmers form a team. Each swimmer swims a 100-meter lap. In this event, how many yards does each team member swim?

3. In 1904, Myer Prinstein won the Olympic Gold Medal in the Men's Triple Jump with a 47-foot effort. Convert the distance of Myer's jump from feet to meters.

4. In ski-jumping events, contestants use either a 70-meter or 90-meter ramp to make jumps. Measured in feet, what is the difference between the lengths of these ramps?

5. The dimensions of the playing field in football are 100 yards long by 50 yards wide. Convert these dimensions to meters.

6. Often, the length of a sailboat is calculated by measuring the boat's water line. Usually, this distance is given in meters. Measured in feet, what is the length of a 12-meter boat? A 6-meter boat?

7. A kilometer equals 1,000 meters. How many miles is a 10-kilometer run? A 5-kilometer run?

8. The Triathalon is a three-part event consisting of a 2-mile swim, a 100-mile bicycle ride, and a 12-mile run. Convert the length of each part of the race from miles to kilometers. Then, calculate how many kilometers a triathlete travels during the entire race.

HANDICAPPING: GRADING SKILLS

In some sports, mathematics can be used to make competition closer between participants with different skill levels. This is done by a process called *handicapping.* Handicapping means awarding points to a less skilled participant to bring that person's score closer to an established standard. Major examples of participant sports in which handicapping is common are bowling and golf. Participant sports are those in which large numbers of people are active. Hunting and fishing are other participant sports. Spectator sports, such as football, baseball, boxing, and horse racing, attract many more observers than participants.

Handicapping usually involves adding points to scores in games, such as bowling, where highest scores win. In golf, where the fewest strokes win each hole, handicapping involves subtracting strokes from a total score for 18 holes, called a "round." The following discussions illustrate scoring in these sports and typical handicapping systems.

Scoring Bowling: Fun With Mathematics

Bowling has its own special scoring system, quite different from that of any other sport or game. A game of bowling, also called a "line," consists of 10 increments called "frames." As each frame is played, the score is added to the previous total until all 10 frames have been played. That sounds simple, but actually, the mathematics is somewhat more complicated.

There are 10 bowling pins in a full rack. If the bowler knocks down all 10 pins on the first try, the result is called a "strike." If the bowler does not strike, but knocks down the remaining pins with a second roll, the result is called a "spare." If one or more pins remain standing after two rolls of the ball, the result is an "open frame."

Scoring is simplest when bowlers are least skilled. In a game with all open frames, the total score is the sum of the number of pins knocked down.

Strikes and spares also are called "marks." This is because marks are made in designated places on bowling scoring sheets to note strikes and spares. Examples of bowling scores are illustrated in Figure 6-4. As shown, an "X" is written for a strike, and a slash (/) is written for a spare. Each requires some extra arithmetic.

Figure 6-4. Strikes and spares are important elements of scoring in bowling. Scores increase rapidly when strikes are made in groups.

DO NOT DESTROY THIS SHEET

	Player	1	2	3	4	5	6	7	8	9	10	Total	
1	Heidi	9	/ 29	X 57	X 77	8 / 96	9 0 105	/ 121	6 3 130	X 150	/ 168	8 168	1
2	Jerry	X 30	X 58	X 77	8 1 86	/ 103	7 2 112	/ 130	8 1 139	/ 159	X 179	/ 179	2
3													3
4													4
5													5
6													6
7													7
8													8
9													9
10													10
11													11
12													12
13													13
14													14
15													15
16													16

Spares. For a spare, the score is 10 pins plus the pinfall on the next ball rolled. If the next ball is a strike, the score for the previous frame is 20. A bowler must make at least one spare to reach a score of 100. This is because the maximum score for an open frame is 9. Ten frames multiplied by 9 is 90. If a bowler knocks down 9 pins on each first try and picks up only one spare, the score would be an even 100.

Knocking down 9 pins with the first ball in each frame and making each spare would produce a score of 190. Thus, the value of spares is obvious.

EXERCISE 6-9

Read each question about bowling scores carefully before calculating the answer.

1. Milt made a spare in the first frame. The next ball Milt rolled knocked down 6 pins. What is Milt's score in the first frame?

2. Miriam made a spare in the first frame. The next ball she rolled knocked down 8 pins. What is Miriam's score in the first frame?

3. Yuri had a score of 120 in the fourth frame. In the fifth frame, Yuri made a spare. The first ball Yuri rolled in the sixth frame knocked down 7 pins. What is Yuri's score in the fifth frame?

4. Ingrid had a score of 104 in the sixth frame. In the seventh frame, Ingrid made a spare. The first ball she rolled in the eighth frame knocked down 9 pins. What is Ingrid's score for the seventh frame?

5. Andre knocked down 8 pins in Frame 1, 9 pins in Frame 2, and made a spare in Frame 3. The first ball he rolled in Frame 4 knocked down 8 pins. Andre then made another spare. The first ball Andre rolled in Frame 5 knocked down 9 pins. Andre then made another spare. The next ball he rolled knocked down 7 pins. What is Andre's total score for the first five frames?

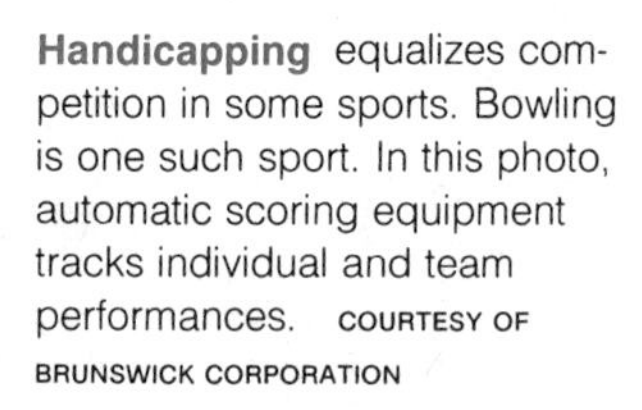

Handicapping equalizes competition in some sports. Bowling is one such sport. In this photo, automatic scoring equipment tracks individual and team performances. COURTESY OF BRUNSWICK CORPORATION

Strikes. For a strike, the score is 10 pins plus the total pinfall on the next two balls rolled. A strike can be scored in several ways, depending upon what a bowler does in succeeding frames.

First, a strike may be preceded by an open frame. If so, the strike frame is scored as 10 pins plus the total pinfall in the following frame. For example, a bowler might get a strike in the first frame. On the first roll in the second frame, 7 pins go down. The second ball knocks down 2 more pins, leaving 1 standing. The score for the first frame is $10 + 7 + 2 = 19$. Since 9 pins were felled in the second frame, the score at the end of that frame is $19 + 9 = 28$.

Second, a strike may be followed by a spare. In that case, the score for the strike frame is $10 + 10 = 20$. If a bowler alternates between strikes and spares for a full game, the score is 200 even. Each frame is scored as 20 pins. A 200 score achieved in this manner is called a "Dutch 200."

Third, a strike might be followed by another strike. Strikes in two successive frames are called a "double." Suppose a bowler starts a game with strikes in the first and second frames. The first frame is scored as 20 pins plus the pinfall of the first roll in the third frame. So, if the bowler knocks down 9 pins in the third frame, the first frame is scored as $10 + 10 + 9 = 29$. The second frame, then, would be scored as 10 plus the total of both balls rolled in the third frame.

Fourth, a bowler may roll three strikes in a row. Three consecutive strikes are called a ''turkey.'' When this happens, the frame in which the first strike is bowled is scored as 10 + 10 + 10 = 30 pins. The second frame would be worth 20 more pins plus the pinfall of the first ball in the fourth frame. If that roll knocked down 9 pins, the score for the second frame would be 30 + 20 + 9 = 59.

EXERCISE 6-10

Read each question about bowling scores carefully before calculating the answer.

1. In the first frame, Ivan scored a strike. On his first roll in the second frame, Ivan knocked down 7 pins. With his second roll, Ivan knocked down 2 pins. What is Ivan's score in the first frame?

2. Harriet made a strike in the first frame and a strike in the second frame. Then, in the third frame, Harriet knocked down 7 pins with her first roll and 2 pins with her second. What is Harriet's score in the first frame? In the second frame?

3. Bill scored a ''turkey.'' On his first roll in the fourth frame, Bill knocked down 5 pins. What is Bill's score in the second frame?

4. Chet scored four strikes in a row. What is Chet's score in the first frame? In the second frame? If Chet made 11 strikes in a row and knocked down 9 pins on his final roll, what would his score be?

5. Ike and Tina have played four frames. Ike knocked down 8 pins in Frame 1, 9 pins in Frame 2, and made a spare in Frame 3. With his first roll in Frame 4, he knocked down 9 pins. Then, with his second roll, he did not knock down any pins. Tina scored a strike in Frame 1. In Frame 2, she knocked down a total of 6 pins. In Frame 3, she knocked down 7 pins. Then, in Frame 4, she knocked down a total of 9 pins. Which player has the best score in Frame 4? How many pins separate the two players?

Bowling Handicaps

Most amateur bowling leagues feature handicapping to make competition as close as possible. Such leagues are called ''handicap leagues.'' Leagues in which all participants are highly skilled may not offer handicaps. Competition without handicaps in bowling or golf is called ''scratch play.''

Bowling handicaps usually are based upon a percentage of the difference between each bowler's average and an arbitrary score. A typical bowling league handicap system might be set as ''two-thirds of 190.'' As a member of this league, your handicap would be figured as follows. Suppose you carry an average of 160. Subtract your average from 190 and multiply by $\frac{2}{3}$ to obtain your handicap:

$$190 - 160 = 30$$
$$30 \times \frac{2}{3} = \frac{60}{3} = 20$$

Thus, your handicap would be 20. This means that 20 pins would be added to your score for each game you play. If you roll a 145 game, you are credited with a 165 game. If you roll a 185, you are credited with a 205. If you maintain your normal average, 160, you earn 180 pins for your team.

Compare this to a bowler who averages 181. This bowler's handicap is figured as follows:

$$190 - 181 = 9$$
$$9 \times \frac{2}{3} = \frac{18}{3} = 6$$

If the 181 bowler rolls his or her average, the handicap score is 187. The result of the handicap system now becomes clear. The 160 bowler, instead of being 21 pins behind the 181 bowler, trails by only 7 pins. This accomplishes three things for a bowling league:

- Closer competition
- Encouragement for the less skilled bowlers
- Challenge for skilled bowlers to improve their scores.

EXERCISE 6-11

Read each question about bowling handicaps carefully. Then calculate the answer. Round off calculations to the nearest whole number. Determine handicaps with the ''two-thirds of 190'' method discussed above.

1. Doris has a 25 handicap. When Doris bowls a 135, what is her adjusted score?

2. Carlos' average score is 140. What is Carlos' handicap? If Carlos bowls a 165, what is his score after the handicap adjustment?

3. Joe's pins per game average is 163. What is Joe's handicap?

4. Randy wants to determine her pins per game average. She has bowled the following scores in her last five games: 145, 170, 110, 125, and 145. Determine Randy's pins per game average.

5. Damian has bowled the following scores: 175, 135, 165, and 150. Herbie has bowled the following scores: 145, 185, 160, and 180. Determine each player's handicap.

6. Damian and Herbie have entered a two-game competition. In the first game, Damian bowled a 175 and Herbie bowled 170. Which player will have the better score after the first handicap adjustment? In the second game, Damian bowled a 150 and Herbie bowled a 140. Now which player will have the better combined score after handicap adjustment?

7. Roberto has bowled games of 160, 166, 182, 174, 147, 132, and 143. Based on these scores, what is Roberto's handicap?

8. Write an algorithm for determining pins per game averages.

Scratch Bowling

In scratch competition, no handicaps are awarded. However, to equalize competition, a limit may be set on the averages of players on individual teams. Typical examples are ''scratch 900'' and ''scratch 950'' leagues.

In a scratch 900 league, the combined averages of five team members must be no greater than 900. That is, the bowlers must average 180 as a group. If one team member has an average of 200, the sum of the averages of the remaining four must not exceed 700 pins.

The limit also applies to substitute bowlers. For example, suppose that a team is right at or near the 900 limit. A 190 bowler would not be allowed to substitute for a 175 bowler. The substitute's higher average would push the team over the league limit.

Scoring Golf: Where Less Is More

Golf is one of the few games in which the lowest score has the highest value. A round of golf usually consists of 18 holes. Golfers also may play nine holes, but scoring and handicapping are based on 18-hole performances.

The typical golf course contains 18 holes of varying lengths and degrees of difficulty. On each hole, golfers begin play from a flat, open section of lawn called a ''tee.'' For this initial club swing, or shot, the ball is placed on a small holder, also called a ''tee.'' The holder raises the ball off the ground an inch or so. Remaining shots are made with the ball resting where it stopped after the previous shot. Golfers try to take as few shots as possible to reach the ''green.'' The green is an area of open lawn in which the hole is located. The golfer's score is the total number of strokes required to move the ball from the tee into the hole.

Because the layout of each hole is different, each is assigned a scoring standard, called ''par.'' Par is expressed as the number of strokes that normally should be required to sink the golf ball in the hole. Most golf holes have par ratings of 3, 4, or 5. Par ratings for 18-hole courses usually range from 70 to 73 strokes. Scores for individual holes are given names, as follows:

Par. When a hole is played in its rated number of strokes, the score is called a ''par.'' A score of par is considered desirable for any hole. A score of par for 9 or 18 holes is considered to be an outstanding performance for all but professional golfers.

Birdie. If a golfer completes a hole with one stroke fewer than par, the score is called a ''birdie.'' The golfer is one stroke under par for the hole.

Eagle. When a hole is completed with two strokes fewer than par, the score is called an ''eagle.'' The golfer is two strokes under par for the hole.

Bogey. A golfer may require one extra stroke to complete a hole. A ''bogey'' is a score one stroke over par. Two strokes over par is a ''double bogey,'' and three strokes over is a ''triple bogey.''

These terms provide a convenient method of identifying the values of scores on holes where values differ. Reporting a score of 5 on the fourth hole has no meaning if the par for that hole is not known. However, if a golfer reported a bogey on the hole, it would be clear that the score was one over par. If the golfer reported a ''bogey 5'' for the hole, it also would be clear that the hole was a par 4.

EXERCISE 6-12

Read each question about golf scores carefully. Then calculate the answer.

1. Sammy scored a double bogey on a par 5. What is Sammy's score for the hole?

2. Lucia scored a birdie on a par 3. What is Lucia's score for the hole?

3. Cleo scored a bogey 6 on the fifth hole. What is par on the fifth hole?

4. Robin scored a triple bogey 6 on the first hole. What is par on the first hole?

5. Howard has finished playing 18 holes of golf. Par for the course is 70. Howard shot 11 pars, 5 bogeys, 1 double bogey, and 1 eagle. What was Howard's final score?

6. Victor and Teresa have finished playing 18 holes of golf. Par for the course is 72. Victor shot 12 bogeys, 1 birdie, and 5 pars. Teresa shot 6 double bogeys, 4 birdies, 5 pars, and 3 bogeys. What is the difference between scores of the two players?

7. The table below shows scores for six golfers. Par for the course is 68. Determine which player shot the best game, second-best game, and third-best game. Also, determine which player's score is closest to par.

Player	Eagles	Birdies	Pars	Bogeys	Double Bogeys
Don Roberts	0	0	12	5	1
Terry Cannon	1	0	10	8	0
Cicero Gomez	0	4	14	0	0
Arnold Gillespi	0	3	5	10	0
Mary Ann Lopez	1	6	6	3	2
Fredette Jones	0	5	8	5	0

8. Eddy has played the same course five times and shot the following scores: 75, 77, 78, 76, and 79. What is Eddy's strokes per game average?

Golf Competition: Stroke Play and Match Play

Competition between or among golfers can be based on either of two systems: stroke play or match play. Both systems utilize the same scoring procedure for individual holes. Most tournaments are conducted under either stroke play or match play rules.

Stroke play. In stroke play, each player is ranked according to the total number of strokes for each 18-hole round. In tournaments, the standings are based upon total scores over one or more rounds. Groupings usually include two to five players. Most of the professional tour events shown on television are scored in this manner.

Match play. Some tournaments, usually amateur championships, are conducted under match play rules. Match play is an elimination form of golf in which scoring is based upon the number of holes won. Competition involves twosomes, with the low scorer on each hole winning that hole. The player who wins the first hole is said to be "1-up." The outcome of a match is determined by the point at which the trailing player is eliminated mathematically.

Golf scores are based on the number of strokes required to sink the ball in the hole. COURTESY OF CARNEGIE LIBRARY OF PITTSBURGH

For example, suppose Player A has won 7 holes and Player B has won 5 as they approach the sixteenth tee. They have tied on three other holes. The leading player is 2-up. Since there are three holes remaining to be played, either player still can win the match.

Player A needs to win just one more hole to end the match. That would give Player A a 3-hole lead with only two holes left to play. The final score of the match would be reported as "3 and 2." This means a lead of 3 holes with 2 remaining to be played.

Player B would have to win all of the final three holes to pull one ahead of Player A and win 8 holes against 7. In this case, the final score would be posted as "1-up" because the match was determined on the final hole.

In another case, if Player C were one hole ahead after 17 holes, Player D could tie the match by winning the final hole. The match then would be extended until one of the players won a hole. This situation is called "sudden death."

On the other hand, if Player C won the final hole, the score would be reported as "2-up."

EXERCISE 6-13

Read each question about golf carefully. Then calculate the answer.

1. Cathy, Monica, and Michi are playing in a three-game, stroke-play tournament. In the first game, Cathy shot a 70, Monica shot a 71, and Michi shot a 68. In the second game, Cathy shot a 71, Monica shot a 68, and Michi shot a 73. In the final game, Cathy shot a 68, Monica shot a 72, and Michi shot a 72. Which player won the tournament?

2. Don, Ted, and Jack played a three-game, stroke-play tournament. Don shot the following scores: 71, 68, and 70. Ted made the following scores: 70, 73, and 69. Jack shot these scores: 67, 73, and 71. Which golfer won the tournament? What is the difference between the first-place contestant's score and the last-place contestant's score?

3. Kent and Lori are playing match-play rules. After 15 holes, Lori is 2-up. If Kent wins the sixteenth, seventeenth, and eighteenth holes, will he win the game?

4. Gail and Hector were playing match-play rules. They were tied going into the last hole. On the eighteenth hole, Hector scored a birdie and Gail scored a bogey. Who won the match?

5. In an 18-hole, match-play tournament, Player A is 2-up on Player B going into the seventeenth hole. If Player B wins the seventeenth and eighteenth holes, will he win the tournament?

6. Four golfers have finished a four-game, stroke-play tournament. Player 1 shot a 68, 73, 71, and 75. Player 2 shot a 69, 74, 71, and 75. Player 3 shot a 75, 78, 67, and 68. Player 4 shot a 73, 72, 73, and 71. Which player won the tournament? Which player had the best strokes per game average?

7. Ai-ling is competing in a four-game, stroke-play tournament. He needs a total score of 282 or less to win the tournament. Ai-ling has turned in the following scores: 71, 70, 69, and 71. Will Ai-ling win?

8. Mona and Allen are on the fifteenth hole in a match-play tournament. Allen is 1-up. If Mona wins the fifteenth and sixteenth holes, Allen wins the seventeenth hole, and Mona wins the eighteenth hole, who will win the tournament? What will the final score be?

Golf Handicapping: An Equalizing Grading System

In organized golf competition, such as that sponsored by golf clubs, handicaps typically are based on 18-hole average scores. Clubs keep records of members' scores. When a player has established an average, his or her handicap can be assigned.

For example, a player might average 85 in playing a par-72 course. That player's handicap would be 13. Simply subtract par from the player's average. Thus, a player with a 91 average would have a 19 handicap on a par-72 course. When making a handicap adjustment to a player's score, the handicap is subtracted from the score. For example, a player with a handicap of 17 who scores 95 in a game, would receive a score of 78.

Some players develop their skills to a point at which they shoot scores consistently at or near par. These players are considered ''scratch golfers.'' In other words, they don't need any handicap.

EXERCISE 6-14

Read each question about golf handicapping carefully before calculating the answer.

1. Jeff has scored an 85. His handicap is 13. What is Jeff's score after the handicap adjustment?

2. Rudy shot the following scores in his last four golf games: 97, 82, 102, and 89. Based on these scores, what is Rudy's strokes per game average?

3. Joji's scores for his last four golf games are 98, 94, 88, and 92. Based on these scores, what is Joji's handicap if par for the course is 69?

4. If Joji scores a 90 in tournament play, what is his score after handicap adjustment?

5. Julio has a handicap of 15. He shot the following scores in a three-game tournament: 86, 90, and 92. Apply Julio's handicap to each game score. What is Julio's adjusted score for all three games?

6. Cisco is playing Reiko in a three-game, stroke-play match. Cisco has a handicap of 20 and shot 93, 94, and 94. Reiko has a 6 handicap and shot 76, 80, and 83. After applying player handicaps to each game, which player has the better score for the three games?

7. Three golfers are competing in a four-game, stroke-play match. Player 1 has an 18 handicap and shot the following scores: 89, 86, 93, and 90. Player 2 has a 16 handicap. He scored 90, 85, 87, and 83. Player 3 has a 24 handicap and she shot these scores: 93, 92, 94, and 90. Which player won the match? What is the difference between the winning and last-place scores?

8. Two golfers are playing a par-70 course in a three-game, stroke-play match. Golfer 1 shot the following scores: 78, 80, and 79. Golfer 2 shot these scores: 98, 92, and 95. Based on scores, determine each player's handicap for this course. After handicap adjustments to all three games, which player has the better score? What is the difference between the winner's adjusted score and the loser's adjusted score?

SUMMING UP

- Mathematics is a key part of most sports and games. The most basic procedures for keeping and reporting scores involve mathematics.
- Mathematics also is necessary for other sports activities, such as presenting information, establishing grading systems, and calculating measurements and conversions.
- Tables are used widely to present sports information. Team standings in a variety of sports are one example. Other uses for tables include individual performances and averages, team statistics, and future schedules.
- Grading systems include formulas for rating athletic performances, such as baseball batting and fielding averages, won-lost records, and others. Most sports grading systems are based on percentages, such as shots or goals scored divided by the number attempted.
- Some sports are scored by relating performances to measurements of time, distance, and/or weight. Track and field, swimming, and weightlifting are examples.
- Conversions have an important role in some sports. Track and field is an example in which events may be measured by either English or metric standards, or both.
- Some sports, such as bowling and golf, utilize handicapping systems. Handicapping is designed to make competition as close as possible among participants with varying levels of skill. A typical handicap either adds to or subtracts from a score to bring it closer to an established standard.

TERMS THAT COUNT

statistics
tabular
batting average
earned run average (ERA)
fielding percentage
pass completion percentage
metric system
English measurements
conversion
handicapping

TESTING YOUR WORKING KNOWLEDGE

Read each question carefully before calculating the answer. Round off calculations to two decimal places.

1. The Blue Jays have four pitchers. Pitcher 1 gave up 111 runs in 312 innings. Pitcher 2 gave up 123 runs in 324 innings. Pitcher 3 gave up 87 runs in 256 innings. Pitcher 4 gave up 45 runs in 190 innings. Which pitcher has the lowest earned run average?

2. In 1983, New York Jets running back, Freeman McNeil, carried the ball 152 times and rushed for 786 yards. What was his yards per carry average?

3. In 1972, Norm Snead, quarterback for the New York Giants, completed 196 of 325 passes. What was Norm Snead's pass completion percentage for 1972?

4. In the 1961–62 basketball season, Wilt Chamberlain scored 4,029 points during 80 games. What was Wilt Chamberlain's points per game average for that season?

5. Linda entered a bowling league that uses the ''two-thirds of 190 method'' to determine handicap. Scores for her first four games were: 145, 132, 167, and 112. What is Linda's pins per game average and handicap?

6. John bowled a 143 in Game 1, a 176 in Game 2, and a 156 in Game 3. Cathy bowled a 175 in Game 1, a 130 in Game 2, and a 142 in Game 3. What is each player's pins per game average? What is each player's handicap? Round off calculations to the nearest whole number.

7. Three golfers were playing a par-68 course in a three-game, stroke-play match. Player 1 shot 89, 90, and 84. Player 2 shot 94, 97, and 91. Player 3 shot 79, 80, and 76. What is each player's handicap for this course? What is each player's three-game score after handicap adjustments? Who won? Round off calculations to the nearest whole number.

PRESENTING INFORMATION

Read each problem carefully and create the table requested.

1. Lewis, Carol, Betty, Dale, and Fred are on the same basketball team and have played in four games this season. Lewis scored 10 points in Game 1, 5 in Game 2, 15 in Game 3, and 13 in Game 4. Carol scored 10 points in Game 1, 18 in Game 2, 2 in Game 3, and 14 in Game 4. Betty scored 8 points in Game 1, 6 in Game 2, 12 in Game 3, and 10 in Game 4. Dale scored 18 points in Game 1, 6 in Game 2, 12 in Game 3, and 6 in Game 4. Fred scored 12 points in Game 1, 14 in Game 2, 8 in Game 3, and 15 in Game 4. Create a table that shows:
   - **a.** How many points each player scored in each game.
   - **b.** Total points scored by each player.
   - **c.** Points per game average for each player. Round off calculations to one decimal place.

2. In the Desert Golf Tournament, Tom Thomas made 4 pars, 3 birdies, 10 bogeys, and 1 eagle; Linda Watson made 11 pars, 4 birdies, 2 double bogeys, and 1 bogey; Gary Weber made 18 pars; Emilia Ramirez made 12 pars, 4 birdies, and 2 double bogeys; Yukio Mashima made 14 pars, 1 double bogey, and 3 birdies; and Olivia Prima made 12 pars, 5 birdies, and 1 double bogey. Par for the course is 70. Create a table that:
   - **a.** Shows how many eagles, pars, birdies, etc. each player scored.
   - **b.** Totals each player's score for 18 holes.
   - **c.** Lists each player in order of finish.

SKILL DRILLS

A. Find the average in each problem. Round off calculations to the nearest whole number.

1. 76, 90, 34, 56, 87
2. 111, 145, 325, 213
3. 99, 101, 78, 45, 123, 45
4. 125, 120, 132, 141, 95
5. 674, 435, 657, 234, 675, 476, 872
6. 234, 213, 243, 222, 254, 203, 231
7. 121, 321, 298, 401, 284, 412
8. 324, 765, 456, 918, 756, 924, 463

B. Find the batting average in each problem. Round off calculations to three decimal places.

1. 8 hits in 23 at-bats
2. 13 hits in 89 at-bats
3. 5 hits in 36 at-bats
4. 25 hits in 134 at-bats
5. 113 hits in 503 at-bats
6. 56 hits in 321 at-bats
7. 34 hits in 154 at-bats
8. 79 hits in 303 at-bats

C. Perform the conversions required in each problem. Round off calculations to two decimal places.

1. 100 miles equal how many kilometers?
2. 230 feet equal how many meters?
3. 640 yards equal how many meters?
4. 2 miles equal how many meters?
5. 35 meters equal how many feet?
6. 400 meters equal how many yards?
7. 10 meters equal how many inches?
8. 5 kilometers equal how many feet?

D. Calculate the following golf scores for a par-72 course.

1. 5 under par
2. 6 over par
3. 2 under par
4. 1 under par
5. 23 over par
6. 17 over par
7. 4 under par
8. 13 over par

7

Mathematics and Your Future

YOUR LEARNING JOB

When you have completed the exercises and assignments for this unit, you should be able to:

- ☐ Define cost, revenue, profit, principal, interest, simple interest, and compound interest.
- ☐ Describe the clearing house function of the Federal Reserve System.
- ☐ Calculate simple interest.
- ☐ Calculate compound interest based on differing frequencies.
- ☐ Explain why interest builds up more rapidly as the frequency of compounding is increased.
- ☐ Explain the value of an individual retirement account in terms of tax benefits.
- ☐ Explain the value of education and/or training in terms of future earnings.

WARM-UP

A. Convert the following fractions to their percentage equivalents.

1. $\frac{1}{10}$
2. $\frac{13}{52}$
3. $\frac{5}{25}$
4. $\frac{9}{100}$
5. $\frac{33}{300}$

B. Convert the following percentages to their decimal equivalents.

1. 9%
2. 25%
3. 33%
4. 7%
5. 13%

C. Compute the following percentages of whole numbers. Round calculations to two decimal places.

1. 10% of 90
2. 15% of 60
3. 6% of 33
4. 113% of 500
5. 106% of 7,300

D. Compute the following percentages of monetary values. Round calculations to the nearest cent.

1. 12% of $234
2. 15% of $3,870
3. 11% of $1,500
4. 9% of $850
5. 6.5% of $1,145

PLANNING FOR YOUR FUTURE

When people talk about planning for the future, they usually mean planning for their financial security. Financial security means having enough money or enough *income* to provide for life's necessities. Income is a regular flow of money that can be used to meet needs or wants. For individuals, incomes can be spent for living expenses and for other purposes, such as entertainment.

Income resulting from labor usually is paid in the form of money. Paychecks and payrolls are discussed in Units 2 and 3. Financial security is something people desire throughout their lives. The key to financial security is the wise use of money. The discussions that follow cover uses of money to produce income—and financial security.

MONEY: A MATTER OF INTEREST

If you went to a swap meet or a garage sale and bought a used chair for $10, the $10 would be the *cost* of the item. Now, say that a friend is visiting at your home and admires the chair. Your friend offers you $20 for the chair, and you agree to sell it. The $20 would be *revenue* from the sale.

The difference between your cost and the revenue you receive ($20 – $10 = $10) is called *profit.* In this case, your profit comes from the sale of a product. Profit also can come from the use of money. One party makes money available for another. The party that owns the money charges the other party for its use. Such charges for the use of money are called *interest.*

Borrowing Money

When you borrow money from a bank or other financial institution, you really are buying the use of that money. You pay for the use of the lender's money by paying back the *principal,* which is the amount borrowed, plus interest.

When the use of money is bought and sold, the charge by the lender is stated as an *interest rate,* or percentage.

For example, say that you want to borow $100 from a bank to purchase a bicycle. The bank may charge an interest rate of 15%. Since 15% of $100 is $15, the bank will loan you the $100 for $115 in payments. This procedure allows you to pay for the money in several payments, such as $11.50 a month for 10 months.

Of the $115 the bank will receive, $100 is the principal and $15 is the profit, or interest.

EXERCISE 7-1

Read each question carefully before calculating the answer. Round calculations to the nearest cent.

1. Carl purchased a used motorcycle for $500. He fixed it up and sold it for $650. How much profit did Carl make?

2. The bank loaned Charro $4,320 at 18% interest per year. In one year, how much, in principal and interest, will Charro need to repay the loan?

3. Rudolph borrowed $500 at 8% interest per year. In one year, how much, in principal and interest, will Rudolph need to repay the loan?

4. Lucia loaned her friend $300 at 11% interest per year for one year. How much money will the friend give to Lucia to repay the loan? How much profit will Lucia make?

5. Gail borrowed $980 at 18% interest per year from her bank. How much will Gail need to repay the loan in one year?

Money as an Investment

You can be a lender, or *investor*, of money as well as a borrower. You can invest, or *deposit*, money in a savings account, for example, and the bank will pay you interest. The longer you allow your money to remain in the bank or other financial institution, the more interest it will earn.

How Banks Make Money

A basic principle of banking involves the setting of interest rates for borrowers and for investors, or depositors. In simple terms, a bank charges higher interest for loans than it pays out on deposits.

For example, you might receive 8.5% interest on a savings account. The bank is paying you 8.5% for use of your money. The bank can lend your money to a borrower and charge 14% interest. The difference between the 14% interest the bank earns and the 8.5% becomes lending profit.

EXERCISE 7-2

Read each question carefully before calculating the answer. Round calculations to the nearest cent.

1. Jerome has $500 in a savings account that earns 6.5% interest per year. How much money will Jerome have in his account after one year?

2. Ms. Billings placed $2,000 in a savings account that earns 5% interest per year. How much money will she have in her account after one year? How much of this money will be interest?

3. Mr. Mitty has $5,000 in a savings account that earns 5.75% interest per year. How much does it cost the bank to use Mr. Mitty's money for one year?

4. Jackie has $6,540 in a savings account that earns 5.5% interest per year. How much money will be in Jackie's account after one year? How much of this money will be profit (interest)?

Banks and the Federal Reserve System

To meet your personal needs, you probably will do business with one bank. This bank, or branch office of a large bank, will usually be located near your home. However, this bank connects you with a vast financial network that deals in billions of dollars every day.

Consider a relatively simple situation: Every two weeks, your employer gives you a check covering your earnings—less taxes and other withholdings. This check is probably written against an account in a bank different from your own. Yet, you expect the convenience of being able to deposit the check in your account. It is up to your bank, then, to get its money from the bank against which the check is written.

Some organized method is needed for the interchange of checks among banks. Some 16,000 commercial banks offer checking accounts in the United States. Many thousands of savings institutions also offer checking service. It simply would not be practical for all of these banks to try to exchange checks directly with each other. Instead, all of the banks use a *clearing house* service provided by one, central bank. This central bank, or main banking source for the entire country, is the *Federal Reserve System,* abbreviated *FED.*

The FED operates 12 regional Federal Reserve Banks in large cities. These regional banks provide clearing house service for checks written on accounts in all member banks. A clearing house receives checks from all banks in which they are cashed. The FED then credits each check to the account of the bank where it was cashed. At the same time, money is taken from the account of the bank on which the check was issued. The clearing house also returns all checks to originating banks. Without this kind of service, the use of checks for buying, selling, or payrolls would not be possible.

The FED also oversees the operation of its member (check writing) banks. Banks deposit some of their funds with the FED to be sure that money is available to meet customer demands. The money kept by the FED is the *reserve* of the depositing banks. Among its responsibilities, the Federal Reserve establishes the amount of reserve that member banks must maintain. This capability, in turn, determines how much money is in circulation in the country. For example, if the FED raises the reserve rate, banks must keep more money on deposit. This leaves less money for banks to lend. If banks can't lend the money, stores can't sell on credit. The entire money market of the country is affected by such decisions.

Other government agencies also stand behind the services you receive from your local bank. A major requirement that affects you is *deposit insurance.* If a bank should lose money and be unable to pay back its depositors, government agencies step in. These agencies insure deposits in banks and savings institutions, usually up to $100,000 per account. Bank deposits are insured by the Federal Deposit Insurance Corporation (FDIC). Savings and loan association deposits are insured by the Federal Savings and Loan Insurance Corporation (FSLIC).

SIMPLE INTEREST

If you borrow money at interest, you agree to pay the lender more than you borrow. For instance, say that you borrow $1,000 for one year at 15%. You would pay the lender $1,000 plus 15% of that amount, or $150, in interest.

When interest is calculated only once, on the original principal of a loan, it is called *simple interest.* Interest rates usually are stated on a yearly basis.

Calculating Interest

If a simple-interest loan is paid back in one year or less, the interest can be calculated by a simple equation: I = PRT. This means:

$$\text{Interest (I)} = \text{Principal (P)} \times \text{Rate (R)} \times \text{Time (T)}$$

For instance, say that you borrowed $1,500 for one year at 18% simple interest. The calculation would be:

$$\text{Interest} = 1{,}500 \times 0.18 \times 1 = 270$$

Your interest would total $270 for the one-year period of the loan. However, you might wish to pay off the loan in less than a year. If the loan is repaid more quickly, less interest is charged, even though the annual rate remains the same. Suppose that you agreed to pay off the loan in nine months instead of 12 months. Nine months is three-quarters of a year, or 0.75 of a year. Again, use the formula:

$$\text{Interest} = 1{,}500 \times 0.18 \times 0.75 = 202.50$$

Your interest is only $202.50 if the loan is repaid in nine months. This is a reduction of $67.50 in interest. To illustrate further, suppose that you borrowed the money for only six months. In this case, the interest is even lower:

$$\text{Interest} = 1{,}500 \times 0.18 \times 0.50 = 135$$

The interest if the loan is repaid over six months is $135. This is only half of the amount paid on a one-year repayment schedule.

EXERCISE 7-3

Read each question carefully before calculating the answer. Round calculations to the nearest cent.

1. Jeff borrowed $800 from the bank at 17% simple annual interest. The loan contract calls for repayment in nine months. How much interest will Jeff be charged?

2. Frances borrows $300 for a six-month period at 16% simple annual interest. How much will she repay the lender?

3. Ron borrowed $750 from his parents and agreed to pay them 12% simple interest at an annual rate. Ron made two equal payments, one after six months and the second at the end of the year. How much did Ron pay his parents each time?

4. Lorraine lent $400 to her friend at 8.5% simple annual interest. The loan is to be paid off in three months. How much interest will Lorraine receive from her friend?

5. Barry borrowed $1,200 for a year at 17% simple interest. His income increased, and he arranged to pay off the loan in nine months instead of a year. How much interest did he save?

Calculating Loan Payments

Many consumer loans are based upon simple interest calculations. Typical examples are loans for automobiles, major appliances, and furniture. These are considered personal loans.

Simple interest, however, does not imply simple payment calculations. This is true especially for loans that are repaid over periods longer than a single year. In practice, loan payments are based on a highly complicated formula. The calculations are so complicated, in fact, that lenders use tables or computer printouts that show how loans are paid off.

The complexity occurs because most consumer loan payments are calculated on a *declining balance* basis. That is, each payment reduces the principal, and interest is calculated on the remaining balance.

Loan payments usually are calculated so that most of the loan interest is paid early in the schedule. For example, if you had a three-year loan, you would pay the bulk of the interest in the first two years. Your early payments would not reduce the principal as rapidly as the last few payments.

Therefore, even though all 36 payments may be the same, they are split differently between principal and interest. Each succeeding payment represents slightly less interest and slightly more principal.

An example of the payment schedule for a 36-month loan is shown in Figure 7-1. Notice how the earned interest decreases slightly with each payment. Also, notice how the principal reduction increases slightly with each payment.

EXERCISE 7-4

Read each question carefully before calculating the answer. All questions are based upon the loan repayment schedule in Figure 7-1. Round all calculations to the nearest cent.

1. During the first year of the loan, how much interest is paid? How much is the principal reduced?
2. How much interest is paid during the second year of the repayment schedule? By how much does this total differ from interest in the first year?
3. What percentage of the loan principal is repaid in the first payment? In the eighteenth payment? In the final payment?
4. How much of the loan principal is repaid during the first year of the loan? During the second year? During the final year?
5. During which month of the payment schedule is the remaining balance reduced to less than 50% of the original principal? How much has been paid in interest at that point? How much interest remains to be paid?

LOAN REPAYMENT SCHEDULE

ITEM: sample
PRINCIPAL: 7,000.00 ANNUAL INTEREST RATE: 12.200
TERM: 36 PAYMENTS PAYMENTS PER YEAR: 12

PMT. NO.	PAYMENT AMOUNT	EARNED INTEREST	PRINCIPAL REDUCTION	REMAINING BALANCE
				7,000.00
1	233.17	71.17	162.00	6,838.00
2	233.17	69.52	163.65	6,674.35
3	233.17	67.86	165.31	6,509.04
4	233.17	66.18	166.99	6,342.05
5	233.17	64.48	168.69	6,173.36
6	233.17	62.76	170.41	6,002.95
7	233.17	61.03	172.14	5,830.81
8	233.17	59.28	173.89	5,656.92
9	233.17	57.51	175.66	5,481.26
10	233.17	55.73	177.44	5,303.82
11	233.17	53.92	179.25	5,124.57
12	233.17	52.10	181.07	4,943.50
13	233.17	50.26	182.91	4,760.59
14	233.17	48.40	184.77	4,575.82
15	233.17	46.52	186.65	4,389.17
16	233.17	44.62	188.55	4,200.62
17	233.17	42.71	190.46	4,010.16
18	233.17	40.77	192.40	3,817.76
19	233.17	38.81	194.36	3,623.40
20	233.17	36.84	196.33	3,427.07
21	233.17	34.84	198.33	3,228.74
22	233.17	32.83	200.34	3,028.40
23	233.17	30.79	202.38	2,826.02
24	233.17	28.73	204.44	2,621.58
25	233.17	26.65	206.52	2,415.06
26	233.17	24.55	208.62	2,206.44
27	233.17	22.43	210.74	1,995.70
28	233.17	20.29	212.88	1,782.82
29	233.17	18.13	215.04	1,567.78
30	233.17	15.94	217.23	1,350.55
31	233.17	13.73	219.44	1,131.11
32	233.17	11.50	221.67	909.44
33	233.17	9.25	223.92	685.52
34	233.17	6.97	226.20	459.32
35	233.17	4.67	228.50	230.82
36	233.17	2.35	230.82	0.00
TOTAL	8,394.12	1,394.12	7,000.00	0.00

Figure 7-1. Loan repayment schedule illustrates how balance of principal is reduced as payments are made.

FINANCE CHARGES

When you borrow money, there may be costs involved in addition to interest charges. These costs may apply to monetary loans and to *credit purchases.* A credit purchase is one in which you receive a product or service and pay over a period of time. Such purchases also are called *installment purchases,* and the payments are called *installments.*

One additional cost is called a *loan origination fee.* A loan origination fee is an extra charge, usually calculated as a percentage of the principal.

Some lenders also require that a borrower purchase a special insurance policy. Such insurance usually is called *credit life insurance.* The policy, usually arranged by the lender, pays off your loan if you die or become disabled.

The sum of interest charges and other loan costs is the *finance charge,* or total cost of borrowing money. Laws require that all charges made by a lender or seller must appear in a loan or installment purchase contract.

The point is that the real cost of borrowing often is greater than the interest charge. A borrower should be aware of the total finance charge before signing a loan or credit purchase agreement.

Contracts are legal documents. Borrowers should determine the total finance charge before signing a loan or credit purchase contract. PHOTO BY CLIFF CREAGER

EXERCISE 7-5

For each question, calculate the finance charge. Round all calculations to the nearest cent.

1. The principal is $1,560. Interest is 12%. The loan is to be repaid in 12 months. There is a charge of $8.50 for credit life insurance.

2. The principal is $400. Interest is 11%. The duration of the loan is nine months.

3. The principal is $2,000. Interest is 18% for one year. There is a loan origination fee of 1%.

4. The principal is $3,640. Interest is 9.75%. The loan is to be paid off in 12 months. A credit life insurance premium of $18.20 is required.

5. The principal is $1,850. Interest is 13.5% for one year. A loan origination fee of 2% is charged, and credit life insurance costs $9.25.

COMPOUND INTEREST

When you deposit or invest your money, you want it to earn as much as possible. For this reason, most financial institutions offer savings accounts with *compound interest.* Compound interest is calculated by adding interest earned to an account balance before calculating interest for the next period. The interest rate is stated as an annual percentage. However, interest may be compounded annually (once a year), quarterly (every three months), monthly, weekly, and even daily. The same formula is used: I = PRT. The difference is that the T (time) is a fraction or percentage of a year. If interest were compounded monthly, for instance, the value of T would be $\frac{1}{12}$. When interest is compounded daily, the T value is $\frac{1}{360}$. Banks and other financial institutions often use 360 days to represent a year. The use of 360 is an example of rounding to simplify calculations.

To illustrate how compounding of interest works, consider an investment of $1,000 at an annual interest rate of 12%. The following examples show what happens to that money under compound interest applications.

Simple interest. Under the formula I = PRT, interest would be added once, at the end of the year:

$$1{,}000 \times 0.12 = 120$$

Thus, the balance in the account at the end of the year would be \$1,000 + \$120 = \$1,120. If the saver left the interest in the account, it would be added to the balance. Assuming that the saver let the interest remain in the account, the interest for the second year would be:

$$1{,}120 \times 0.12 = 134.40$$

The account balance at the end of the second year would be \$1,254.40. The \$1,000 investment would have earned \$245.40. In effect, this amounts to compounding annually. This is because the second year's interest is calculated on a principal that includes the first year's interest.

Interest compounded monthly. At the end of the first month, interest would be calculated as follows:

$$I = 1{,}000 \times 0.12 \times \frac{1}{12}$$

$$I = \frac{120}{12} = 10$$

Thus, \$10 would be added to the account at the end of the first month, leaving a new balance of \$1,010. At the end of the second month, the same process would occur. The table in Figure 7-2 shows how the account would grow during the first year under monthly compounding.

The account balance at the end of the first year is \$1,126.84, which means that the investment earned \$126.84. This is \$6.84 more than was earned by compounding annually. The difference in earnings increases as time goes on. At the end of two years, under monthly compounding, the account balance is \$1,269.74. This is \$15.34 greater than the balance under the annual compounding method described above.

Month	Starting Balance	Interest Calculation	Ending Balance
1st	$1,000	1,000 × 0.01 = 10.00	$1,010
2nd	1,010	1,010 × 0.01 = 10.10	1,020.10
3rd	1,020.10	1,020.10 × 0.01 = 10.20	1,030.30
4th	1,030.30	1,030.30 × 0.01 = 10.30	1,040.60
5th	1,040.60	1,040.60 × 0.01 = 10.41	1,051.01
6th	1,051.01	1,051.01 × 0.01 = 10.51	1,061.52
7th	1,061.52	1,061.52 × 0.01 = 10.62	1,072.14
8th	1,072.14	1,072.14 × 0.01 = 10.72	1,082.86
9th	1,082.86	1,082.86 × 0.01 = 10.83	1,093.69
10th	1,093.69	1,093.69 × 0.01 = 10.94	1,104.63
11th	1,104.63	1,104.63 × 0.01 = 11.05	1,115.68
12th	1,115.68	1,115.68 × 0.01 = 11.16	1,126.84

Figure 7-2. Monthly compounding involves multiplying the balance by one-twelfth of the annual interest rate each month. Each month's interest is added to the balance prior to the next month's calculation.

EXERCISE 7-6

For each question, calculate the ending balance after one year. Round all calculations to three decimals. Round the answer to the nearest cent.

1. Carlos has $500 deposited in a savings account that earns 5.5% interest annually. Interest is compounded monthly.

2. Jill has $3,500 deposited in a savings account that earns 6.5% interest annually. Interest is compounded monthly.

3. Benito has $4,670 deposited in a savings account that earns 8% interest annually. Interest is compounded monthly.

4. Kuang-fu has $713 deposited in a savings account that earns 5% interest annually. Interest is compounded monthly.

5. Saburo has $6,900 deposited in a savings account that earns 7% interest annually. Interest is compounded monthly.

Long-Term Investments

The advantage of frequent compounding shows up more dramatically when an investment is allowed to grow over longer periods. For this illustration, \$20,000 is being invested for a period of 10 years. In Case A, the investment will earn 12% interest compounded annually, with all earnings remaining in the account. In Case B, the 12% interest rate will be compounded monthly.

Case A involves a simple algorithm:

Algorithm	Numerical
1. Multiply balance by 0.12.	$20{,}000 \times 0.12 = 2{,}400$
2. Add result to balance.	$20{,}000 + 2{,}400 = 22{,}400$

These two steps produce the interest and new balance at the end of the first year in Case A. Repeating those steps nine more times will give the balance after 10 years: \$62,116.98.

However, there is an easier way to calculate compound interest. Multiplying the balance by the interest rate and then adding the result to the balance carried forward is a two-step procedure. It is possible to perform these functions in a single procedure: Simply multiply the balance by 1 + the interest rate. Multiplying by 1 is the same as adding the balance. In the case above, the algorithm would be expressed in the following step:

Algorithm	Numerical
Multiply balance by 1.12.	$20{,}000 \times 1.12 = 22{,}400$

The same procedure applies to compounding at more frequent intervals. For Case B, the equation I = PRT involves multiplying the principal by the interest rate, 12%, and by the time, which is one month, or one-twelfth of a year. This can be done in the following steps:

$$20{,}000 \times 0.12 = 2{,}400$$

$$2{,}400 \times \frac{1}{12} = 2{,}400 \div 12 = 200$$

The interest for the first month, therefore, is $200. Adding the $200 to the original balance of $20,000 gives a new balance of $20,200 after one month. You can see that many calculations would be necessary to calculate compounded interest over a 10-year period. You would have to repeat these calculations 120 times. Obviously, a simpler method is desirable.

Instead, multiplying the annual interest rate by the time factor will give you a constant interest rate to be calculated monthly. In the case of 12% annual interest compounded monthly, the interest rate factor is calculated this way:

$$R \times T = 0.12 \times \frac{1}{12} = \frac{0.12}{12} = 0.01$$

Thus, multiplying the balance by 0.01, or 1%, produces the interest each month. Further, using the factor of 1 to carry forward each previous balance enables you to complete each month's computation in one simple step. Just multiply the previous balance by 1.01. The progression is shown in Figure 7-3.

Notice that the balance at the end of the first year is $22,536.49. This is $136.49 more than the $22,400 balance calculated on 12% interest compounded annually. The difference

Month	Beginning Balance	Monthly Calculation	Ending Balance
1st	$20,000	20,000 × 1.01	$20,200
2nd	20,200	20,200 × 1.01	20,402
3rd	20,402	20,402 × 1.01	20,606.02
4th	20,606.02	20,606.02 × 1.01	20,812.08
5th	20,812.08	20,812.08 × 1.01	21,020.20
6th	21,020.20	21,020.20 × 1.01	21,230.40
7th	21,230.40	21,230.40 × 1.01	21,442.70
8th	21,442.70	21,442.70 × 1.01	21,657.13
9th	21,657.13	21,657.13 × 1.01	21,873.70
10th	21,873.70	21,873.70 × 1.01	22,092.44
11th	22,092.44	22,092.44 × 1.01	22,313.36
12th	22,313.36	22,313.36 × 1.01	22,536.49

Figure 7-3. An easy method for monthly compounding is to multiply the balance by a factor of 1 plus the interest rate. This eliminates the necessity of adding the balance to each calculation.

in earnings becomes more pronounced as time goes on. The difference after five years, for example, is $1,450.41. With monthly compounding, the balance at the end of 10 years would be $66,007.74, an improvement of $3,890.78 over annual compounding.

EXERCISE 7-7

Read each question carefully before calculating the answer. Round all calculations to three decimals. Round the answer to the nearest cent.

1. Bill has two savings accounts. Account 1 earns 5.5% interest per year. Interest is compounded quarterly. Account 2 earns 6% interest per year. Interest is compounded annually. Bill has $1,000 deposited in each account. What will be the ending balance of each account after one year?

2. Eddy has $5,000 in a savings account that earns 6.5% interest per year, compounded annually. Madelline has $5,000 in a savings account that earns 6% interest per year and in which interest is compounded quarterly. What will be the difference between Eddy's ending balance and Madelline's ending balance after one year?

3. Benji has $500 in a savings account that earns 5% interest per year and in which interest is compounded quarterly. Nancy has $500 in a savings account that earns 5% interest per year and in which interest is compounded monthly. In comparison with Benji's account, how much more will Nancy have in her account after one year?

4. Paco has $10,000 in a savings account that earns 6% interest, compounded annually. Rhett also has $10,000 in a savings account. However, Rhett's account earns 5.75% interest, compounded monthly. After one year, who will have the higher ending balance, Rhett or Paco? How much more money will that person have in his account?

5. Lawanda has $13,500 in a savings account that earns 6% interest, compounded quarterly. If she allows all interest to remain in the account, how much will she have after two years?

INVESTMENTS AND YOUR FUTURE SECURITY

The effect of compounding interest is proof of the saying that the best time to start saving is when you are young. The longer your savings earn interest, the greater their value. There are several sources of retirement income available to most working citizens. The following discussion describes the most flexible type of retirement savings program and underscores the value of compounding interest.

Individual Retirement Account (IRA)

To illustrate, suppose that you opened an *individual retirement account (IRA)* at the age of 22. An individual retirement account is a special investment program that offers substantial income tax benefits as you save for retirement. IRA investors save on taxes in three ways:

- Money invested in an IRA in any year becomes an income tax deduction for that year up to a certain amount. In the mid-1980s, each individual working person could deduct up to $2,000 of IRA deposits annually.

- Interest earned on an IRA account is not taxed at the time it is earned. IRA account holders pay no income tax on those monies until they are withdrawn.

- Individuals typically have lower income levels after retirement. Thus, the tax rate on funds withdrawn from an IRA is significantly lower than the rate paid on earned income.

Suppose that, starting at age 22, you invested $2,000 in an IRA each year during your working life. Also, suppose that your IRA account earned 12% interest, compounded annually. Under these circumstances, you would be a millionaire by the age of 58. Further, if you elected to retire at the age of 65, your account balance would exceed $2 million ($2,421,609.62).

Consider that, in 43 years of investing, your total contribution to the IRA would be only $86,000. More than $2,335,000 would be interest. This money growth is illustrated by the chart in Figure 7-4. For simplicity, this chart uses simple interest compounded annually. If interest were compounded more frequently, earnings would be higher.

Year	Interest Earned
1st	$ 240.00
5th	1,524.68
10th	4,211.69
20th	17,292.58
30th	57,919.84
40th	184,101.94
43rd	259,459.83

Figure 7-4. Chart shows how an IRA account balance grows when interest is compounded.

In the first year, your original $2,000 investment would earn $240. Assuming that you retired at age 65, interest accumulated on that first $2,000 investment would total $259,459.83. You can use a calculator to check these figures. Just use the interest compounding algorithm described above. Multiply the balance by 1.12 to calculate the new balance, including interest, for each year.

The point is that compounded interest starts building slowly, but after a period of time, your savings will grow rapidly. Thus, the earlier a person begins saving, the greater the financial security that person can enjoy later in life.

EXERCISE 7-8

Read each question carefully before calculating the answer. Because these problems require large numbers of calculations, you should use a calculator if possible. If no calculator is available, try working one or two questions by hand. Round all calculations to three decimals. Round answers to the nearest cent.

1. Conrad has deposited $5,000 in an IRA that earns 10% interest per year and in which interest is compounded annually. How much will Conrad's $5,000 be worth in 20 years?

2. Melba has deposited $2,000 in an IRA that earns 9% interest per year and in which interest is compounded annually. What will be the ending balance in Melba's account after 25 years?

3. Charlotte has deposited $500 in an IRA that earns 11% interest per year and in which interest is compounded annually. What will be the balance of her account in 5 years? 10 years? 15 years? 20 years?

4. On her fiftieth birthday, Shiba placed $2,000 in an IRA that earned 8% interest per year and in which interest was compounded annually. At the age of 54, Shiba deposited an additional $3,500. No more money was deposited. How much did Shiba have in her account when she reached the age of 65?

5. At 22, Rebecca decided to open an IRA that earned 12% interest per year and in which interest was compounded annually. Her first deposit totaled $1,000. After that, she made a $1,000 deposit every 12 months for 14 years. How much did she have in her account when she reached the age of 36?

Other Retirement Plans

There are a number of ways an individual can plan for financial security during retirement years. Depending on an individual's circumstances, two or more methods of retirement planning may be available. Common sources of retirement income include:

- Social security
- Employer pension plans
- Annuities, including life insurance plans.

Social security. The federal government operates a program of old age, survivors', and disability insurance that pays benefits to millions of Americans. *Social security* funds come from payroll deductions and other contributions by millions of workers and employers. Social security old age benefits are intended to supplement other retirement savings.

Pensions. A *pension* is a program in which an employer continues to pay employees a portion of wages after retirement. Some companies administer their own pension plans. Many such plans are based on *profit sharing* plans. Profit sharing means that the company contributes a portion of its profit to employee benefits. Other employers utilize plans offered by insurance companies and other financial institutions. Pension plan funds usually are invested in a manner that offers both a reasonable rate of return and protection against loss.

Annuities. Another method of providing for retirement income is through an *annuity.* An annuity is a savings program designed to pay out predetermined amounts at *maturity.* Maturity is the point at which an investment is fully developed and payout begins. Typical annuity programs are life insurance policies in which cash values increase as premiums are paid. This type of life insurance is called ordinary life insurance, straight life insurance, or whole life insurance. Life insurance that does *not* build up cash values is called term life insurance. Because no cash value is involved, term insurance is cheaper to purchase than whole life.

EXERCISE 7-9

Read each question carefully before calculating the answer. Round the answer to two decimal places.

1. Barbara makes $38 monthly payments on a life insurance policy worth $150,000. How much does this policy cost Barbara annually? How much money will Barbara pay to her insurance company over a 25-year period?

2. Norman is examining three life insurance policies, each worth $50,000. Policy A requires payments of $235 a year. Policy B requires payments of $55 a quarter. Policy C requires payments of $19 a month. Which policy will cost Norman the least amount of money per year?

3. Jordan Manufacturing offers the following pension plan to employees: Persons who work for 20 years are entitled to 55% of yearly salary upon retirement. Persons who work for 25 years are entitled to 65% of yearly salary upon retirement. Persons who work for 30 years are entitled to 75% of yearly salary upon retirement. Bernice has worked for Jordan Manufacturing for 25 years and is planning to retire. Her yearly salary is $25,000. What will her monthly retirement income be?

4. Payroll clerks deduct 7.05% from each employee paycheck to cover social security payments. How much would be deducted from a $1,190 paycheck? A $2,345 paycheck? A $719.20 paycheck?

Education as an Investment

The discussions throughout this unit are largely about managing income. Income, of course, is money you earn. However, before you can plan for and manage earnings, you need to receive actual income. To realize income, you must first be qualified to hold a job. In turn, you have to prepare for a job that will produce enough income to satisfy your future needs. Job preparation, then, involves education and training.

The thing that complicates your educational investment is you, yourself. Your educational investment must build upon and develop your natural abilities and skills. You should seek a position that will prove satisfying. After all, you will work for 40 years or more. That, in itself, is a big investment. Before you make this investment, therefore, you should ask some important questions: What are the prospects for continuing future employment in each field you consider? What investment will you have to make in education or training to qualify for the job? What are the prospects for growth or advancement? What are wage levels? What will the prospective wages buy? Will you be satisfied in that field in 10, 20, or 30 years?

These questions are hard. The arithmetic can be simple: Suppose a job requires two years of education and training after high school. Your costs to qualify are two years of time and $7,000 in tuition and expenses. At entry-level pay, your time might be valued at $18,000. Thus, job preparation represents an investment of $25,000. ($18,000 + $7,000 = $25,000)

Now, suppose you estimate that this training will increase your earnings by an average of $3,000 per year. If you go to work at age 20 and retire at 65, your added earnings would be $135,000. ($3,000 × 45 = $135,000)

During your working life, your investment would be paid back more than five times. Actually, job preparation through education can lead to even greater differences in earnings.

SUMMING UP

- Planning for the future usually involves working toward the attainment of financial security for retirement. Financial security means adequate income or capital to provide for life's necessities.
- The use of money can be bought and sold. If you borrow money, you pay back the principal plus interest. The amount of interest collected by a lender of money is profit. The interest rate is the percentage of the principal charged to the borrower.
- When you invest money, you are selling the use of that money to a person or organization. As an investor, or depositor, you receive interest as payment for the use of your money.
- Banks and other financial institutions accept deposits, pay interest to depositors, and lend their money to borrowers. Financial institutions make a profit by charging higher interest rates to borrowers than they pay out to depositors.
- The Federal Reserve System provides clearing house service for banks throughout the country. Bank deposits are protected by government insurance.
- There are two basic ways of calculating interest charges. Simple interest is calculated once. Compound interest is added to the balance carried forward at specified times. The formula for computing interest is: I = PRT (interest equals principal × interest × time).
- Compounding interest involves calculating interest more frequently and adding interest to the balance each time. Since interest is added to the balance, interest is being paid on interest as well as on the balance. The more frequently interest is compounded, the faster a balance will grow.
- Interest can be compounded in a single step of multiplication. Multiplying by 1 carries forward the value of the balance. Adding the decimal expressing interest rate × time to 1 provides the correct multiplier. Thus, if a 10% interest rate were compounded quarterly, the multiplication factor would be 1 plus $\frac{1}{4}$ of the annual interest rate, or 1.025.
- A finance charge is the total cost of obtaining a loan. Finance charges can include interest, loan origination fees, credit life insurance, and other costs.
- An investment in education or training will pay you back in your future earnings.

TESTING YOUR WORKING KNOWLEDGE

Read each question carefully before calculating the answer. Round calculations to three decimal places. Round answers to the nearest cent.

1. Dorman borrowed $689 at 12% simple annual interest. The duration of the loan was 12 months. What were Dorman's monthly payments? How much was the finance charge?

2. Tracy has deposited $2,500 in a savings account that earns 6% interest per year. If interest is compounded monthly, how much will she have in her account after one year? If interest is compounded annually, how much will she have in her account after one year? What is the difference?

3. Dale borrowed $10,000 at 11% interest per year. He paid off the loan in 12 months. In addition, to receive the loan, Dale had to pay a loan origination fee of $100 and a credit life insurance fee of $5 per month until the loan was paid off. What was the finance charge on Dale's loan? What were Dale's *total* monthly payments?

4. Bridgette began working for the Furniture Mart when she was 22. The Furniture Mart offered the following pension plan: Persons who work for 20 years receive 60% of annual salary upon retirement. Persons who work for 25 years receive 65% of annual salary upon retirement. Persons who work for 30 years receive 70% of annual salary upon retirement. Bridgette worked for the Furniture Mart for 25 years. Her monthly retirement income was $1,300. What was Bridgette's annual salary before retirement?

5. Dolly opened an IRA that earned 12% interest per year, compounded annually. Her initial deposit was $2,000. After that, she made a $2,000 deposit every 12 months for 9 years. How much did she have in her account after 10 years? Round the answer to the nearest dollar.

TERMS THAT COUNT

income
cost
revenue
profit
interest
principal
interest rate
investor
deposit
clearing house
Federal Reserve System (FED)
reserve
deposit insurance
simple interest
declining balance
credit purchase
installment purchase
installment
loan origination fee
credit life insurance
finance charge
compound interest
individual retirement account (IRA)
social security
pension
profit sharing
annuity
maturity

PRESENTING INFORMATION

Read each of the following questions carefully. Then present the information in the format required.

1. Ricardo has placed $2,000 in a savings account that earns 10% simple interest, compounded annually. Draw a line graph that shows how Ricardo's money will grow during a five-year period

2. Jenine has borrowed $2,500 at 13% simple annual interest. The duration of the loan is 12 months. In addition, Jenine paid a $250 loan origination fee. Draw a pie chart that relates the principal and the total finance charge to the total of payments for the loan.

3. Teressa has to decide what to do when she completes high school. One of her choices is to become a receptionist. She feels that, with this choice, she could receive on-job training and be promoted into office administration. But Teressa would prefer to become a dental hygienist. This career would require two additional years of education. Schooling would cost about $5,000, plus a loss of two years of wages. In office work, Teressa would average $10,000 per year in earnings for three years. Then, she would average $12,500 for three more years. After that, she thinks she would earn $15,000 for four years. Thus, her earnings over 10 years would be $127,500. As a dental hygienist, Teressa would expect to earn $20,000 during her first two years, $25,000 thereafter for the next six years. Remember, her two years of education involve a cost of $25,000 ($5,000 tuition + $20,000 lost earnings = $25,000). Teressa's earnings during eight working years would be $190,000 (six years at $25,000 = $150,000 and two years at $20,000 = $40,000). Thus, in 10 years after high school graduation, Teressa would earn $190,000 if she goes for extra education. If she goes right to work after high school, she will earn $127,500. To help Teressa appreciate this comparison draw a bar chart showing the total earnings for these two options.

SKILL DRILLS

In each of the following loan problems, use the simple interest formula, I = PRT, to calculate total interest paid. Round calculations to the nearest cent.

1. The principal is $1,400. Interest is 12%. Duration of the loan is 12 months.

2. The principal is $2,800. Interest is 13%. Duration of the loan is 12 months.

3. The principal is $1,650. Interest is 18%. Duration of the loan is nine months.

4. The principal is $360. Interest is 11%. Duration of the loan is three months.

5. The principal is $2,500. Interest is 10%. Duration of the loan is one year.

6. The principal is $1,500. Interest is 15%. Duration of the loan is six months.

7. The principal is $3,600. Interest is 13%. Duration of the loan is 12 months.

8. The principal is $3,000. Interest is 10%. Duration of the loan is 10 months.

Mathematics and Money

YOUR LEARNING JOB

When you have completed the exercises and assignments for this unit, you should be able to:

- ☐ Describe ways in which mathematics is used in retail business.
- ☐ Explain why the price of merchandise rises when handling of the items increases.
- ☐ Make change in a consumer purchasing situation.
- ☐ Maintain a running total of purchase transactions.
- ☐ Compute sales tax on a retail purchase.
- ☐ Use a sales tax table to determine sales tax.
- ☐ Describe factors that affect retail merchandise markup.

WARM-UP

A. Find the sum in each problem.

1. 45.039 + 4.201 + 78.12 + 0.34 =
2. 3,412 + 3.487 + 51.09 + 73.016 =
3. 92.8347 + 102.4721 + 540.3928 + 5,632.922 =
4. 0.3971 + 1.28743 + 65.002 + 543 + 5,398 + 0.4723 =
5. 0.9876 + 45.1234 + 891.009 + 9.5471 + 56.0398 =

B. Find the difference in each problem.

1. 73.981 − 25.253 =
2. 89.0192 − 3.8794 =
3. 1,203.09 − 341.00342 =
4. 482.93276 − 95.888565 =
5. 0.91635 − 0.906748 =

C. Find the product in each problem.

1. 4.039 × 0.25 =
2. 100.1 × 0.33 =
3. 2,918 × 0.065 =
4. 435.95 × 0.07 =
5. 2.1013 × 1.09 =

D. Find the quotient in each problem. Round calculations to two decimal places.

1. 23.45 ÷ 1.15 =
2. 54.43 ÷ .99 =
3. 3.41 ÷ 2.19 =
4. 114.01 ÷ 32.9 =
5. 12.34 ÷ 2.11 =

RETAILING AND WHOLESALING: HOW PRODUCTS REACH USERS

As a student, your major involvement in the world of business may be as a consumer. A consumer is one of two parties involved in a sales transaction. The consumer buys something. The other party, of course, is the seller.

Most consumer purchases are made at *retail* businesses. *Retailing* is the activity of selling products or services in small quantities, usually directly to consumers (the public). Retailers also are called *merchants.*

Retail merchants are in the business of selling products directly to consumers. PHOTO BY LISA SCHWABER-BARZILAY

Retailers obtain goods in quantity, usually from *wholesale* suppliers. A *wholesaler* is an individual or an organization that buys and sells goods in large quantities. Wholesalers then sell and distribute products to retailers. These types of customers buy in quantity. In between the wholesaler and the consumer, a lot of arithmetic takes place.

One type of arithmetic is used to figure costs of items to be sold. Wholesalers buy in greater quantities than retailers. Retailers, in turn, buy greater quantities than consumers. At each level, the arithmetic of buying and selling is the same. That is, the amount of a purchase is determined through multiplication. The cost of a single item is multiplied by the number of units purchased:

$$\text{Unit cost} \times \text{Units purchased} = \text{Purchase price}$$

Typically, the larger the purchase, the smaller the price per unit will be. For example, you, as a consumer, might buy a sweater for $15. The store might have bought a dozen of the same type of sweater. The store might have paid $10.50 each for one dozen (12) sweaters. As another option, a store might be able to buy six dozen (72) sweaters for $9 each. Using the arithmetic of buying and selling, the transactions work out as follows:

$$\$10.50 \times 12 = \$126$$
$$\$9 \times 72 = \$648$$

The arithmetic of buying and selling can produce some simple lessons. One important lesson is: The more money that is spent on a total purchase, the smaller the unit cost should be. Thus, the wholesaler would buy more sweaters at smaller prices for each. For example, a wholesaler may be able to buy 50 dozen (600) sweaters at $7.25 each. If 100 dozen (1,200) sweaters are purchased, the unit price may be $6. The arithmetic for these purchases would be:

$$\$7.25 \times 600 = \$4{,}350$$
$$\$6 \times 1{,}200 = \$7{,}200$$

Again, notice that the larger the amount of a total purchase, the lower the unit cost will be. The differences in unit price make it possible for companies to be in business. A retailer has to receive more for a sweater than it costs. The difference provides the profit that enables a business to exist. Business profits from merchandise sales are described later in this chapter.

EXERCISE 8-1

Read each question carefully before calculating the answer. Round each answer to the nearest cent.

1. The Copy House purchases paper for its copy machine for $4 per ream. How much does the company spend on an order of 150 reams?

2. Ralph manages an auto parts store. He purchases spark plugs for $1.05 per unit. How much does an order for 3,600 spark plugs cost?

3. Corner Market purchases large quantities of soft drinks for $1.20 per six-pack. How much does an order for 180 six-packs of cola cost?

4. Dean's Stationers purchases ink pens for $3.85 per dozen. How much does an order for 15 dozen cost?

5. AOR Industries purchases paper clips at a price of 35 cents per box of 100. How much does an order for 575 boxes cost?

6. The Copy House purchases paper for its copy machine at $2.50 per ream. How much does the company spend on an order of 150 reams?

7. Connie buys cardboard containers for packaging goods. The containers cost 35 cents each. What does an order for 225 containers cost?

8. Lori purchases hamburger buns for 85 cents per package of eight. How much does it cost to purchase 113 packages?

MERCHANDISING

The buying and selling of consumer goods is called *merchandising*. *Merchandise* is the product or products sold by a merchant. Merchants use mathematics in many areas of their business, in addition to figuring amounts of purchases. Merchandising activities requiring mathematical calculations may include:

- Setting prices for merchandise
- Making change for customers
- Totaling sales
- Calculating and collecting sales tax
- Computing payrolls and other business expenses
- Calculating profit or loss.

FACTORS AFFECTING PRICES

You are aware that wholesale prices usually are lower than retail prices. You also realize that many products are less expensive when they are purchased in large quantities. The reasons are quite simple.

Think of product distribution as a channel in which two factors affect prices. These factors are:

- Handling
- Quantity.

Each time a product is handled, there is a cost factor added to its price. Handling includes packaging, transportation, and store display of merchandise. People are paid to perform each of these tasks.

For example, you might order a product from a mail order company. The order blank probably has a space for adding a shipping and handling charge to the total cost of your order. Such costs are included in the prices of goods sold in retail stores.

The other price factor, quantity, is related closely to handling. Soft drinks, for example, usually cost less by the six-pack or the case than by the individual can. One reason is that the larger quantities require less handling than individual containers.

Say that the price of a single soft drink can is 30 cents, and the price of a six-pack is $1.35. If you purchase a six-pack, the cost

Handling of merchandise is a major factor in its eventual selling price. Minimizing handling helps hold costs down.

of each can is $1.35 ÷ 6, which equals 23 cents. If you save 7 cents per can, your total savings for a six-pack is 42 cents.

If an employee in a retail store places four six-packs of soft drinks on a shelf, the employee is moving four items. If those soft drink cans were sold individually, the employee would have to handle 24 items. It is quicker and easier to handle fewer items. This greater efficiency means lower labor costs for the store owner, because that employee can accomplish more work in a given time. Part of this saving is passed on to the consumer in the form of a lower price.

EXERCISE 8-2

Read each question carefully before calculating the answer. Round each answer to the nearest cent.

1. Ink pens sell for 45 cents each or for $4.20 per dozen. How much does a consumer save by purchasing a dozen pens?

2. A six-pack of cola sells for $1.35. A case of cola, or 4 six-packs, sells for $4.75. How much less will each can of cola cost when purchased by the case?

3. Mi-ling is shopping for spiral ring notebooks. The school store sells packages of four notebooks for $9.40 per package. A local store sells notebooks for $2.76 each. Mi-ling needs eight notebooks. At which store will Mi-ling spend the lesser amount of money? How much money will she save?

4. Guitar strings sell for $1.35 per string. Sets of six strings sell for $7.49. How much does a customer save by purchasing a set of strings? What is the unit price?

5. Hand-made coffee mugs sell for $1.50 each. The same mugs also are sold in sets of 10 for $13.75. How much do consumers save by purchasing a set of 10 mugs? What is the unit price?

6. Bob's Self Serve Station sells gasoline for $1.10 per gallon. Janice's Full Serve Station sells gasoline for $1.25 per gallon. When purchasing 15 gallons, how much can a customer save at Bob's Self Serve?

7. Raffle tickets for the Booster Club can be purchased in groups of 5, 10, or 25. Five tickets sell for $2.50. Ten tickets sell for $4.50. Twenty-five tickets sell for $10. Determine the unit price for each group.

8. Ice cream bars sell for 35 cents each. Boxes containing a dozen sell for $2.58. What is the unit price when purchased by the dozen? At that rate, how much would four dozen ice cream bars cost?

Making change is a required skill for anyone who works for a retail business.

MAKING CHANGE

If you go to work in a retail store, one of the first skills you must master is making change. Most customers do not present the exact change for a purchase.

For example, say that you are a cashier in a restaurant. A luncheon customer might come to your counter with a bill for $4.75. The customer hands you a $5 bill. Your job is to determine the difference between $4.75 and $5. This difference is the customer's change. You can use two approaches. First, you can subtract $4.75 from $5. The difference is 25 cents. The second approach is to start with $4.75 and determine how much has to be added to reach $5. The answer is the same, 25 cents.

That example is simple enough. Most transactions involve numbers that are less obvious. You might be given a $20 bill for purchases totaling $13.72. Somewhat more effort is required to calculate the change due: $6.28. However, you still have two methods available, subtraction or addition.

Addition and subtraction really are mirror images of the same process. Either method can be used to figure change. Change can be defined as the difference between the purchase price of an item and the *amount tendered.* The amount tendered is the amount of money you receive from a customer.

In the first example, the solution is easy. Adding 25 cents to $4.75 makes $5. In the second example, the change is more than a dollar. The idea is to add the amount of change to reach the next whole dollar amount. Then add the number of dollars to equal the amount tendered. Thus, adding 28 cents to $13.72 gives $14. Adding $6 to $14 gives $20.

EXERCISE 8-3

For each question, think of yourself as a cashier. Read each question carefully. Then calculate the sum of the transaction and determine the correct amount of change, if any, to return to the customer. No sales tax is involved in these problems.

1. A customer places 4 cans of soup, 2 pounds of apples, and 3 six-packs of cola on the counter. Soup sells for 49 cents per can. Apples are 70 cents per pound. Cola is on sale for $1.35 per six-pack. The customer pays with a $10 bill.

2. Tickets for a rock concert are $12.50 for reserved seating and $7.50 for festival seating. A customer purchases six tickets, three for festival seating and three for reserved seating. The customer pays with three $20 bills.

3. Snow skis cost $227 and bindings cost $128. A customer purchases three sets of skis and bindings and gives you twenty-two $50 bills.

4. A man rents eight tuxedos. Each set rents for $38. A set consists of trousers, coat, shirt, and tie. The man also rents eight pairs of shoes for $7 each. In addition, the store offers a 15% discount on rentals of eight tuxedos or more. You are given ten $20 bills, ten $10 bills, one $5 bill, and one $1 bill.

5. Tokens for video games sell for 25 cents each or 5 for $1. A girl asks for 38 tokens and gives you one $5 bill and three $1 bills.

6. You are selling raffle tickets for 75 cents each or five tickets for $3.25. Your neighbor purchases 15 tickets and gives you two $5 bills.

7. A young couple place three record albums and four cassette tapes on the counter. Albums sell for $6.98. Tapes sell for $5.98. The couple give you two $20 bills and one $5 bill.

8. A customer puts 15.7 gallons of unleaded gasoline in her car and gives you a $20 bill and a $1 bill. Unleaded gasoline sells for $1.30 per gallon.

TRANSACTION ACCUMULATION

Retail sales transactions usually involve more than a single item. It usually is necessary to keep a running total of transactions. Cash registers and computer sales terminals perform this task in most business establishments. However, salespersons in some circumstances may need to keep a running account of transactions prior to entering them in a register.

One example might be a yardage store. A customer might wish to buy material from several bolts of cloth. Each purchase may involve a different length of material. Each purchase also may involve a different price per yard. In such a situation, the salesperson may keep a running total on a pad of paper as a courtesy to the customer. Calculations could be performed with pencil and paper, or on a hand-held calculator.

Say that the customer buys material from four bolts. For each transaction, the salesperson would multiply the length of the material by the price per yard. A running total could be kept, so the customer could be informed of the cumulative price at any point.

A similar situation might occur at a lumber yard or building materials center. A customer might wish to buy a number of different items from various locations in the yard or center. The customer might request a salesperson to keep a running total so the customer could stay within a budget.

In such cases, writing the information also helps the salesperson. Having the information in writing makes it easier to enter data in a register or terminal when the sale is completed.

EXERCISE 8-4

Read each question carefully before calculating the answer. Round calculations to the nearest cent. (No sales tax is involved in these problems.)

1. Bob and Carrol have $25. They need to buy three cans of paint and several paint brushes. Paint costs $6.12 per can. Brushes cost $1.10 each. How much money will they have left after purchasing the paint? How many brushes will they be able to purchase?

2. Matilda wants to purchase 3 yards each of four different fabrics. She has $55.50. Fabric A costs $6 per yard. Fabric B costs $5.50 per yard. Fabric C costs $7 per yard. Fabric D costs $6.25 per yard. First, Matilda pays for 3 yards of Fabric A. How much money does she have left? Then, she has the salesperson cut 3 yards of both Fabric B and Fabric C. Now how much money does she have left? How many yards of Fabric D can she purchase?

3. A customer purchases seven 8-foot studs, two sheets of plywood, one box of nails, and one hammer. The box of nails costs $2.49, and the hammer costs $6.75. Plywood costs $6.11 per sheet, and studs are $1.49 each. Create a list that shows the cost of each item, subtotals for cost of plywood and studs, and the final total.

4. Ms. Masami had $200 with which to purchase clothes. In the first store, she bought two skirts, one for $15 and one for $19. In the second store, she purchased two pairs of shoes, one for $21.50 and the other for $28.95. How much money had she spent? How much money did she have left? In the third store, Ms. Masami found four blouses. The first cost $22; another, $17.50; and the other two, $15 each. Did Ms. Masami have enough money to purchase all four blouses? If so, how much money will she have left over? If not, how much more money will she need?

5. You work as a cashier in a record and tape store. A customer comes in with a $100 gift certificate and selects five cassette tapes and three record albums. Each cassette sells for $5.98, and each album sells for $7.65. What percentage of the customer's gift certificate has been used? A week later, the same customer purchases four more cassette tapes. The following week the customer comes in once more and buys three record albums. What percentage of the gift certificate has been used?

SALES TAXES

Another skill needed by salespersons in most states is the ability to calculate *sales tax*. Forty-five states and the District of Columbia have sales taxes. Some local governments—counties, cities, and towns—also have sales taxes. A sales tax is a tax imposed on retail business transactions. Sales taxes range from a low of 2% in Oklahoma to a high of 8% in New York City. States that do not have sales taxes are Alaska, Delaware, Montana, New Hampshire, and Oregon. Retail businesses collect sales taxes from their customers and pay the money to governmental agencies.

Most sales tax computations are simple. The salesperson multiplies the price of an item by the decimal value of the sales tax. The product is the amount of tax. The tax then is added to the price of the item.

For example, say that you worked in a clothing store. The sales tax in your community is 5%. If a customer purchases a shirt for $8.50, you would calculate the sales tax this way:

$$8.50 \times 0.05 = 0.43$$

The sales tax is 43 cents. This amount of tax is added to the price of the shirt to obtain the total price to the customer:

$$8.50 + 0.43 = 8.93$$

Say that you work in an auto parts store, and the sales tax in your state is $6\frac{1}{2}$%. How would you figure the total amount due from a customer on purchases totaling $27.50?

The same two steps are involved. First, calculate the sales tax by multiplying the combined price of the items purchased by 0.065. Then, add the tax to the total price of the merchandise:

$$27.50 \times 0.065 = 1.79$$
$$27.50 + 1.79 = 29.29$$

EXERCISE 8-5

Each question contains a transaction amount and a sales tax rate. Calculate the sales tax amount and the total price for each transaction. Round calculations to the nearest cent.

1. $23 at 6.5% sales tax
2. $5,670 at 4.5% sales tax
3. $75,000 at 7.5% sales tax
4. $450 at 4% sales tax
5. $2,500 at 3% sales tax
6. $254.32 at 6.5% sales tax
7. $24.68 at 6% sales tax
8. $660.99 at 5.5% sales tax
9. $34,750 at 5% sales tax
10. $5,438.98 at 3.5% sales tax
11. $12,323.09 at 4% sales tax
12. $108,000 at 7.5% sales tax
13. $35,789.32 at 5% sales tax
14. $67,819.83 at 4.5% sales tax
15. $254.43 at 6.5% sales tax
16. $9,990.85 at 3% sales tax
17. $325,938.10 at 5.5% sales tax
18. $962.76 at 8% sales tax

Tax Tables

Many retail businesses use *tax tables* for fast and accurate determination of sales tax amounts on purchases. A sales tax table usually consists of two columns of information. On the left are dollar amounts of purchases. On the right are the amounts of sales tax to be charged.

The sales tax column is rounded to the penny. The left-hand column contains a range of amounts for which each sales tax amount is added. The table in Figure 8-1 is an example of sales tax progression for a community in which the sales tax rate is 6.5%.

Sales Tax Exemptions

State and local sales tax laws can differ widely. In many places, certain types of consumer goods are *exempt* from the sales tax. Exempt means free from, or not subject to, a regulation. A major example is food. However, in some states, the food exemption does not apply to food prepared and consumed in public eating establishments.

Figure 8-1. Sales tax tables aid cashiers in adding the proper amount of tax to purchases.

6½% TAX SCHEDULE

STATE OF CALIFORNIA—BOARD OF EQUALIZATION

SALES AND USE TAX REGULATIONS

APPENDIX H TO REGULATION 1700, REIMBURSEMENT FOR SALES TAX

STATE OF CALIFORNIA 6½% SALES TAX REIMBURSEMENT SCHEDULE Civil Code Section 1

Transaction	Tax	Transaction	Tax	Transaction	Tax	Transaction	Tax	Transaction	Tax	Transaction	Tax	Transaction
.01– .10	.00	14.24–14.38	.93	28.54–28.69	1.86	42.85–42.99	2.79	57.16–57.30	3.72	71.47–71.61	4.65	85.77–85.92
.11– .20	.01	14.39–14.53	.94	28.70–28.84	1.87	43.00–43.15	2.80	57.31–57.46	3.73	71.62–71.76	4.66	85.93–86.07
.21– .35	.02	14.54–14.69	.95	28.85–28.99	1.88	43.16–43.30	2.81	57.47–57.61	3.74	71.77–71.92	4.67	86.08–86.23
.36– .51	.03	14.70–14.84	.96	29.00–29.15	1.89	43.31–43.46	2.82	57.62–57.76	3.75	71.93–72.07	4.68	86.24–86.38
.52– .67	.04	14.85–14.99	.97	29.16–29.30	1.90	43.47–43.61	2.83	57.77–57.92	3.76	72.08–72.23	4.69	86.39–86.53
.68– .83	.05	15.00–15.15	.98	29.31–29.46	1.91	43.62–43.76	2.84	57.93–58.07	3.77	72.24–72.38	4.70	86.54–86.69
.84– .99	.06	15.16–15.30	.99	29.47–29.61	1.92	43.77–43.92	2.85	58.08–58.23	3.78	72.39–72.53	4.71	86.70–86.84
1.00– 1.15	.07	15.31–15.46	1.00	29.62–29.76	1.93	43.93–44.07	2.86	58.24–58.38	3.79	72.54–72.69	4.72	86.85–86.99
1.16– 1.30	.08	15.47–15.61	1.01	29.77–29.92	1.94	44.08–44.23	2.87	58.39–58.53	3.80	72.70–72.84	4.73	87.00–87.15
1.31– 1.46	.09	15.62–15.76	1.02	29.93–30.07	1.95	44.24–44.38	2.88	58.54–58.69	3.81	72.85–72.99	4.74	87.16–87.30
1.47– 1.61	.10	15.77–15.92	1.03	30.08–30.23	1.96	44.39–44.53	2.89	58.70–58.84	3.82	73.00–73.15	4.75	87.31–87.46
1.62– 1.76	.11	15.93–16.07	1.04	30.24–30.38	1.97	44.54–44.69	2.90	58.85–58.99	3.83	73.16–73.30	4.76	87.47–87.61
1.77– 1.92	.12	16.08–16.23	1.05	30.39–30.53	1.98	44.70–44.84	2.91	59.00–59.15	3.84	73.31–73.46	4.77	87.62–87.76
1.93– 2.07	.13	16.24–16.38	1.06	30.54–30.69	1.99	44.85–44.99	2.92	59.16–59.30	3.85	73.47–73.61	4.78	87.77–87.92
2.08– 2.23	.14	16.39–16.53	1.07	30.70–30.84	2.00	45.00–45.15	2.93	59.31–59.46	3.86	73.62–73.76	4.79	87.93–88.07
2.24– 2.38	.15	16.54–16.69	1.08	30.85–30.99	2.01	45.16–45.30	2.94	59.47–59.61	3.87	73.77–73.92	4.80	88.08–88.23
2.39– 2.53	.16	16.70–16.84	1.09	31.00–31.15	2.02	45.31–45.46	2.95	59.62–59.76	3.88	73.93–74.07	4.81	88.24–88.38
2.54– 2.69	.17	16.85–16.99	1.10	31.16–31.30	2.03	45.47–45.61	2.96	59.77–59.92	3.89	74.08–74.23	4.82	88.39–88.53
2.70– 2.84	.18	17.00–17.15	1.11	31.31–31.46	2.04	45.62–45.76	2.97	59.93–60.07	3.90	74.24–74.38	4.83	88.54–88.69
2.85– 2.99	.19	17.16–17.30	1.12	31.47–31.61	2.05	45.77–45.92	2.98	60.08–60.23	3.91	74.39–74.53	4.84	88.70–88.84
3.00– 3.15	.20	17.31–17.46	1.13	31.62–31.76	2.06	45.93–46.07	2.99	60.24–60.38	3.92	74.54–74.69	4.85	88.85–88.99
3.16– 3.30	.21	17.47–17.61	1.14	31.77–31.92	2.07	46.08–46.23	3.00	60.39–60.53	3.93	74.70–74.84	4.86	89.00–89.15
3.31– 3.46	.22	17.62–17.76	1.15	31.93–32.07	2.08	46.24–46.38	3.01	60.54–60.69	3.94	74.85–74.99	4.87	89.16–89.30
3.47– 3.61	.23	17.77–17.92	1.16	32.08–32.23	2.09	46.39–46.53	3.02	60.70–60.84	3.95	75.00–75.15	4.88	89.31–89.46
3.62– 3.76	.24	17.93–18.07	1.17	32.24–32.38	2.10	46.54–46.69	3.03	60.85–60.99	3.96	75.16–75.30	4.89	89.47–89.61
3.77– 3.92	.25	18.08–18.23	1.18	32.39–32.53	2.11	46.70–46.84	3.04	61.00–61.15	3.97	75.31–75.46	4.90	89.62–89.76
3.93– 4.07	.26	18.24–18.38	1.19	32.54–32.69	2.12	46.85–46.99	3.05	61.16–61.30	3.98	75.47–75.61	4.91	89.77–89.92
4.08– 4.23	.27	18.39–18.53	1.20	32.70–32.84	2.13	47.00–47.15	3.06	61.31–61.46	3.99	75.62–75.76	4.92	89.93–90.07
4.24– 4.38	.28	18.54–18.69	1.21	32.85–32.99	2.14	47.16–47.30	3.07	61.47–61.61	4.00	75.77–75.92	4.93	90.08–90.23
4.39– 4.53	.29	18.70–18.84	1.22	33.00–33.15	2.15	47.31–47.46	3.08	61.62–61.76	4.01	75.93–76.07	4.94	90.24–90.38
4.54– 4.69	.30	18.85–18.99	1.23	33.16–33.30	2.16	47.47–47.61	3.09	61.77–61.92	4.02	76.08–76.23	4.95	90.39–90.53
4.70– 4.84	.31	19.00–19.15	1.24	33.31–33.46	2.17	47.62–47.76	3.10	61.93–62.07	4.03	76.24–76.38	4.96	90.54–90.69
4.85– 4.99	.32	19.16–19.30	1.25	33.47–33.61	2.18	47.77–47.92	3.11	62.08–62.23	4.04	76.39–76.53	4.97	90.70–90.84
5.00– 5.15	.33	19.31–19.46	1.26	33.62–33.76	2.19	47.93–48.07	3.12	62.24–62.38	4.05	76.54–76.69	4.98	90.85–90.99
5.16– 5.30	.34	19.47–19.61	1.27	33.77–33.92	2.20	48.08–48.23	3.13	62.39–62.53	4.06	76.70–76.84	4.99	91.00–91.15
5.31– 5.46	.35	19.62–19.76	1.28	33.93–34.07	2.21	48.24–48.38	3.14	62.54–62.69	4.07	76.85–76.99	5.00	91.16–91.30
5.47– 5.61	.36	19.77–19.92	1.29	34.08–34.23	2.22	48.39–48.53	3.15	62.70–62.84	4.08	77.00–77.15	5.01	91.31–91.46
5.62– 5.76	.37	19.93–20.07	1.30	34.24–34.38	2.23	48.54–48.69	3.16	62.85–62.99	4.09	77.16–77.30	5.02	91.47–91.61
5.77– 5.92	.38	20.08–20.23	1.31	34.39–34.53	2.24	48.70–48.84	3.17	63.00–63.15	4.10	77.31–77.46	5.03	91.62–91.76
5.93– 6.07	.39	20.24–20.38	1.32	34.54–34.69	2.25	48.85–48.99	3.18	63.16–63.30	4.11	77.47–77.61	5.04	91.77–91.92
6.08– 6.23	.40	20.39–20.53	1.33	34.70–34.84	2.26	49.00–49.15	3.19	63.31–63.46	4.12	77.62–77.76	5.05	91.93–92.07
6.24– 6.38	.41	20.54–20.69	1.34	34.85–34.99	2.27	49.16–49.30	3.20	63.47–63.61	4.13	77.77–77.92	5.06	92.08–92.23
6.39– 6.53	.42	20.70–20.84	1.35	35.00–35.15	2.28	49.31–49.46	3.21	63.62–63.76	4.14	77.93–78.07	5.07	92.24–92.38
6.54– 6.69	.43	20.85–20.99	1.36	35.16–35.30	2.29	49.47–49.61	3.22	63.77–63.92	4.15	78.08–78.23	5.08	92.39–92.53
6.70– 6.84	.44	21.00–21.15	1.37	35.31–35.46	2.30	49.62–49.76	3.23	63.93–64.07	4.16	78.24–78.38	5.09	92.54–92.69
6.85– 6.99	.45	21.16–21.30	1.38	35.47–35.61	2.31	49.77–49.92	3.24	64.08–64.23	4.17	78.39–78.53	5.10	92.70–92.84
7.00– 7.15	.46	21.31–21.46	1.39	35.62–35.76	2.32	49.93–50.07	3.25	64.24–64.38	4.18	78.54–78.69	5.11	92.85–92.99
7.16– 7.30	.47	21.47–21.61	1.40	35.77–35.92	2.33	50.08–50.23	3.26	64.39–64.53	4.19	78.70–78.84	5.12	93.00–93.15
7.31– 7.46	.48	21.62–21.76	1.41	35.93–36.07	2.34	50.24–50.38	3.27	64.54–64.69	4.20	78.85–78.99	5.13	93.16–93.30
7.47– 7.61	.49	21.77–21.92	1.42	36.08–36.23	2.35	50.39–50.53	3.28	64.70–64.84	4.21	79.00–79.15	5.14	93.31–93.46
7.62– 7.76	.50	21.93–22.07	1.43	36.24–36.38	2.36	50.54–50.69	3.29	64.85–64.99	4.22	79.16–79.30	5.15	93.47–93.61
7.77– 7.92	.51	22.08–22.23	1.44	36.39–36.53	2.37	50.70–50.84	3.30	65.00–65.15	4.23	79.31–79.46	5.16	93.62–93.76
7.93– 8.07	.52	22.24–22.38	1.45	36.54–36.69	2.38	50.85–50.99	3.31	65.16–65.30	4.24	79.47–79.61	5.17	93.77–93.92
8.08– 8.23	.53	22.39–22.53	1.46	36.70–36.84	2.39	51.00–51.15	3.32	65.31–65.46	4.25	79.62–79.76	5.18	93.93–94.07
8.24– 8.38	.54	22.54–22.69	1.47	36.85–36.99	2.40	51.16–51.30	3.33	65.47–65.61	4.26	79.77–79.92	5.19	94.08–94.23
8.39– 8.53	.55	22.70–22.84	1.48	37.00–37.15	2.41	51.31–51.46	3.34	65.62–65.76	4.27	79.93–80.07	5.20	94.24–94.38
8.54– 8.69	.56	22.85–22.99	1.49	37.16–37.30	2.42	51.47–51.61	3.35	65.77–65.92	4.28	80.08–80.23	5.21	94.39–94.53
8.70– 8.84	.57	23.00–23.15	1.50	37.31–37.46	2.43	51.62–51.76	3.36	65.93–66.07	4.29	80.24–80.38	5.22	94.54–94.69
8.85– 8.99	.58	23.16–23.30	1.51	37.47–37.61	2.44	51.77–51.92	3.37	66.08–66.23	4.30	80.39–80.53	5.23	94.70–94.84
9.00– 9.15	.59	23.31–23.46	1.52	37.62–37.76	2.45	51.93–52.07	3.38	66.24–66.38	4.31	80.54–80.69	5.24	94.85–94.99
9.16– 9.30	.60	23.47–23.61	1.53	37.77–37.92	2.46	52.08–52.23	3.39	66.39–66.53	4.32	80.70–80.84	5.25	95.00–95.15
9.31– 9.46	.61	23.62–23.76	1.54	37.93–38.07	2.47	52.24–52.38	3.40	66.54–66.69	4.33	80.85–80.99	5.26	95.16–95.30
9.47– 9.61	.62	23.77–23.92	1.55	38.08–38.23	2.48	52.39–52.53	3.41	66.70–66.84	4.34	81.00–81.15	5.27	95.31–95.46
9.62– 9.76	.63	23.93–24.07	1.56	38.24–38.38	2.49	52.54–52.69	3.42	66.85–66.99	4.35	81.16–81.30	5.28	95.47–95.61
9.77– 9.92	.64	24.08–24.23	1.57	38.39–38.53	2.50	52.70–52.84	3.43	67.00–67.15	4.36	81.31–81.46	5.29	95.62–95.76
9.93–10.07	.65	24.24–24.38	1.58	38.54–38.69	2.51	52.85–52.99	3.44	67.16–67.30	4.37	81.47–81.61	5.30	95.77–95.92
10.08–10.23	.66	24.39–24.53	1.59	38.70–38.84	2.52	53.00–53.15	3.45	67.31–67.46	4.38	81.62–81.76	5.31	95.93–96.07
10.24–10.38	.67	24.54–24.69	1.60	38.85–38.99	2.53	53.16–53.30	3.46	67.47–67.61	4.39	81.77–81.92	5.32	96.08–96.23
10.39–10.53	.68	24.70–24.84	1.61	39.00–39.15	2.54	53.31–53.46	3.47	67.62–67.76	4.40	81.93–82.07	5.33	96.24–96.38
10.54–10.69	.69	24.85–24.99	1.62	39.16–39.30	2.55	53.47–53.61	3.48	67.77–67.92	4.41	82.08–82.23	5.34	96.39–96.53
10.70–10.84	.70	25.00–25.15	1.63	39.31–39.46	2.56	53.62–53.76	3.49	67.93–68.07	4.42	82.24–82.38	5.35	96.54–96.69
10.85–10.99	.71	25.16–25.30	1.64	39.47–39.61	2.57	53.77–53.92	3.50	68.08–68.23	4.43	82.39–82.53	5.36	96.70–96.84
11.00–11.15	.72	25.31–25.46	1.65	39.62–39.76	2.58	53.93–54.07	3.51	68.24–68.38	4.44	82.54–82.69	5.37	96.85–96.99

Supermarket and grocery store personnel often must separate taxable items and nontaxable items. For example, a shopper might have tissues, magazines, soap, and toothpaste in a shopping cart. Sales tax would be charged on these items. The shopper also may be purchasing meat, vegetables, fruit juice, and a frozen pie. These food items would not be subject to sales tax.

Modern supermarkets have computerized cash registers that add sales tax automatically to taxable items. The operator merely depresses a special key each time the price of a taxable item is entered. The register keeps a running total of those items. When a subtotal is reached, the appropriate amount of sales tax is added. Figure 8-2 shows a supermarket receipt with sales tax included.

In markets equipped with label-reading devices, information on sales tax is fed directly into a computer. Most packages of food

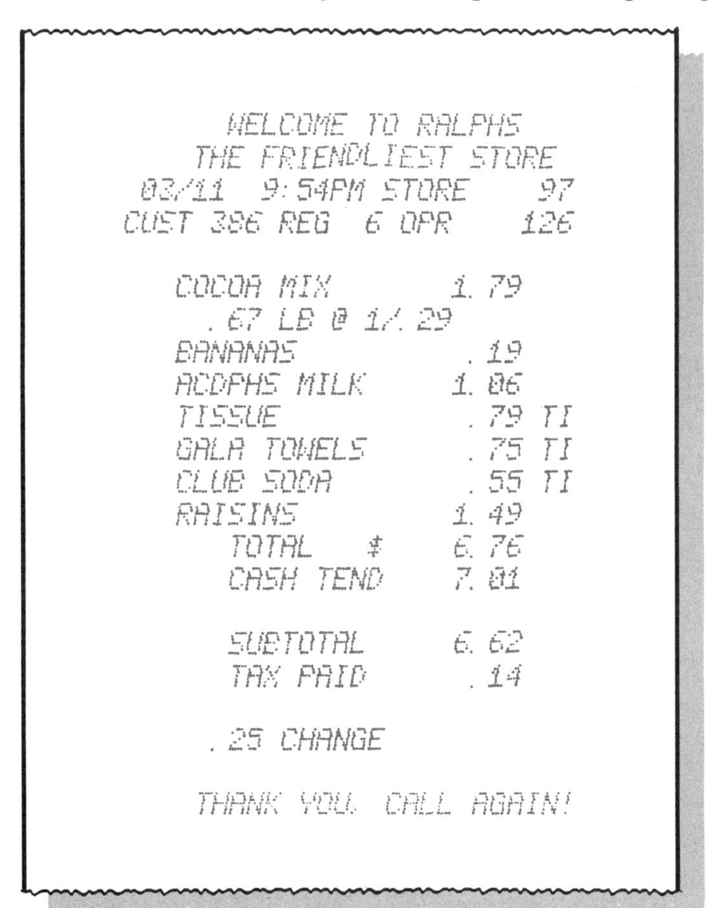
WELCOME TO RALPHS
THE FRIENDLIEST STORE
03/11 9:54PM STORE 97
CUST 386 REG 6 OPR 126

COCOA MIX 1.79
.67 LB @ 1/.29
BANANAS .19
ACDPHS MILK 1.06
TISSUE .79 TI
GALA TOWELS .75 TI
CLUB SODA .55 TI
RAISINS 1.49
TOTAL $ 6.76
CASH TEND 7.01

SUBTOTAL 6.62
TAX PAID .14

.25 CHANGE

THANK YOU CALL AGAIN!

Figure 8-2. Computer-generated supermarket receipt contains information of value to both store and customer.

Figure 8-3. Bar codes can be read directly into computers. Data sensed from these labels are used to record and total supermarket purchases.

and merchandise have special labels that can be read automatically by computers operating cash registers. The special label consists of a series of lines and blank spaces that form a code called a *bar code.* The bar code identifies the package and its product. The bar code is passed over a beam of light at the checkstand. The information is read and transmitted to a computer in the store that operates the cash registers. The bar code "tells" the computer if the item is taxable. With this equipment and labeling, sales can be handled quickly at checkstands. The computer processes the information as rapidly as the checker picks up and moves items across the bar code reader. A bar code is shown in Figure 8-3.

EXERCISE 8-6

Read each question and calculate the total purchase price. Note that, in this exercise, only milk, meats, and fresh fruits and vegetables are nontaxable. The sales tax rate is 4.5%. Round calculations to the nearest cent.

1. Felix went to the market and bought four ball point pens priced at 98 cents each, a notebook for $1.99, and a ream of paper for $4.

2. Luisa placed the following items on the checkout counter: four cans of soup for $0.49 per can, two gallons of milk priced at $1.99 per gallon, and two dozen eggs that cost $1.15 per dozen.

3. Gabriel is purchasing the following items: 2 gallons of milk at $1.92 per gallon, 2 six-packs of cola for $2.45 per six-pack, a loaf of bread for $0.99, 3 packages of hot dogs for $1.79 per package, and 4 packages of buns at $0.55 each.

4. Wilma bought three pounds of hamburger, which was selling for $2 per pound, a jar of mayonnaise for $1.24, a head of lettuce for 52 cents, a pound of tomatoes for 89 cents, and two packages of hamburger buns for 55 cents a package.

5. Jill and Sue purchased two colas for 35 cents apiece, a package of peanuts for 89 cents, three candy bars priced at 35 cents each, and one pint of milk for 35 cents.

6. Glen bought 8 spark plugs, 5 quarts of oil, an oil filter, and an air filter. Each spark plug cost $1.29. A quart of oil sold for $1.15. The air filter cost $10. The oil filter was on sale for $5.67.

7. Bruce handed the following items to the salesperson: three pairs of designer jeans, two long-sleeve shirts, three belts, and a pair of leather dress shoes. The shoes cost $45. The belts sold for $3.50 each. The long sleeve shirts were $18 each. Each pair of jeans cost $28.

8. Ching-yu purchased a jar of spaghetti sauce, a package of spaghetti, three quarts of milk, and a loaf of bread. The spaghetti cost $0.92, and the spaghetti sauce cost $2.65. The milk sold for $0.57 per quart. Bread was $1.22 per loaf.

REGISTERS THAT DO ALL THE WORK

Many businesses have electronic cash registers that perform a number of functions. In some cases, the salesperson need not even worry about prices.

One example is the type of register used in some fast-food restaurants. These new registers have picture keys that identify each item ordered by a customer. For example, there might be a key for a regular hamburger. Another key might indicate a large order of french fries. Still another might be used to order a particular soft drink. As each product key is pressed, the register enters descriptions and prices automatically.

These computer-operated registers also deliver orders electronically to the restaurant's food preparation area. This type of instant communication offers increased accuracy as well as speed. Avoided are the errors that often result from hastily written (or shouted) orders that are easily misunderstood.

The customer also benefits. In addition to fast and accurate service, the computer-operated register also provides an itemized receipt. Typically, the receipt lists, by name, each item ordered, as well as its price.

The business may benefit in still another way. Computers may be used to keep track of the *inventory*, or stock, of food items and supplies. This continuous inventory helps the business to operate smoothly.

This type of computer operation is useful only for businesses that sell a limited number of products. However, these operations make it possible for such businesses to do an almost unlimited volume of sales.

The Cashier as Banker

Regardless of the type of equipment used, a cashier must be accurate in completing each transaction. A register may calculate change due on a purchase, but the cashier must count the change correctly from the cash drawer.

In a very real sense, a cashier is a banker. When a cashier goes on duty, he or she brings a *bank* to the register. A bank is a cash drawer containing a certain number of bills and coins in various denominations. The bank is designed to enable the cashier to make change during a normal shift.

The cash drawer is the bank from which a cashier makes change. The money in the drawer must balance with the register at the end of the shift. PHOTO BY CLIFF CREAGER

At the end of a shift, the register should contain a cash amount equal to the bank plus all sales receipts. The cashier is responsible for this amount of money. Mistakes cost you money. Thus, it pays to develop math skills for this type of job.

EXERCISE 8-7

Read each question carefully before calculating the answer.

1. When Benji started his shift, he had $450 in the cash register. During his shift, he rang up $275.36 worth of sales. How much money was in his register, or bank, at the end of his shift?

2. Luis started his shift with $350 in his register. His sales total for the day was $1,239.67. How much money should be in the register at the end of his shift?

3. Leonore had $1,547.87 in her register at the end of her shift. Her sales total for the day was $1,317.87. What amount was in her bank at the start of her shift?

4. Joanne had $1,231.43 in her register at the end of her shift. She started with $375.25. What amount of sales were processed through her register?

5. Fujio rang up $758.93 worth of sales during his shift. At quitting time, he had $1,012.56 in his register. At the start of his shift, Fujio had $275 in his bank. By what amount was Fujio's register off at the end of the day?

MARKUP AND PROFIT

Retail businesses charge more for merchandise than they pay for it. This difference in prices, called a *margin,* represents the income of a retailer. The difference between retail prices and merchant costs is called *markup.*

The operating margin of a retail business must cover the expenses of the business. This margin also contains the profit earned by the business.

A number of factors affect markup decisions. Common markup factors include:

- Sales volume
- Costs of handling merchandise
- Prevailing market prices
- Demand for merchandise.

Sales volume. Large stores and retail chains may sell great quantities of certain items of merchandise. These large retailers usually can offer lower prices than smaller competitors. One reason is that large retailers typically buy merchandise at lower cost because of greater volumes.

Costs of handling merchandise. Another advantage generally enjoyed by large retailers is greater efficiency in the handling and distribution of merchandise. As volume grows, the per-unit cost of handling and selling merchandise becomes reduced. This

Lower prices can be charged by stores that buy merchandise in large quantities.

efficiency can be reflected in smaller margins between wholesale and retail prices.

Prevailing market prices. Merchants must be competitive to be successful. Their prices must be in line with competing stores offering similar merchandise. Higher levels of service can attract some buyers, but price remains the greatest single concern of most consumers. Even the highest-priced retail stores may have special sale events to boost volume or to reduce their inventories.

Demand for merchandise. A basic rule of the marketplace involves the relationship between supply and demand. Say that a store has 20 units of a certain item in stock. If 25 customers want to buy one of those items at the stated price, demand is greater than supply. This condition tends to force the price upward.

On the other hand, if only 15 customers want one of the items at the stated price, supply is greater than demand. In this condition, the price will tend to be forced downward.

The point is that as the price of merchandise decreases, demand increases. If prices are too high, demand will not equal supply.

Sometimes, a price is just right for market conditions. When this occurs, supply and demand are equal. The number of consumers willing to buy an item is equal to the supply of that item offered for sale. This is called the *market price.*

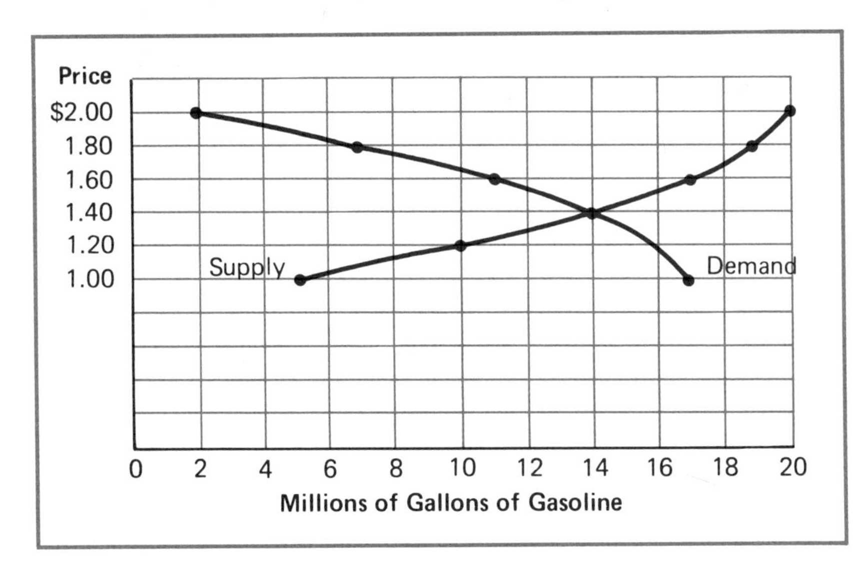

Figure 8-4. Market price is the point at which the demand for a product equals its supply. People are willing to buy all of a product that is offered for sale at the market price.

EXERCISE 8-8

Read each question carefully before calculating the answer. Round calculations to the nearest cent.

1. Po-Ling sells a gallon of milk for $2.05. He purchases milk from the local dairy for $1.35 a gallon. How much does Po-Ling mark up the price of milk?

2. Jensen sells designer jeans. His markup is 65%. If Jensen purchases jeans for $10, what price does he charge?

3. Cecilia sells used musical instruments. She buys a used saxophone for $560. She determines selling price by adding $25 for rent, $14 for advertising, and $135 for profit to her cost. How much does she ask for the saxophone? By what percentage does she mark up this item?

4. The Ticket Agency purchases tickets to rock concerts for $7.50. They sell tickets for $15 apiece. By what percentage does The Ticket Agency mark up tickets?

5. Jeanie purchases purebred dogs for $125 and sells them to pet lovers. Last year, she sold 100 dogs and earned $25,000. What was the average price at which she sold each dog?

6. Dandy's Auto Parts purchases spark plugs for $1.05 each. The price is marked up 22%. What price does Dandy's Auto Parts charge for spark plugs?

7. Alice's Restaurant purchases a six-pack of cola for $1.13. Her markup is 20%. How much does Alice charge for one six-pack of cola? What is her price margin?

8. Charlotte buys record albums for $3.55 apiece. Her markup is 125%. How much does she sell them for?

SUMMING UP

- Retail businesses account for most consumer purchasing transactions. Retail merchants use mathematics in almost every aspect of their business.
- Major factors that affect prices are the amount of handling required to show and sell merchandise and the volume of sales.
- One of the first skills most retail employees must develop is the ability to make change. A related skill is maintaining a running total of purchases by a customer.
- In most states, sales tax must be computed and added to many retail purchases. Sales taxes are based on percentages of retail prices.
- Sales tax tables can be used for greater speed and accuracy in determining sales tax amounts.
- Some retail purchases, such as food items, are exempt from sales taxes in many states. This exemption frequently does not apply to food purchased and consumed in restaurants.
- Electronic cash registers can perform most of the computations involving transaction accumulation and sales taxes. Bar codes carry information on prices and on whether the item is taxable. This information is transmitted electronically to computers that operate registers. Some point-of-sale terminals, such as those in many fast-food restaurants, have keyboards with pictures of products. These registers are simpler and faster to operate than those with standard keyboards.
- Markup is the process of setting retail prices for merchandise. A retailer must cover all expenses of doing business, and realize profit as well, from the margin between wholesale and retail prices.

TERMS THAT COUNT

retail
retailing
merchant
wholesale
wholesaler
merchandising
merchandise
amount tendered
sales tax
tax table
exempt
bar code
inventory
bank
margin
markup
market price

TESTING YOUR WORKING KNOWLEDGE

Read each question carefully before calculating the answer. Round calculations to the nearest cent, or two decimal places.

1. The Tire House purchased 2,500 tires for $55,000, not including sales tax. What was the unit price?

2. The student store at school sells packages of three spiral ring notebooks for $7.95 per package. A local store, Benard's, sells the same notebooks for $3.31 each. Which store offers the better unit price?

3. The Book Nook marks up each book 55%. Sales tax is 4%. If the Book Nook buys a book for $3.50, how much will it cost a customer? If the store purchases a book for $1.99, how much will it cost? The Book Nook buys a book for $4.60, marks up the price, and then applies a 10% discount. How much will a customer pay for the book? (Note: Sales tax is calculated on the *final* sales price.)

4. Janice received a gift certificate for 100 free game tokens at a local video arcade. On her first trip, Janice used 13 tokens. On her second trip, she used 19. On her third, she used 17. How many free tokens did Janice have left? On her fourth trip, Janice took a friend and used up 47 tokens. How many tokens did she have left after the fourth trip?

5. A customer places 6 cans of soup, 3 pounds of apples, 2 gallons of milk, and 2 six-packs of cola on the counter. Soup sells for 49 cents per can. Apples are 68 cents per pound. Cola is on sale for $1.35 per six-pack. Milk sells for $1.98 per gallon. The store charges 5.5% sales tax (milk and apples are non-taxable items). If the customer pays with a $20 bill, how much change should he or she receive?

6. Tickets for a rock concert are $25 for front-row seats, $15 for balcony seats, and $12.50 for standard seating. A customer purchases eight tickets. Half of them are for standard seating, two are for front-row seats, and two are for balcony seats. The sales tax rate is 6%. The customer pays with seven $20 bills. How much change should the customer receive?

7. Snow skis cost $197 a pair and bindings cost $128. A customer purchases two pairs of skis with bindings and gives you thirty-three $20 bills. How much change should the customer receive?

8. Larry is a florist. He purchases red roses for $1.50 each and sells them for $2.25 each. By what percentage rate does Larry mark up the roses? If Larry charges 6.5% sales tax, how much does he charge for one dozen roses?

9. Charley sells video games for home computers. He purchases 100 games for $1,200 and marks up the unit price $4. What percentage rate of markup does this represent?

10. Randy had $1,558.93 worth of sales receipts at the end of his shift. He had $1,908.93 in his cash register. What amount was in the register when Randy started his shift?

11. A young woman purchased a computer system in the store where you work. The computer cost $1,950. The printer cost $545, and the display screen cost $455. The woman paid you the exact amount in $50 bills. How many bills did she give you?

12. Jeff has just purchased an electric guitar and has given the cashier twelve $50 bills and eight $20 bills. The cashier returned $10 in change. What was the selling price of the guitar?

PRESENTING INFORMATION

Read each of the following questions carefully. Then, present the information in the format required.

1. Karen's Deli charges $3.50 plus 5.5% sales tax for a super submarine sandwich. Karen places $0.50 worth of meat, $0.35 worth of cheese, and $0.45 worth of bread on each sandwich. She also places $0.25 worth of miscellaneous products, such as mayonnaise, mustard, lettuce, tomato, and pickles, on each super sub. Draw a pie chart that shows how each cost-area, such as miscellaneous or sales tax, compares with the total sale price of a super sub. (Note: Include the price margin.)

2. Sales tax rates vary from state to state. For example, Alaska does not charge sales tax. Arizona charges 5% sales tax. Connecticut charges 7.5% sales tax. New Mexico charges 3.5% sales tax. Pennsylvania charges 6% sales tax. Draw a vertical bar chart showing the cost of a $7,000 car, including sales tax, in each state listed above.

SKILL DRILLS

Calculate the sales tax and the total price of merchandise including sales tax for each of the following sales transactions. Round calculations to the nearest cent.

1. $13 at 4.5% sales tax
2. $2,432 at 6.5% sales tax
3. $75,356 at 2.5% sales tax
4. $450 at 3% sales tax
5. $9,223 at 6% sales tax
6. $221.39 at 7.5% sales tax
7. $754.18 at 6.5% sales tax
8. $12.45 at 3.5% sales tax
9. $19.23 at 7% sales tax
10. $485.78 at 3.5% sales tax
11. $9,871.52 at 2% sales tax
12. $12,123.67 at 3.7% sales tax
13. $2,175.99 at 6.5% sales tax
14. $18,450.13 at 7.5% sales tax
15. $31,436.71 at 2.5% sales tax
16. $10.35 at 5.5% sales tax
17. $8.99 at 7.5% sales tax
18. $561.49 at 4.5% sales tax

Mathematics and Your Body

YOUR LEARNING JOB

When you have completed the exercises and assignments in this unit, you should be able to:

- ☐ Explain why body measurements are important in health care.
- ☐ Describe how mathematics is used to monitor such indicators as heartbeat and pulse rate.
- ☐ Calculate proper aerobic exercise rates.
- ☐ Calculate proper dosages of medications, following established guidelines.
- ☐ Calculate an individual's caloric needs based upon activity levels.
- ☐ Explain how data are transformed into useful information in the health care field.

WARM-UP

A. Find the sum in each problem.

1. 3,424 + 345 + 12,393 + 54,325 =
2. 436 + 23,142 + 76,402 + 4,398 + 658 =
3. 5,493 + 1,426 + 9,487 + 132,483 + 83,002 =
4. 6,283 + 871 + 45,396 + 10,498 + 352,958 + 1,232,948 =
5. 6,546,983 + 4,392,580 + 1,283 + 34,926 + 915,347 =

B. Find the difference in each problem.

1. 45,587 − 32,897 =
2. 143,008 − 25,579 =
3. 573,846 − 325,836.8 =
4. 2,573,957.08 − 1,976,879.56 =
5. 8,843,807 − 45,382.945 =

C. Find the product in each problem.

1. 56.7 × 56.6 =
2. 194 × 45.93 =
3. 5,943 × 712 =
4. 825 × 356.75 =
5. 4,398 × 8,192 =

D. Find the quotient in each problem. Round calculations to two decimal places.

1. 4,573 ÷ 25.5 =
2. 367 ÷ 2.46 =
3. 8,549 ÷ 756 =
4. 2,950 ÷ 7.243 =
5. 45,768 ÷ 5,840 =

MEASUREMENTS AND YOUR BODY

Mathematics plays a major role in your physical well-being and in the maintenance of good health. Mathematics also is used extensively in the treatment of disease.

Think about how you describe—and how others describe—your body. Aside from the color of your skin, hair, and eyes, your physical description consists primarily of numbers. The two measurements most frequently considered are your height, expressed in feet and inches, and weight, in pounds. In addition, other bodily measurements are used frequently, such as when you buy clothes. Your chest, waist, and other measurements are important. Your feet must be measured so shoes and other footwear will fit properly and be comfortable. Your head measurement is necessary for the fitting of hats.

Bodily measurements can be taken in English or metric terms. For example, a person who stands 6 feet is 1.83 meters or 183 centimeters tall. Simply multiply feet by 0.305. A person who measures 6 feet 6 $\frac{3}{4}$ inches would be exactly 2 meters tall. For this conversion, it is necessary to convert the English measurement to decimal inches. Then, total inches are multiplied by 0.0254 to obtain meters:

$$6 \text{ (feet)} \times 12 \text{ (inches per foot)} = 72 \text{ (inches)}$$
$$72 + 6\frac{3}{4} = 78\frac{3}{4} \text{ or } 78.75$$
$$78.75 \times 0.0254 = 2.00$$

Similarly, weight in pounds (lb) can be converted to kilograms (kg). Multiply pounds by 0.4536. A 150-lb person also weighs 68.04 kg. A 200-pound football player could be listed in a program as weighing 90.72 kg.

$$150 \text{ (pounds)} \times 0.4536 = 68.04 \text{ (kilograms)}$$
$$200 \times 0.4536 = 90.72$$

The point is that numbers are used to provide a precise description of your physical self. This unit discusses some ways in which mathematics is used in monitoring and maintaining the condition of your body.

EXERCISE 9-1

Read each question carefully before performing the conversions. Round answers to two decimal places.

1. Ted's waist is 32.5 inches around. What is the distance around Ted's waist in meters?

2. Carrol's legs are 33 inches long. How long are her legs in centimeters?

3. Ramona weighs 112 pounds. What is her weight in kilograms?

4. Craig is 5 feet 2 inches tall. What is Craig's height in meters?

5. Cecilia's arms are 28 inches long. What is the length of her arms in centimeters?

6. Rueven's feet are 10 $\frac{1}{4}$ inches long. How long are Rueven's feet in centimeters?

7. Lowanda weighs 104 pounds. What is her weight in kilograms?

8. On the average, an African elephant weighs 9,000 pounds and stands 10 feet high at the shoulders. What are these measurements in kilograms and meters?

MATHEMATICS AND YOUR HEART

Your heart is one of the essential *organs* of your body. An organ is a bodily part that performs a function. The heart also is one of the hardest-working bodily organs.

Your heart is a pump that moves blood throughout your body. The blood is pumped through a series of vessels. *Blood vessels* carry blood to different parts of the body. Vessels that carry blood from the heart to parts of the body are called *arteries.* Another set of vessels return blood from body parts to the heart. The vessels that return blood to the heart are called *veins.* This movement of blood from the heart to the body parts and back again is called *circulation.* The entire system for moving blood in the body is called the *circulatory system.*

Body measurements are meaningful mathematical notations in your everyday life.

Your blood delivers oxygen and *nutrients,* or food, to all parts of your body. The blood also picks up waste matter from body parts and delivers it to organs that eliminate it from the body.

You can feel your heart pumping, or beating. Your heart pumps slightly more than two ounces of blood each time the heart muscle contracts. You also can feel your blood moving through some parts of your body. The blood moves as your heart expands (rests and fills with blood) and contracts (pumps). The heart produces a *pulsing* movement of the blood. A pulse is a beating or throbbing. The rate of your heartbeat can be determined by measuring your *pulse rate.* Pulse rate usually is measured by placing a finger against an artery on the thumb side of the inside of the wrist. A watch is used to count off a minute. The pulse rate is the number of pulses, or beats, that occur within a minute.

A typical heartbeat at rest is 75 beats per minute. At this rate, your heart would beat 4,500 times an hour, and 108,000 times in a 24-hour period. In a year, your heart would beat 39.4 million times. If you live to be 80, your heart will beat more than 3 billion times.

At 2 ounces per heartbeat, your heart will have pumped about 49 million gallons of blood in 80 years. Check the accuracy of the preceding statements and then answer the following questions.

EXERCISE 9-2

Read each question carefully before calculating the answer. Round calculations to two decimal places.

1. On the average, the heart pumps 2 ounces of blood per heartbeat. If your heart beats 75 times per minute, how many ounces of blood pass through the heart during one hour?

2. Juan's pulse measures 73 beats per minute. At this rate, how many times will his heart beat during a three-day period?

3. At rest, Maurie's heartbeat averages 79 beats per minute. His heart pumps 2.05 ounces of blood per beat. How many minutes will it take Maurie's heart to pump 16,195 ounces of blood?

4. On three separate occasions, Julie measured her pulse rate. The first time it was 76 beats per minute. The second time it was 74. The third time it was 80. Determine her average pulse rate.

5. If your pulse rate is 78 beats per minute, how many times will your heart beat in 36 hours? 72 hours? 30 days?

Heartbeat and Blood Supply

The volume of blood in a person's body depends upon the surface area of the body. Average blood volume for men is about $12\frac{2}{3}$ pints, or 5.99 liters. Women, on average, have about 7.5% less blood per square meter of body surface.

How rapidly do you think your blood flows through your body? Can you guess how long it takes for your heart to pump your blood through your entire body?

Mathematics makes such questions easy to answer. Say that your normal heartbeat rate is 75 at rest, and that your heart pumps 2.3 ounces of blood on each beat. Also say that your body's blood supply is 6.2 quarts (12.4 pints).

The first step is to convert your total blood volume to ounces. Remember, there are 32 ounces in a quart and 16 ounces in a pint:

$$12.4 \times 16 = 198.4 \text{ ounces}$$

Next, divide that volume by the volume of blood moved by each heartbeat, 2.3 ounces, to determine the number of beats needed:

$$198.4 \div 2.3 = \frac{198.4}{2.3} = 86.26$$

Finally, divide the number of beats by the average number of beats per minute to determine the time required:

$$86.26 \div 75 = \frac{86.26}{75} = 1.15$$

The answer, 1.15 minutes, is approximately 69 seconds. This is the time required by your heart to pump your blood through your body at rest. When you are active, however, your heart beats more rapidly.

Suppose you jog around your school's running track for 15 minutes as part of your physical education class. This exercise increases your heartbeat to 110 beats per minute. Assume that each beat still pumps 2.3 ounces of blood. How long does it take for your heart to pump your blood through your entire body under these conditions? How many times will your blood be circulated during your 15-minute jogging exercise?

An algorithm for finding the solution is the same as the one discussed earlier:

1. Divide total volume of blood by volume pumped per heartbeat. Result is number of beats required to circulate blood.
2. Divide number of beats required to circulate blood by heartbeat rate per minute. Result is time required to circulate blood.

The calculations are as follows:

$$198.4 \div 2.3 = \frac{198.4}{2.3} = 86.26$$

$$86.26 \div 110 = \frac{86.26}{110} = 0.78$$

The result, 0.78 minutes, equals approximately 46.8 seconds.

EXERCISE 9-3

Read each question carefully before answering the question. Round calculations to two decimal places.

1. Martha's pulse rate is 74 beats per minute and her heart pumps 2.1 ounces of blood per heartbeat. Her body holds 195.2 ounces of blood. How long does it take for her heart to circulate her blood through her entire body?

2. Art's pulse rate is 78 beats per minute and his heart pumps 2.35 ounces of blood per heartbeat. His body holds 201.6 ounces of blood. How long does it take for Art's heart to circulate his blood?

3. Betty's pulse rate is 72 beats per minute and her heart pumps 2 ounces of blood per heartbeat. She has 6 quarts of blood in her body. How long does it take for Betty's heart to circulate her blood?

4. When Hisako plays football, his pulse rate averages 140 beats per minute. Hisako has 6.3 quarts of blood in his body and his heart pumps 2.15 ounces of blood per heartbeat. When Hisako plays football, how many times is his blood circulated in 20 minutes?

5. Joey can run a mile in 6.42 minutes. While running, his pulse rate averages 146 beats per minute. Joey's heart pumps 2.25 ounces of blood during each beat and Joey has 12.8 pints of blood in his body. At 115 beats per minute, how many minutes will it take to circulate all of the blood in Joey's body? How many times will his blood be circulated when he runs a mile?

Strenuous activities, such as hiking or running, increase the rate at which a person's heart beats.
COURTESY OF VIRGINIA STATE TRAVEL SERVICE

MATHEMATICS AND EXERCISE

Regular exercise is vital to the maintenance of good health. The two general approaches to exercise are:

- Aerobic
- Isometric.

Exercise involves straining, or putting to heavy use, your circulatory and *respiratory systems*. The respiratory system delivers oxygen to circulating blood. Oxygen is necessary for *combustion,* or burning, to take place. Your body constantly burns fuel, the food you eat, converting it to energy. To do this efficiently, your body must have an adequate supply of oxygen.

Respiration means breathing. Mechanisms within your lungs draw oxygen from the air you breathe and feed the oxygen into your blood.

At the same time, your lungs remove gases from your blood. These gases are the result of the burning of fuel within your body. Thus, your respiratory system takes in oxygen and puts out gases.

Of the two general types of exercise, *aerobic* exercise is designed to benefit your circulatory and respiratory systems. Aerobic exercise involves movement of the body. Examples of aerobic exercise are walking, running, and swimming. Aerobic exercises help condition the heart and lungs by increasing the efficiency of oxygen intake by the body.

Isometric exercise is designed primarily for building physical strength. Weight lifting is an example of isometric exercise.

Mathematics is an important element in aerobic exercise. Proper heart rate, or pulse, for obtaining maximum aerobic benefit is calculated mathematically.

Reaching and Maintaining an Aerobic Condition

Aerobic exercise must be carried on for a period of time to be effective. When your heartbeat, or pulse rate, reaches a certain point, an *aerobic condition* exists. This condition, or heart rate, should be maintained for 12 minutes or more for full physical benefits.

Your age determines your proper aerobic pulse rate. The mathematical formula consists of two simple steps. Start with the value of 220. The first step is to subtract your age. The remainder is your maximum aerobic heart rate. If you are 15 years of age, your maximum aerobic heart rate is:

$$220 - 15 = 205$$

The second step is to multiply your maximum aerobic heart rate by 70 percent. The product is your proper aerobic heart rate, or pulse rate, during exercise. At age 15, therefore, your aerobic pulse rate is:

$$205 \times 0.70 = 143.50$$

The idea is to maintain a pulse rate between 70% and 80% of your maximum aerobic heart rate. Thus, the 15-year-old individual would have an aerobic pulse rate between 143.50 and 164.

EXERCISE 9-4

Read each question carefully before calculating the answer. Round calculations to two decimal places.

1. April is 25 years old. What is her maximum aerobic heart rate?

2. Kenneth is 12 years old. What is his maximum aerobic heart rate?

3. Raymond is 36 years old. What is his maximum aerobic heart rate?

4. Sarah's maximum aerobic heart rate is 175. How old is Sarah?

5. Carlos has a maximum aerobic heart rate of 145. How old is Carlos?

6. Barbara is 16 years old. What is the range (70% to 80%) for her proper aerobic heart rate?

7. Ron is 67 years old. What is the range for his proper aerobic heart rate?

8. Greg is 29 years old. What is the range for his proper aerobic pulse rate?

The Time Factor

Exercise should be continued for 12 minutes or more after the aerobic pulse rate has been achieved. Time also must be allowed for the heart rate to increase from a normal condition to the aerobic rate. The type of exercise determines how long this takes. For example, running is more strenuous than walking. Therefore, less time is required to reach aerobic pulse rate when running. In general, a total exercise period of 20 minutes or more is recommended.

Aerobic exercise causes the heart to beat more often during a given time period. For example, suppose a 20-year-old person has a normal pulse rate of 68 beats a minute. In a 40-minute period of normal activity, that person's heart would beat 2,720 times. How many times would that person's heart beat during a two-mile aerobic hike that required 40 minutes?

First, subtract the person's age from 220. Second, multiply the remainder by 0.70. Third, muliply the product by 40:

$$220 - 20 = 200$$
$$200 \times 0.70 = 140$$
$$140 \times 40 = 5{,}600$$

Thus, the hiker's heart would beat approximately 5,600 times during the 40-minute exercise period. That represents an increase of 2,880 beats over normal.

Aerobic exercise helps condition the respiratory and circulatory systems. PHOTO BY LIANE ENKELIS

EXERCISE 9-5

Read each question carefully before calculating the answer. Round calculations to two decimal places.

1. David is a 35-year-old jogger and his normal pulse rate is 73 beats per minute. On the average, how many times does his heart beat during a 35-minute jog?

2. Elise is a 24-year-old basketball player and her normal pulse rate is 69 beats per minute. On the average, how many times does her heart beat during a 12-minute quarter?

3. Eduardo is a 28-year-old soccer player. His normal pulse rate is 70 beats per minute. On the average, how many times does Eduardo's heart beat during 50 minutes of normal activity? During a 50-minute soccer game? What is the difference?

4. Otis is a 14-year-old runner. On the average, how many times does his heart beat during a 2-hour-and-10-minute marathon?

5. Marion races motorcycles. She is 19 years old. Her normal pulse rate is 72 beats per minute. On the average, how many times does her heart beat during 40 minutes of normal activity? During a 40-minute motorcross race? What is the difference?

6. Howard is a 47-year-old tennis player. His normal pulse rate is 74 beats per minute. What is Howard's proper aerobic pulse rate during a 30-minute tennis match?

7. Carolyn enjoys ice skating. She is 53 years old. Her body contains 6 quarts of blood and her heart pumps 2.1 ounces of blood with each beat. On the average, how many times does her heart beat during 20 minutes of skating? At that pulse rate, how long does it take for her heart to circulate her blood completely?

8. Allen plays ice hockey and is 38 years old. His body holds 6.15 quarts of blood and his heart pumps 2 ounces of blood during each beat. On the average, how many times does Allen's blood circulate during a 25-minute hockey game?

MATHEMATICS AND MEDICATION

Mathematics has a tremendous impact in the area of drug and medication dosages. Several general factors must be considered in determining the proper dosage of a drug for an individual. These factors include:

- Age
- Body weight
- Percentage of body fat
- General level of activity.

Correct dosage is important for two reasons. First, too small a dose will produce little or no effect. Second, too large a dose may produce undesirable, or even *toxic,* side effects. Toxic means poisonous.

Age	Portion of Adult Dose
20 years	Full dose
10	$\frac{1}{2}$
5	$\frac{1}{4}$
$2\frac{1}{2}$	$\frac{1}{8}$
1	$\frac{1}{12}$

Figure 9-1. Traditional dosage chart for medications is based upon age.

Measuring Dosage by Age

Most drugs are packaged with directions for adult dosages. The age factor is very important in the use of any form of medication, even aspirin. A traditional rule of thumb is represented in the chart in Figure 9-1.

Another formula for children's dosages is the following, suggested by New York physician W. A. Bastedo, M.D. Bastedo's Rule is:

$$\text{Adult dose} \times (\text{age} + 3) \div 30 = \text{child's dose}$$

To illustrate these two rules in practice, consider this problem:

> How close are the dosages given by the two rules for a 10-year-old child if an adult dose of a certain medication is 4 teaspoons?

The first rule states that a 10-year-old should receive one-half the adult dosage. In this case, the dose would be 2 teaspoons. The following equation applies under Bastedo's Rule:

$$4 \times (10 + 3) \div 30 = 4 \times \frac{13}{30} = \frac{52}{30} = 1.73$$

The suggested dosages are quite close. The difference is 0.27 teaspoons, or a little more than one-fourth of a teaspoon.

Try this exercise for further comparison of dosages under these two rules. Compare the dosages of a medicine with an adult dosage of 30 milliliters (ml), which is approximately 2 tablespoons. Use both methods to determine dosages for a 5-year-old, a 10-year-old, and a 15-year-old. Can you find an age for which the two methods suggest identical dosages?

EXERCISE 9-6

Read each question carefully before calculating the answer. Round calculations to two decimal places.

1. Bobby is five years old and is taking medicine for a cold. The adult dosage is 2 tablespoons of medicine every four hours. Using the traditional proportion method, what dosage should Bobby be given every four hours?

2. Janice is 10 years old and is taking medicine for the flu. The adult dosage is two pills every six hours. Using the traditional proportion method, how many pills should Janice be given every six hours?

3. Harriet is 12 years old and is taking cough syrup for a cold. The adult dosage is 4 tablespoons. Using Dr. Bastedo's Rule, what dosage should Harriet be given?

4. Joe is $2\frac{1}{2}$ years old and has a fever. His doctor has recommended that Joe's mother give Joe aspirin every four hours. The adult dosage is two every four hours. What dosage should Joe's mother give Joe? Use the traditional proportion method.

5. Susan is five years old and is not feeling well. The doctor has instructed Susan's mother to give the child a special medicine. The adult dosage is $\frac{1}{2}$ ounce every eight hours. Using the proportion method, determine the dosage Susan should receive.

6. Landus has a fever and his mother is giving him aspirin. Landus is nine years old. The adult dosage is two aspirin every four hours. Using Dr. Bastedo's Rule, what dosage should Landus' mother be giving him?

7. Nestor has a cold and is taking aspirin. His doctor used the traditional proportion method to determine the dosage. The doctor said Nestor should get one aspirin every four hours. The adult dosage is two aspirin every four hours. How old is Nestor?

8. Kamitaka is 10 years old and has a cold. His mother is giving him cough syrup to soothe his throat. The adult dosage is 3 teaspoons every three hours. Determine what dosage Kamitaka should be taking. Use both Dr. Bastedo's Rule and the traditional proportion method. Then, find the difference between the two methods.

MATHEMATICS AND DIET

The food you eat is composed primarily of three nutrients that produce the energy your body needs to function. Nutrients are substances that are utilized by the body to promote growth and to maintain life. Energy-producing nutrients are:

- Proteins
- Fats
- Carbohydrates.

A balanced diet should contain these three nutrients in a balanced proportion. Nutrition experts today suggest the following percentages in a normal diet:

Nutrient	Percentage Intake
Proteins	12%
Fats	30%
Carbohydrates	58%

Current research estimates describe the average American diet as containing more than 42% fats instead of the recommended 30%. A hamburger is a good example. An ordinary hamburger is two-thirds fat and only one-third protein. Lean means containing relatively little or no fat. One disadvantage to a diet high in fats is a dramatic increase in the probability of heart disease.

Calories are a measure of the energy needed to perform an activity. Taking part in an athletic game consumes more calories than watching the game.

Counting Calories

You may have heard people on diets talk about the calories they consume. A *calorie* is the amount of heat required to raise the temperature of 1 gram of water 1 degree centigrade [1.8 degrees F]. This formula applies at 1 *atmosphere* of pressure. An atmosphere is a unit of pressure equal to the pressure of air at sea level, approximately 14.7 pounds per square inch (psi).

In terms of food, the complete utilization of a gram of protein or carbohydrate yields about four calories. The complete utilization of a gram of fat yields about nine calories.

Calories provide the energy the body needs to function. The more active you are, the more calories you need. The table in Figure 9-2 offers examples of typical caloric needs. The numbers represent calories needed per hour, based upon body weight expressed in kilograms. You can determine your weight in metric terms by dividing your weight in pounds by 2.2. Thus, if you

Activity	Caloric Needs per Kilogram per Hour
Sleeping	0.93
Sitting at rest	1.43
Standing relaxed	1.50
Walking	2.00
Light exercise	2.43
Moderate exercise	4.14

Figure 9-2. Increasing activity levels bring a need for greater caloric intake. The body burns calories to produce energy.

weigh 120 pounds, your metric weight is 120 ÷ 2.2 = 54.5 kilograms. If you weigh 160 pounds, your metric weight is 72.7 kilograms.

$$120 \div 2.2 = 54.5 \text{ kilograms}$$
$$160 \div 2.2 = 72.7 \text{ kilograms}$$

Thus, a 120-pound [54.5-kilogram] person would need 132.4 calories per hour during light exercise. The computation is:

$$54.5 \times 2.43 = 132.4$$

Calories that are not burned up by your body to produce needed energy are stored by the body. Thus, excess calories cause weight gain. This is why people on diets often keep track of their caloric intake.

Measuring Caloric Needs

Suppose a 120-pound [54.5-kilogram] person kept a record of daily activity for a normal work week. The record shows that an average day for this person includes the following activities:

- 8 hours sleep
- 3 hours moderate exercise
- 1 hour walking
- 7 hours sitting at rest
- 2 hours standing relaxed
- 3 hours light exercise.

This person's daily caloric need can be calculated by using information from the Activity/Calorie Chart in Figure 9-2. The solution involves computing the number of calories required for each activity and then adding these totals. To determine caloric need for each activity, multiply weight (rounded to 55) × caloric need per hour for the activity × number of hours:

Sleep	55 × 0.93 × 8 = 409.2
Moderate exercise	55 × 4.14 × 3 = 683.1
Walking	55 × 2.00 × 1 = 110.0
Sitting at rest	55 × 1.43 × 7 = 550.6
Standing relaxed	55 × 1.50 × 2 = 165.0
Light exercise	55 × 2.43 × 3 = 401.0
Total caloric need:	2,318.9

EXERCISE 9-7

Read each question carefully. Then determine the person's daily caloric needs. Refer to the Activity/Calorie Chart in Figure 9-2 to find data about caloric needs per kilogram per hour.

1. Joshua sleeps 10 hours per day and spends three hours performing moderate exercises. He also walks for one hour, stands relaxed for five hours, and sits relaxed for five hours. He weighs 110 pounds.
2. Kim performs light exercises six hours per day. In addition, she sleeps nine hours, walks for two hours, sits relaxed for five hours, and stands relaxed for two hours. Kim weighs 95 pounds.
3. Kimo sits relaxed for seven hours per day and stands relaxed for two hours per day. Before work, he performs moderate exercises for one hour and light exercises for three hours. After work, Kimo walks for one hour and sleeps for 10 hours. Kimo weighs 215 pounds.
4. At work, Leon spends eight hours a day performing moderate exercise. To relax after work, he exercises lightly for two hours, walks for one hour, and then sits and relaxes for four hours. At night, he sleeps for nine hours. Leon weighs 186 pounds.
5. I-chen performs light exercises eight hours per day. In addition, she walks for one hour, sits at rest for three hours, and stands at rest for three hours. At night she gets nine hours of sleep. I-chen weighs 105 pounds.

Diet and Caloric Intake

The chart in Figure 9-3 shows the caloric content of a variety of foods. Using tables such as this, you can plan meals that will provide your body with the appropriate number of calories. First, calculate your caloric needs as shown above.

You also can check the caloric intake of some of your favorite meals. Compare these figures to your caloric needs on a typical day.

Finally, you can calculate the percentages of protein, fat, and carbohydrate in each food item. This enables you to plan daily meals that will provide the proper proportions of these substances: 12% protein, 30% fat, 58% carbohydrate.

Proper diet is a vital part of maintaining good health. Caloric intake should be measured according to your level of activity.

FRUIT AND FRUIT JUICES

	SIZE OF PORTION	CALORIES
Apple	1 medium	76
Avocado	½ peeled	279
Banana	1 medium	88
Cantaloupe	½ medium	37
Fruit mix	½ cup . . . approx.	100
Grapefruit	½ medium	75
Grapefruit juice	½ cup	65
Lemon and lime	1 medium	20
Olives, green	2 medium	15
Orange	1 medium	70
Orange juice	½ cup fresh	58
Peach	1 medium	46
Pear	1 medium	95
Pineapple juice	½ cup	65
Prune juice	½ cup	85
Prunes	4 or 5 with liquid	121
Raisins	¼ cup	215
Strawberries	1 cup raw	54

BREADS AND CEREALS

	SIZE OF PORTION	CALORIES
Biscuit	1 medium	129
Bread, white	1 slice	63
Bread, whole wheat	1 slice	55
Cereal cooked	½ cup	74
Cereal, ready to eat	1 cup	100
Flour—all purpose	1 cup sifted	401
Noodles	½ cup cooked	54
Pancake	1 thin 4" diam.	59
Rice	½ c. cooked	101
Spaghetti	½ c. cooked	109
Rolls, plain	1 medium	118
Rolls, sweet	1 medium	178
Waffle	1 medium	216

SALADS WITH DRESSING

	SIZE OF PORTION	CALORIES
Carrot and raisin salad	½ cup	100
Chicken and celery salad	½ cup	250
Coleslaw	½ cup	51
Cottage cheese, pineapple	½ cup, 1 slice	240
Gelatin, cabbage, carrot	½ cup	40
Potato salad	½ cup	200
Tossed salad, mixed veg.	½ cup	50

MEAT, EGGS, CHEESE

	SIZE OF PORTION	CALORIES
Bacon, crisp	2 strips	97
Beef—chuck	3 oz.-avg. serv.	265
Cheese, American	1" cube (1 oz.)	113
Cheese, Cottage	½ cup	107
Chicken, fried	3 oz. meat— avg. serv.	169
Chicken	½ broiler	332
Egg, boiled or poached	1 medium	77
Egg, scrambled	1 medium	106
Halibut, baked	avg. serving	228
Tuna, canned	3 oz.	247
Frankfurter	1 medium	124
Ham, baked	3 oz. (no bone)	339
Ham, boiled	2 oz.	172
Hamburger	3 oz.	316
Luncheon meat	2 med slices	165
Steak, round	3 oz. (no bone)	197
Steak, sirloin	3 oz. (no bone)	257

POPULAR MAIN DISHES

	SIZE OF PORTION	CALORIES
Chicken pie	1 pie, 3¾" diam.	350
Chop Suey—American	1 cup	400
Macaroni and cheese	1 cup	464
Beef and vegetable stew	1 cup	252
Pizza pie	4" diameter	185
Soup, creamed tomato	1 cup	201
Soup, vegetable	1 cup	82
Spaghetti, Italian style	1 cup + 1 oz. cheese	331

VEGETABLES

	SIZE OF PORTION	CALORIES
Asparagus	½ c. spears	21
Beans, dry	½ c. cooked plain	115
Beans, lima	½ cup cooked	76
Beets	½ cup	34
Broccoli	½ cup	22
Cabbage	½ cup	20
Carrots, cooked	½ cup, diced	22
Carrots, raw	one 5½" long	21
Cauliflower, cooked	½ cup	15
Celery, raw	three 5" stalks	9
Corn, cooked	½ cup	70
Greens	½ cup	20
Lettuce, head	¼ of lb.	17
Peas	½ cup cooked	55
Potato, baked	1 med. size	97
Potato, mashed	½ cup	120
Potato, french fried	8 avg. size	157
Radishes	4 small	4
Sweet potatoes, baked	1 avg. size 5"x 2"	252
Sweet potatoes, candied	1 small	314
Tomatoes, fresh	1 medium, raw	30
Tomatoes, canned	½ cup	23

BEVERAGES

	SIZE OF PORTION	CALORIES
Carbonated, cola type	8 oz. or 1 cup	107
Cocoa, with whole milk	1 cup	236
Coffee, black	1 cup	–
Milk, whole	1 cup	166
Milk, skim or buttermilk	1 cup	86
Tea, plain	1 cup	–

FATS

	SIZE OF PORTION	CALORIES
Butter or margarine	1 tbsp.	100
Cream, heavy (40%)	1 tbsp.	49
Cream, coffee (20%)	1 tbsp.	30

SOME MAJOR CULPRITS

	SIZE OF PORTION	CALORIES
Candy bar, milk chocolate	1 bar, 1.6 oz.	250
Chocolate candy + nuts	1 piece (1 oz.)	151
Chocolate cream	1 oz.	110
Jam, jellies	1 tbsp.	55
Malted milk	1 cup	281
Nuts, cashew	1 oz.	164
Peanuts	10 kernels or 1 tbsp.	50
Pecans	1 tbsp. chopped	52
Soda with ice cream	1 large glass	325
Sundae, chocolate	½ cup ice cream, 2 tbsp. sauce	330
Whipping cream	1 heaping tbsp.	49
Sugar	1 tbsp	48

Figure 9-3. Different foods contain varying amounts of calories. Tables such as this help in meal planning.

EXERCISE 9-8

Read each question carefully before calculating the answer. Refer to the chart in Figure 9-3 for the number of calories per food item.

1. On Monday, Jim ate the following foods for breakfast: two cups of whole milk, two scrambled eggs, four strips of crisp bacon, and two medium-sized biscuits with one tablespoon of butter. How many calories were in Jim's breakfast?

2. Jane ate the following foods for lunch: one cup of whole milk, one 1.6-ounce milk chocolate bar, one 3-ounce hamburger, and one medium-sized banana. How many calories did she consume during lunch?

3. Kermit had two cups of vegetable soup, one medium-sized baked potato, and a 3-ounce sirloin steak for dinner. He also drank 16 ounces of cola. How many calories did Kermit take in during dinner?

4. Geraldine ate the following foods yesterday: For breakfast, she had one scrambled egg, one cup of fresh orange juice, one medium-sized peach, and two crisp strips of bacon. For lunch, she ate a 3-ounce hamburger and drank 16 ounces of cola. In the evening, she had one-half cup of potato salad, one-half cup of cooked corn, one cup of whole milk, and an average-sized serving of baked halibut. In addition, she ate one medium-sized apple and a 1-ounce chocolate candy bar with nuts as a snack during the day. How many calories were in the foods she ate?

5. Adrian ate the following foods on Friday: For breakfast, he had two cups of ready-to-eat cereal, two cups of whole milk, one-half cup of grapefruit juice, and two scrambled eggs. He ate two cups of Italian style spaghetti with cheese and drank two cups of whole milk for lunch. At dinner, he had one cup of whole milk, a 3-ounce round steak, one cup of tossed green salad with dressing, and one-half cup of mashed potatoes. For dessert he had a chocolate sundae with one-half cup of ice cream and two tablespoons of sauce. How many calories did Adrian take in during the day?

PRESENTING DATA FOR HEALTH

One of the most valuable uses of mathematics in the health-care field is the gathering and analyzing of statistical data. Improved information-gathering techniques have made it possible to identify public health trends. From these trends, government and private agencies can determine research and treatment needs.

As with other forms of statistics, these data can be presented in graph or chart form to increase their impact. The graph in Figure 9-4 shows the death rate from rheumatic fever in the United States per 100,000 population. Rheumatism is a disease that causes painful swelling in muscles and joints. Rheumatic fever also can attack such vital muscles as the heart, which can lead to death. The graph presents several important items of information. One item is the downward trend in deaths caused by rheumatic fever. See if you can determine some others in the following exercise.

EXERCISE 9-9

Read each question carefully. Refer to the graph in Figure 9-4 to determine each answer.

1. In which year was the number of deaths from rheumatic fever the highest?

2. Which decade shows the least amount of change in the number of deaths caused by rheumatic fever?

3. Which decade shows the greatest amount of change in the number of deaths caused by rheumatic fever?

4. In which year was the number of deaths caused by rheumatic fever the lowest?

5. Since 1910, have the number of deaths caused by rheumatic fever increased or decreased?

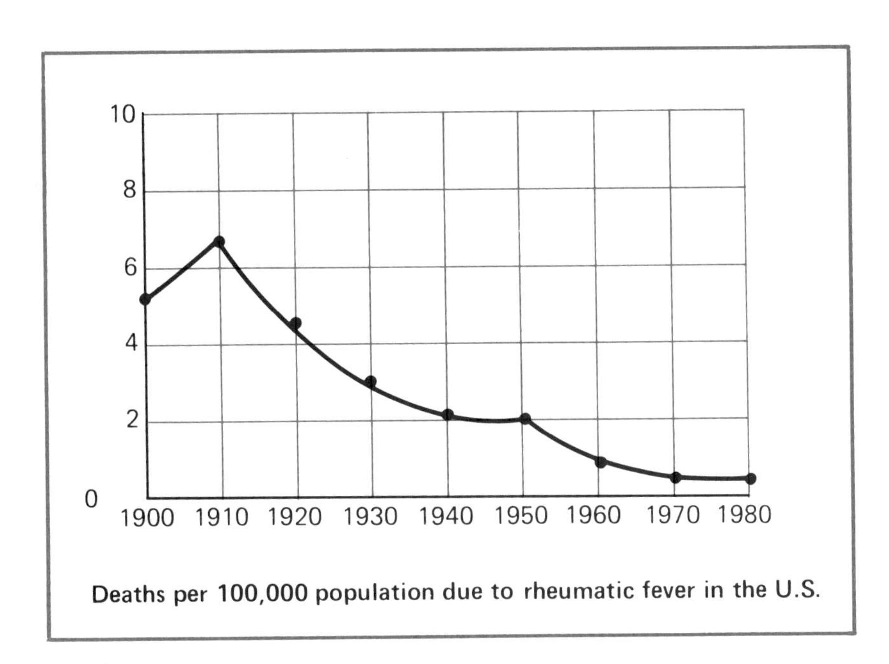

Figure 9-4. Graphs are used to present statistical data on public health trends. This graph shows the declining occurrence of deaths due to rheumatic fever in the United States.

Determining Probabilities

The graph in Figure 9-5 depicts the rate of heart disease death rate per 100,000 population in the United States. The information this graph presents is of immediate concern today. This is because the graph shows that the various kinds of heart disease represent an increasing problem.

The graph in Figure 9-5 offers an opportunity to test your skill in determining *probabilities.* A probability is the likelihood of something happening. For example, the probability of a person dying of heart disease in 1930 was approximately 200 in 100,000 population. If you wanted to determine your statistical chances of dying from heart disease in 1930, you would perform this calculation:

$$\frac{200}{100{,}000} = \frac{2}{1{,}000} = 0.002$$

Thus, the probability that you might die of heart disease was 0.2%, or one-fifth of one percent. There are other probabilities that can be determined from the information in the chart. The following exercise will test your skills in this area.

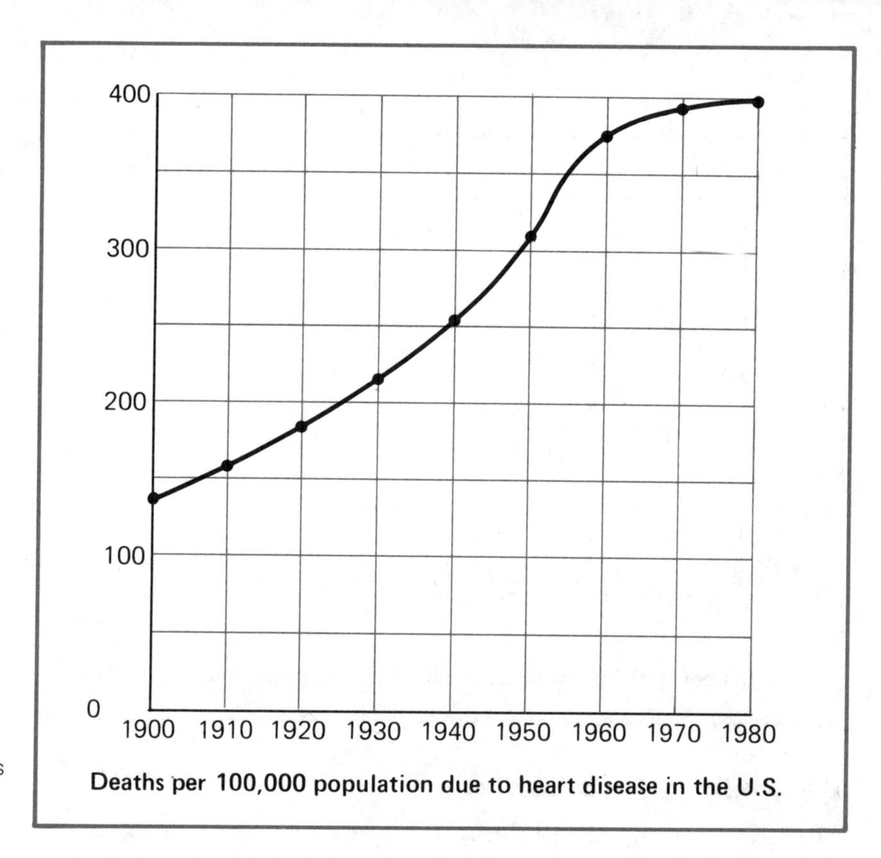

Figure 9-5. Graphs such as this one can be used to determine the probability, or statistical chance, of an occurrence. The graph indicates an increasing public health problem of major concern.

EXERCISE 9-10

Read each question carefully. Refer to the graph in Figure 9-5 to determine each answer.

1. Were there more deaths resulting from heart disease in 1940 or in 1970?
2. Since 1900, has the number of deaths related to heart disease increased or decreased?
3. During which decade did the number of deaths related to heart disease change the most?
4. In 1910, what was the probability that a person living in the United States would die from heart disease?
5. Determine the probability that a person living in 1980 would die from heart disease.

SUMMING UP

- Mathematics is used in many ways in monitoring, evaluating, and treating the human body.
- As a special language, mathematics is used to describe a person's physical characteristics. Measurements such as height and weight may be related to your health. Other measurements help you to select clothes that fit properly. Your physical characteristics can be expressed in English or metric measurements.
- One of the hardest working organs in your body is your heart. This muscular pump moves blood throughout your body. The speed at which your heart beats, or your pulse rate, is measured in beats per minute. Your level of activity affects your heart rate.
- Mathematics can be used to plan exercise programs. Exercises that condition the circulatory and respiratory systems are called aerobic. An aerobic exercise involves movement of the body.
- To determine your aerobic pulse rate, subtract your age from 220 and multiply the remainder by 0.70. For maximum benefit, aerobic exercise should be performed for at least 12 minutes once the aerobic pulse rate is achieved.
- Mathematics also is important for determining proper dosages of medications. Dosage levels usually are based on such factors as age, body weight, percentage of body fat, and general activity level.
- Activity level also is a determining factor in a person's caloric needs. A calorie is a unit of heat energy. Calories are contained in food, which is the fuel the body burns. Mathematics is used to plan and control caloric intake.
- Health data, or statistics, can be shown in graphic form to present meaningful information about populations. Charts and graphs can be used to show trends and to evaluate the results of public health programs.

TERMS THAT COUNT

organ
blood vessel
artery
vein
circulation
circulatory system
nutrient
pulse
pulse rate
respiratory system
combustion
respiration
aerobic
isometric
aerobic condition
toxic
calorie
atmosphere
probability

TESTING YOUR WORKING KNOWLEDGE

Read each question carefully before calculating the answer. Round calculations to two decimal places.

1. John is $5\frac{1}{4}$ feet tall and weighs 125 pounds. What are his height and weight when measured in meters and kilograms?

2. At rest, Lori's pulse rate averages 72 beats per minute. Her heart pumps 2 ounces of blood per beat. How many minutes will it take Lori's heart to pump 225 quarts of blood?

3. Ed's pulse rate is 75 beats per minute and his heart pumps 2.15 ounces of blood per beat. He has 6 quarts of blood in his body. How long does it take for Ed's heart to circulate his blood completely?

4. Benji is 32 years old. What is his proper aerobic heart rate?

5. Harold enjoys roller skating. He is 33 years old and his body holds 6.3 quarts of blood. His heart pumps 2.05 ounces of blood with each beat. How many times does his heart beat during 10 minutes of skating? At that pulse rate, how long does it take for Harold's heart to circulate his blood completely?

6. Ramona is five years old and has a cold. Her mother is giving her cough syrup to soothe her throat. The adult dosage is 4 teaspoons every three hours. Determine what dosage Ramona should be taking. Use both Dr. Bastedo's Rule and the traditional proportion method. Then, find the difference between the two methods.

7. At work, Larry spends six hours per day performing moderate exercise and two hours standing at rest. To relax after work, he exercises lightly for two hours and walks for one hour. Then, he sits and relaxes for three hours. At night, he sleeps for 10 hours. Larry weighs 157 pounds. What are Larry's daily caloric needs? Refer to Figure 9-2.

8. Ann ate the following foods on Monday: For breakfast she had one cup of ready-to-eat cereal, two cups of whole milk, four crisp strips of bacon, and two scrambled eggs. She ate one 3-ounce hamburger and one medium-sized apple and drank two cups of whole milk for lunch. At dinner, she had one cup of whole milk, one 3-ounce sirloin steak, one cup of potato salad, and one-half cup of cooked peas. In addition, she had one large glass of soda with ice cream for dessert. How many calories did Ann take in during the day? Refer to Figure 9-3.

9. Robert is $6\frac{1}{4}$ feet tall and weighs 225 pounds. What are his height and weight in meters and kilograms?

10. Eugene's head measurement is 24 inches. What is this in centimeters?

11. The range of Martha's proper aerobic heart rate is between 115.5 and 132 beats per minute. What is Martha's age?

12. Emily is a 17-year-old tennis player. Her normal pulse rate is 68 beats per minute. Approximately how many times will Emily's heart beat during a 30-minute tennis match?

13. Mario's doctor prescribed $\frac{1}{4}$ teaspoon of medicine every four hours. The doctor uses the traditional proportion method to determine dosage. The adult dosage is 2 teaspoons every four hours. How old is Mario?

14. Karl is 49 years old. What is his proper aerobic heart range?

15. Todd and Terry are identical twins. However, Todd weighs 4.54 kilograms more than Terry who weighs 120 pounds. How many pounds does Todd weigh?

PRESENTING INFORMATION

Read each question carefully. Then present the information in the format required.

1. A balanced diet contains three nutrients—proteins, fats, and carbohydrates—in balanced portions. Draw a pie chart that shows the combination of these nutrients that constitutes a balanced diet. The information you will need to draw your pie chart is given earlier in this chapter.

2. Carlos consumed 2,700 calories on Monday; 1,940 calories on Tuesday; and 3,060 calories on Wednesday. Then, he took in 2,670 calories on Thursday; 2,600 on Friday; 2,800 on Saturday; and 2,500 on Sunday. Draw a line graph that shows how Carlos' calorie intake changed for the week. On which day did Carlos' calorie intake change the most? On which day did it change the least?

SKILL DRILLS

A. Find the sum in each problem.

1. 354 + 845 + 65,427 + 52,876 =
2. 3,245 + 11,845 + 58,326 + 8,436 + 333 =
3. 3,214 + 5,334 + 4,522 + 235,667 + 76,886 =
4. 3,647 + 744 + 47,754 + 47,747 + 745,528 + 8,332,652 =
5. 2,414,432 + 3,325,666 + 6,745 + 54.653 + 915.547 =

B. Find the difference in each problem.

1. 45.587 − 2.325 =
2. 121.008 − 22.526 =
3. 528,436 − 176.5828 =
4. 634,678.11 − 536,879.76 =
5. 8,324.876 − 7,321.989 =

C. Find the product in each problem.

1. 12.4 × 32.7 =
2. 876 × 15.03 =
3. 2,745 × 0.755 =
4. 3,452 × 91.23 =
5. 7,327 × 8.102 =

D. Find the quotient in each problem. Round calculations to two decimal places.

1. 363 ÷ 17.68 =
2. 823 ÷ 5.96 =
3. 8.438 ÷ 698 =
4. 1.546 ÷ 3.604 =
5. 45.254 ÷ 3.622 =

P A R T

Mathematics, Products, and Travel

IN THIS PART

Horsepower is a word you hear frequently in discussions of automobiles. The term actually represents a measure of the capacity of a horse to perform work. Horsepower is one of the values you will learn to develop mathematically.

People build machines to perform tasks that are beyond the physical abilities of humans. Mathematics is used to measure the amount of energy needed to perform work. As you will see, mathematics also is used to measure a machine's capacity to apply energy.

Manufacturing businesses rely heavily upon some basic mathematical calculations in decision making. The calculations have to do with determining costs. The basic mathematics is similar, whether the business is a giant corporation or a one-person shop.

The management of inventory involves the same basic mathematical considerations, regardless of the size of a business. Mathematics is used to maintain proper inventory levels. Too much inventory reduces profit. Too little can mean lost sales.

Inventory also has to be moved. You will explore a variety of transportation systems designed to solve special problems.

Mathematics and Machines

YOUR LEARNING JOB

When you have completed the exercises and assignments for this unit, you should be able to:

- ☐ Explain the terms energy, power, and work.
- ☐ Calculate mechanical energy in terms of foot-pounds.
- ☐ Calculate mechanical power in terms of horsepower.
- ☐ Grade horsepower requirements for work to be done.
- ☐ Calculate power transmission with belts.
- ☐ Calculate power transmission with gears.
- ☐ Explain the function of an idler gear.
- ☐ Calculate mechanical advantage of levers.
- ☐ Calculate mechanical advantage of gears.
- ☐ Calculate mechanical advantage of hydraulic systems.

WARM-UP

A. Find the sum in each problem.

1. 3,645.9 + 452.87 + 23,172 + 56.901 =
2. 45,983 + 823,472 + 54,334 + 728,902 + 82,387 =
3. 21,472 + 8,372.83 + 643,285.3 + 45,473.12 =
4. 1,283,845 + 4,327,543 + 483,694 + 8,837,556 =
5. 8.983 + 34.2193 + 187.04 + 73.0074 + 54.21734 =

B. Find the difference in each problem.

1. 6,584.9 – 4,382.032 =
2. 45,832 – 45.02 =
3. 674.34 – 435.003 =
4. 8,795.8 – 643.948 =
5. 2,745.03 – 2,740.917 =

C. Find the product in each problem. Round calculations to two decimal places.

1. 45.82 × 36.47 =
2. 243.8 × 35.09 =
3. 65.012 × 4.86 =
4. 72.94 × 109.3 =
5. 6,748 × 4.904 =

D. Find the quotient in each problem. Round calculations to two decimal places.

1. 53.78 ÷ 4.57 =
2. 635 ÷ 35.98 =
3. 3,192 ÷ 54.08 =
4. 6,832.9 ÷ 465 =
5. 34,621.34 ÷ 3,465.6 =

ENERGY, POWER, AND WORK

Many significant events in history have involved, or resulted from, advances in the way *work* is done. Physical work includes the transfer of *energy*, or the application of force, and the movement or change that results from this application. Energy is the capacity to perform work.

An engine converts the energy of combustion, or burning, into mechanical force and motion.

Think of work as moving a chair from one side of a room to the other. You must expend energy to lift the chair, carry it a

certain distance, and place it back on the floor. As discussed in Unit 9, your body generates energy by burning calories.

However, no matter how much energy you have, your body is limited in the way it can apply that energy. You might be able to carry 20 light chairs across the room without becoming tired. However, you may not be able to lift a really heavy chair. The ability to apply energy is called *power.*

Humans are limited in their capacity to perform physical work; they have limited power. To overcome this limitation, people have invented *machines* to do work. A machine is a device designed to transmit or modify the application of power, force, or motion. Machines can be designed to perform work very rapidly. Large machines can lift or move very heavy objects, far beyond the capacity of humans. Machines can be operated manually, mechanically, electrically, or electronically. This unit deals with some common applications of energy and power to perform work.

MEASURING MECHANICAL ENERGY

Energy involved in motion is called *mechanical energy.* If you roll a bowling ball down an alley, the moving ball contains energy. That energy will have an effect when the ball strikes one or more bowling pins. The amount of energy depends upon the weight of the ball and the speed at which it travels. A 16-pound ball rolled at the same speed as a 10-pound ball will exert more force against the pins.

As a measure of mechanical energy, work usually is measured in *foot-pounds.* One foot-pound is the work required to lift one pound a distance of one foot. The equation for measuring work is expressed this way:

$$\text{Work} = \text{Weight (pounds)} \times \text{Distance (feet)}$$

Example: How much work is accomplished when a 150-pound person climbs a 12-foot ladder?

Solution:

$$\text{Work (W)} = 150 \text{ pounds} \times 12 \text{ feet} = 1{,}800 \text{ foot-pounds}$$

EXERCISE 10-1

For each question, determine the amount of mechanical energy, or work, that is accomplished. Round calculations to two decimal places.

1. A 50-pound box is lifted 100 feet.
2. A 25-pound bale of hay is lifted 2.5 yards.
3. A 125-pound person climbs a 75-foot telephone pole.
4. A 250-pound machine is lifted 4 feet.
5. A 2-ton truck is lifted 5 feet.
6. A 145-pound man carries a 35-pound box up a 25-foot-high staircase.
7. A 95-pound person carries a 15-pound dog up a 15-foot-high staircase.
8. A 468.25-pound machine is lifted 20 yards.

MEASURING POWER

When work is measured according to the rate at which it is performed, the resulting value is called power. Mechanical power is measured in terms of *horsepower.* One horsepower is the work required to lift 550 pounds a distance of one foot in one second.

The definition of horsepower is credited to James Watt, the inventor of the steam engine. Watt determined, through experimentation, the capability of a healthy work horse. He calculated this capacity as equal to lifting 33,000 pounds 1 foot in one minute. This is converted into seconds by dividing by 60:

$$33{,}000 \div 60 = 550$$

Think about an engine rated at 3 horsepower. Theoretically, a 3-horsepower engine could lift 1,650 pounds a distance of 1 foot in one second. Or, it could lift 550 pounds 3 feet in one second.

Power of man, beast, or machine is needed to unload cargo from ships. The steam engine made it possible to perform this job in less time and with less human effort.

Or, it could lift 550 pounds 1 foot in one-third of a second. This concept is expressed by the following equations:

$$\text{Horsepower (hp)} = \frac{\text{Weight (lbs.)} \times \text{Distance (ft.)}}{\text{Time (sec.)} \times 550}$$

$$\text{Hp} = \frac{550 \times 3}{1 \times 550} = \frac{1{,}650}{550} = 3$$

$$\text{Hp} = \frac{1{,}650 \times 1}{1 \times 550} = \frac{1{,}650}{550} = 3$$

$$\text{Hp} = \frac{550 \times 1}{\frac{1}{3} \times 550} = \frac{550 \times 3}{550} = 3$$

EXERCISE 10-2

Read each question carefully before calculating the answer. Round calculations to two decimal places.

1. How long would it take a 5-horsepower engine to lift 550 pounds 1 foot?

2. How long would it take a 20-horsepower engine to lift 11,000 pounds 1 foot?

3. How high will a $\frac{1}{2}$-horsepower engine lift 550 pounds in one second?

4. If a $\frac{1}{2}$-horsepower engine can lift a motorcycle 1 foot in one second, how heavy is the motorcycle?

5. How long would it take a $\frac{1}{4}$-horsepower engine to lift a 550-pound machine 1 foot?

6. Raymond is using a 1-horsepower engine to lift a 50-pound sandbag 1 foot. How long will it take?

7. How long would it take a 25-horsepower engine to lift 6,875 pounds 1 foot?

8. Mona used a 1-horsepower engine to lift a machine 1 foot in $\frac{1}{4}$ of a second. What was the weight of the machine?

Measuring Horsepower Requirements

If you know the amount of work to be done in any situation, you can calculate the power requirement of an *engine* or a *motor*. The terms engine and motor frequently are used with the same meaning. However, an engine generally changes the energy of *combustion*, or burning, into mechanical force and motion. A motor usually is a rotating device that changes electrical energy into mechanical energy.

For example, say that you want to buy a gasoline-engined forklift to be used in a lumber yard. You may have two absolute requirements: maximum weight and height. The third requirement—time—may be flexible.

Suppose that the heaviest load to be handled by the forklift weighs 3,000 pounds. Also, suppose that the forklift may have to raise this load to a storage shelf 12 feet above the ground. To complete the equation, you need determine only how rapidly you want the forklift to operate.

Based on warehouse operation, you may want the forklift to be able to perform its maximum lift in approximately 10 seconds. You now have all the elements needed to complete the equation and calculate horsepower requirements:

$$Hp = \frac{3{,}000 \times 12}{10 \times 550} = \frac{36{,}000}{5{,}500} = \frac{360}{55} = 6.55$$

EXERCISE 10-3

Read each question carefully. From the information given, determine the horsepower requirement. Round calculations to two decimal places.

1. Ignacio needs an engine that can lift 50-pound bales of hay 25 feet in 15 seconds.

2. Julie needs to lift 250 pounds 38 feet in 24 seconds.

3. Ron needs a crane that can lift a 500-pound rock from an 8-foot hole. He wants to perform this task in 38 seconds.

4. Omni Construction needs a forklift that can lift up to 800 pounds of material 25 feet. Management wants to lift the materials in 15 seconds.

5. Ed is building an elevator that can lift 4,000 pounds 100 feet in 32 seconds.

6. Jane is building a service elevator to carry groceries from her garage upstairs to the kitchen. It must lift 150 pounds 15 feet in 10 seconds.

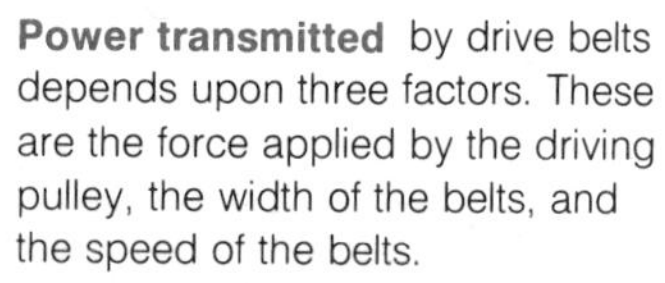

Power transmitted by drive belts depends upon three factors. These are the force applied by the driving pulley, the width of the belts, and the speed of the belts.
PHOTO BY CLIFF CREAGER

Transmitting Power: Driving Belts

Power produced by an engine or motor frequently is transmitted away from the source to perform work. A common method of transmitting power is with a belt, which then drives another machine or mechanical device. The amount of power transmitted by a belt depends upon the force of the pull and the width and speed of the belt. The following equation expresses the horsepower transmitted by a belt:

$$Hp = \frac{F \times W \times S}{550}$$

In the above equation:

F = pull in pounds per inch of belt width
W = width of belt
S = speed of belt in feet per second.

Example: Suppose a belt 10 inches wide has a pull of 75 pounds per inch of width. If the belt travels 2 feet per second, how much horsepower is transmitted?

Solution:

$$Hp = \frac{75 \times 10 \times 2}{550} = 2.73$$

EXERCISE 10-4

Read each question carefully before calculating the answer. Round calculations to two decimal places.

1. How many horsepower are transmitted by a 5-inch-wide belt traveling at 13 feet per second with a pull of 50 pounds per inch?

2. How many horsepower are transmitted by a 3-inch-wide belt traveling at 25 feet per second with 120 pounds of pull per inch?

3. Julian has built a go-cart that is powered by a lawn mower engine. The engine drives a 5-inch-wide belt at 14 feet per second and creates a pull of 25 pounds per inch. How much horsepower is transmitted to the wheels?

4. Isako has a lawn mower that uses a 2-inch-wide belt to drive the cutting blade. The belt moves 25 feet per second with a pull of 15 pounds per inch. How much horsepower is transmitted to the cutting blade?

5. The forklift at City Builders uses a 12-inch belt to raise the forks. The belt turns at 0.5 feet per second and has a pull of 150 pounds per inch. How much horsepower is transmitted to the lifting forks?

Belts and Pulleys

Typically, a belt is driven by a device called a *pulley.* A pulley is a wheel with a *grooved* rim. A groove is a depression or channel. A belt fits into the groove of a pulley and is held in place by the raised ridges of the pulley's rim.

The belt, in turn, wraps around one or more driven pulleys. For example, an automotive engine may drive one belt that turns pulleys to operate an alternator and an air conditioning compressor.

The speed of a belt can be calculated by determining the *circumference* of the pulley and the pulley's speed. The circumference is the distance around the outer edge of a round object,

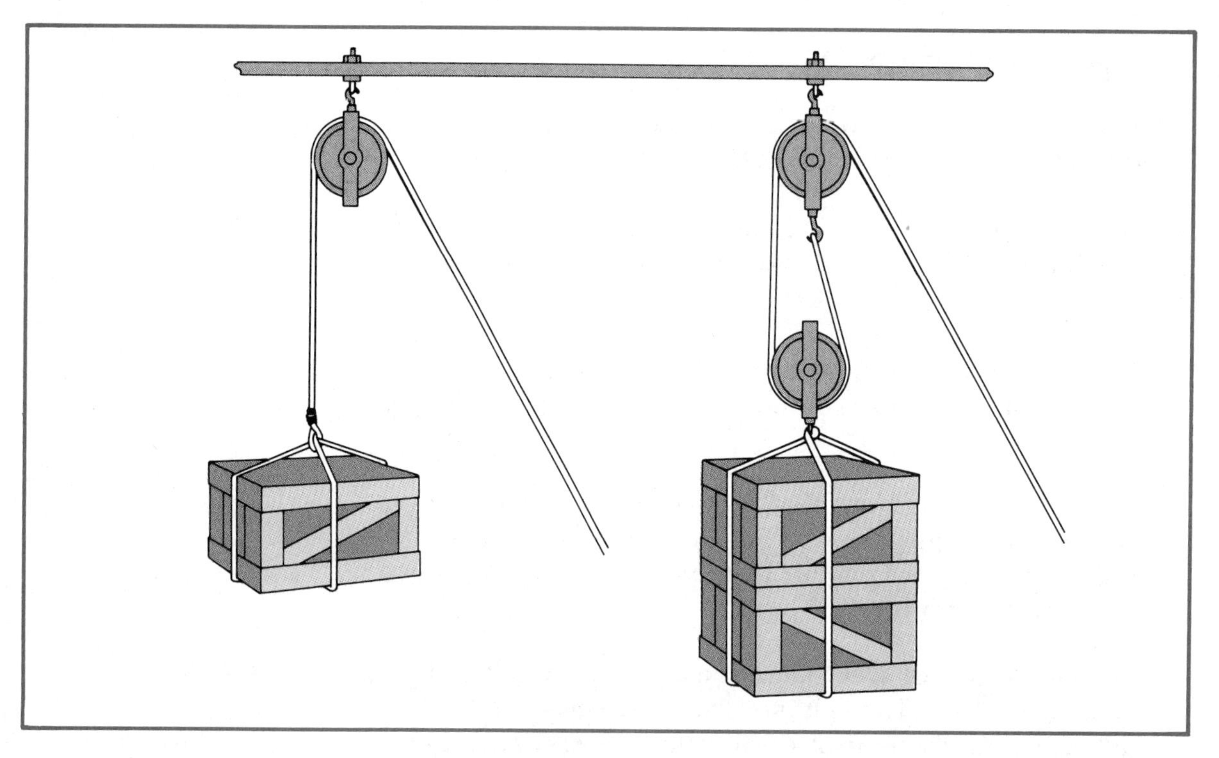

Belt-and-pulley assemblies are one common method of transferring power.

or circle. The important circumference on the pulley is the distance around the surface inside the groove on which the belt rests. The pulley's speed is expressed as *revolutions per minute (rpm)*. For each complete revolution of the pulley, the belt travels a distance equal to the pulley's circumference.

Note: When a horsepower computation is based on distance traveled per minute, the denominator of the equation becomes 33,000. Earlier equations used 550 to express the work performed per second. Multiply by 60 to determine the constant for work per minute: 33,000.

Example: Suppose a belt 5 inches wide runs over a pulley 2 feet in circumference. The pulley is turning at a rate of 150 rpm, and the pull of the belt is 30 pounds per inch. Determine the horsepower transmitted.

Solution:

$$\text{Hp} = \frac{30 \times 5 \times (150 \times 2)}{33{,}000} = \frac{45{,}000}{33{,}000} = 1.36$$

EXERCISE 10-5

Read each question carefully. Then, determine the horsepower transmitted by the belt. Round calculations to two decimal places.

1. The pulley that drives the air conditioning in Terry's car is 1.5 feet in circumference and turns at 1,000 rpm. The pull is 22 pounds per inch on a belt that is $\frac{1}{2}$ inch wide.

2. The circumference of the pulley that drives the cutting blade on a lawnmower is 0.75 feet. The belt is 2 inches wide and has a pull of 45 pounds per inch. The pulley turns at 225 rpm.

3. The belt that drives a 100-yard assembly line is 12 inches wide and has a pull of 110 pounds per inch. The pulley that drives the belt rotates at 100 rpm and has a circumference of 2.5 feet.

4. The belt that drives the power steering on an automobile is 0.76 inches wide and has a pull of 25 pounds per inch. The pulley that drives the belt is 1.25 feet in circumference and rotates at 1,100 rpm.

5. A cement mixer is turned by a 2-inch belt that has a pull of 50 pounds per inch. The pulley that drives the mixer is 6.6 feet in circumference and rotates at 25 rpm.

Transmitting Power: Gears

Belts provide an efficient means of transmitting power for many jobs. However, a belt may be subject to stretching, wearing, and slippage. For many mechanical jobs, a more efficient method of transmitting power is through *gears.* A gear, basically, is a wheel with teeth around its circumference. The teeth of a driving (powered) gear *mesh,* or fit together, with the teeth of a driven gear. When two or more gears are connected, power is said to be transmitted through a *gear train.*

Think of a gear train as two pulleys with teeth, moved together to eliminate the need for a belt.

The general formula for two meshed gears is: Multiply the number of teeth on one gear by its speed. This is equal to the number of teeth on the second gear multiplied by its speed. The formula can be expressed in the following equation:

$$T \times S = t \times s$$

Example: A 30-tooth gear turning at 200 rpm drives a larger gear with 40 teeth. What is the speed of the larger gear?

Solution:

$$40 \times s = 30 \times 200$$
$$40s = 6{,}000$$
$$s = \frac{6{,}000}{40}$$
$$s = 150$$

EXERCISE 10-6

Read each question carefully before calculating the answer. Round calculations to two decimal places.

1. A 25-tooth gear rotating at 150 rpm is driving a 50-tooth gear. What is the speed of the 50-tooth gear?

2. A 75-tooth gear is spinning at 75 rpm. It is driving a 150-tooth gear. What is the speed of the larger gear?

3. A 125-tooth gear is driven by a 15-tooth gear that rotates at 225 rpm. Determine the speed of the 125-tooth gear.

4. A gear rotating at 25 rpm has 45 teeth and drives a 20-tooth gear. At what speed is the 20-tooth gear rotating?

5. A 35-tooth gear spinning at 225 rpm drives a 45-tooth gear. What is the speed of the larger gear?

6. A 10-tooth gear drives a 25-tooth gear at 100 rpm. What is the speed of the smaller gear?

Driving Gear Requirements

Using the same equation, you can determine the number of teeth in a driving or driven gear.

Example: A driven gear with 24 teeth rotates at 400 rpm. The driving gear with which it meshes must rotate at 300 rpm. How many teeth must the driving gear have?

Solution:

$$t \times 300 = 24 \times 400$$

$$300t = 9{,}600$$

$$t = \frac{9{,}600}{300}$$

$$t = 32$$

Gears are a highly efficient mechanical method for the transmission of power.

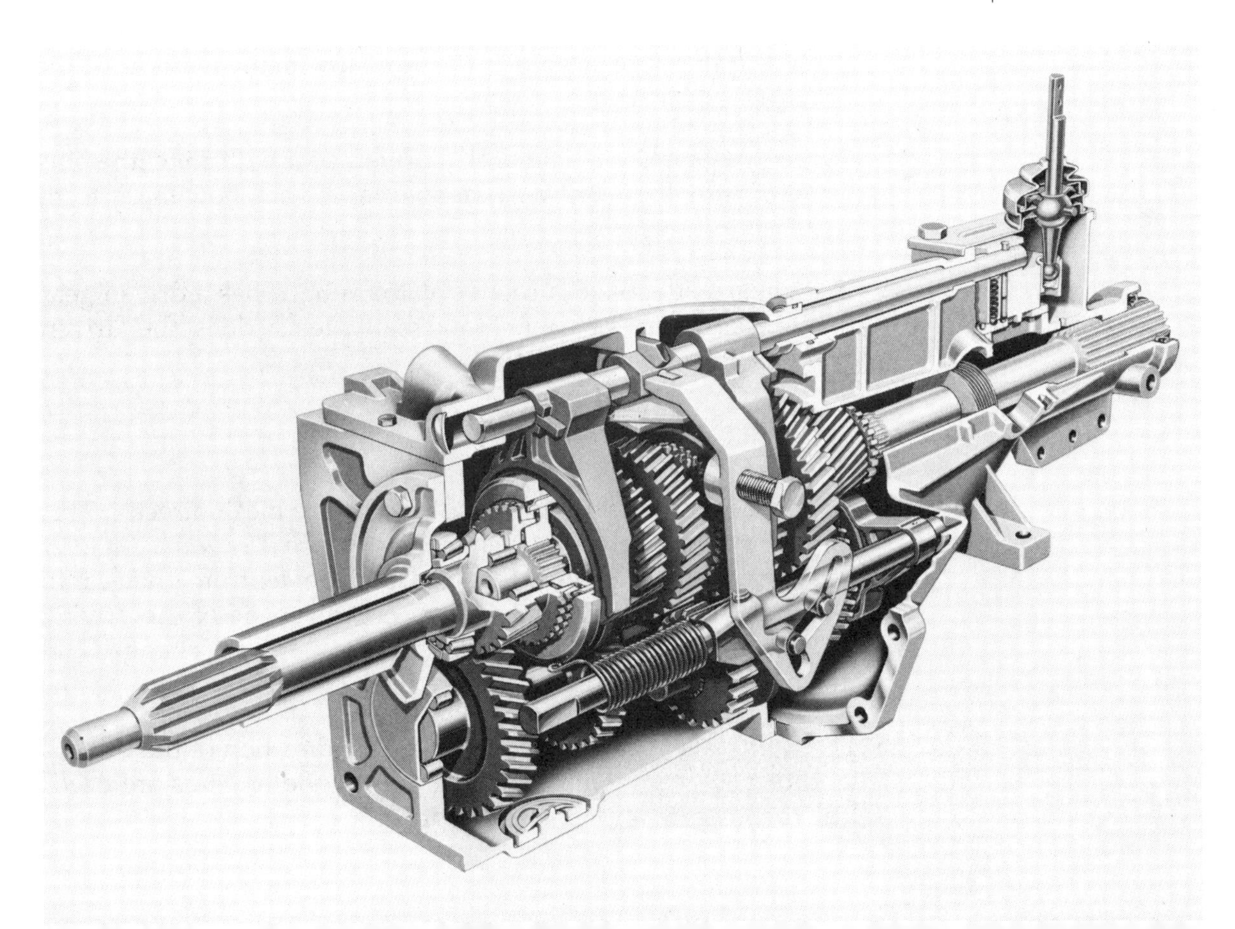

EXERCISE 10-7

Read each question carefully before calculating the answer.

1. A 45-tooth gear rotating at 50 rpm must drive another gear at 150 rpm. How many teeth should the driven gear have?

2. A 50-tooth gear is spinning at 25 rpm. It is driving another gear at 50 rpm. How many teeth does the driven gear have?

3. A gear that is driven by a 45-tooth gear rotating at 125 rpm is spinning at 75 rpm. How many teeth does it have?

4. A gear rotating at 25 rpm is driving a 20-tooth gear at 85 rpm. How many teeth does the driving gear have?

5. A 35-tooth gear spinning at 220 rpm is being driven by a gear rotating at 100 rpm. How many teeth does the driving gear have?

6. A driving gear with 10 teeth rotates at 100 rpm. It meshes with a gear that must rotate at 25 rpm. How many teeth must the driven gear have?

7. To power a hand-held tool, a driven gear with 55 teeth spins at 65 rpm. If the driving gear spins at 25 rpm, how many teeth does it have?

8. A gear with 20 teeth rotates at 150 rpm. It has a driven gear that spins at 50 rpm. How many teeth does the driven gear have?

Idler Gears

In some cases, gears may be too far apart to mesh directly. This problem can be solved by using one or more *idler gears.* An idler gear is a free-turning gear that transmits power from one gear to another. An idler gear does not affect the speed of the driving or driven gears. The formula $T \times S = t \times s$ is not changed, no matter how many idler gears are used.

If one idler gear is used, the turning direction of the driving and driven gears will be the same. If two idler gears are used, the turning directions will be opposite.

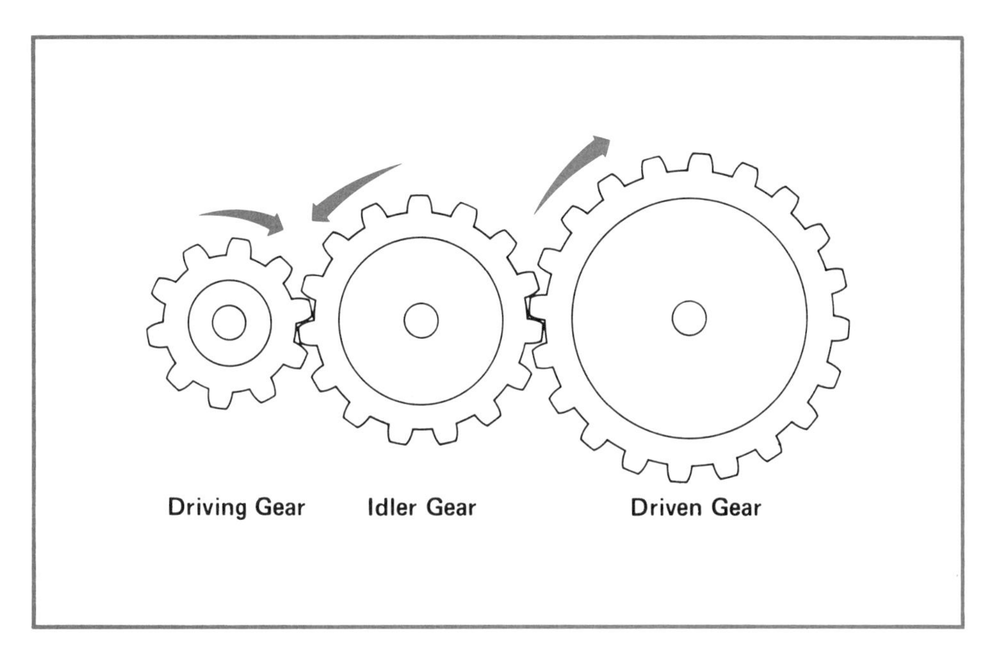

Figure 10-1. One idler gear causes a driven gear to rotate in the same direction as the driving gear is rotating.

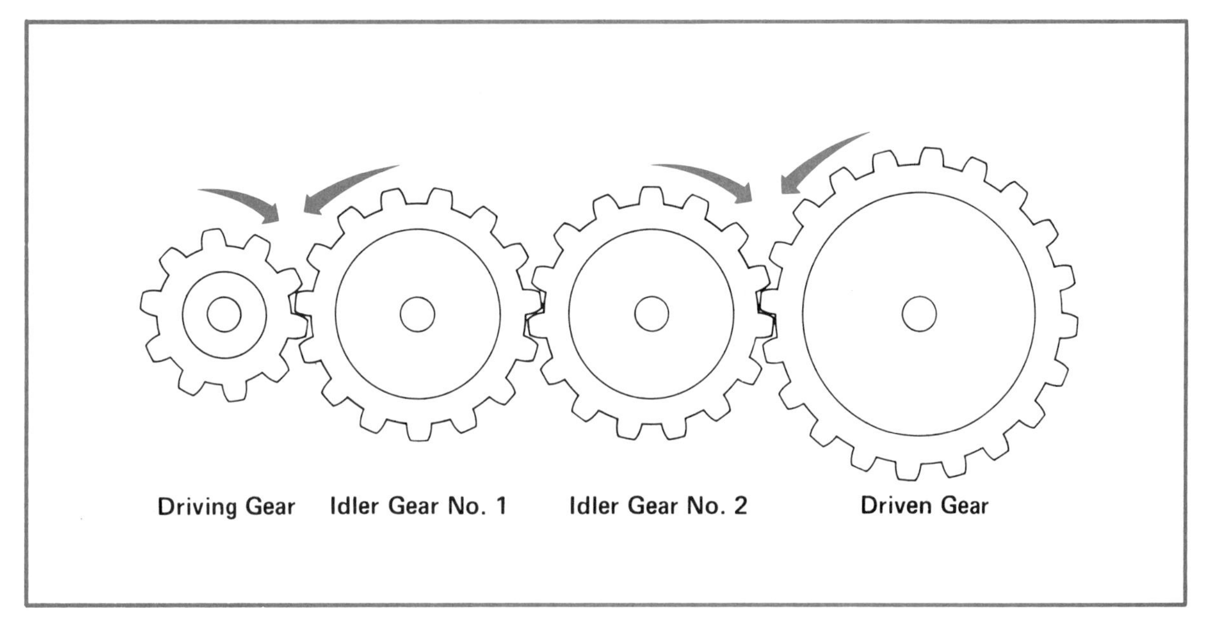

Figure 10-2. Two idler gears cause a driven gear to rotate in the opposite direction from that of the driving gear.

Thus, the number and sizes of idler gears are determined by the distance power is transmitted, and the desired direction of rotation. When driving and driven gears are to turn in the same direction, an odd number of idler gears is used. If driving and driven gears are to turn in opposite directions, an even number of idler gears is used. (See Figures 10-1 and 10-2.)

EXERCISE 10-8

Read each question carefully before calculating the answer. Round calculations to two decimal places.

1. Three gears are being used to power a machine. The driving gear has 25 teeth. The idler gear has 36 teeth. The driven gear has 15 teeth. If the driving gear spins at 150 rpm, how fast is the driven gear turning?

2. A 50-tooth driving gear that is rotating at 75 rpm meshes with a 75-tooth idler gear that meshes with a 10-tooth driven gear. How fast is the driven gear rotating?

3. A 20-tooth gear that rotates at 125 rpm drives one idler gear. The idler turns the driven gear at 50 rpm. Does the driven gear turn in the same or opposite direction as the driving gear? How many teeth are on the driven gear?

4. A driven gear with 80 teeth rotates at 200 rpm. There are two idler gears between the driving and driven gear. If the driving gear has 100 teeth, at what speed does it rotate? Does the driven gear turn in the same direction as the driving gear?

5. A 60-tooth driving gear rotates at 120 rpm and drives a 30-tooth gear. How fast is the driven gear turning? Three idler gears are placed between the driving gear and the driven gear. Will the driving and driven gears turn in the same direction? Will the addition of three idler gears change the speed of the driven gear?

MECHANICAL ADVANTAGE: MULTIPLYING FORCE

At the beginning of this unit, work is described as being the use, or application, of energy and its result. The concept of work can be expressed in an equation:

$$\text{Work} = \text{Force} \times \text{Distance}$$

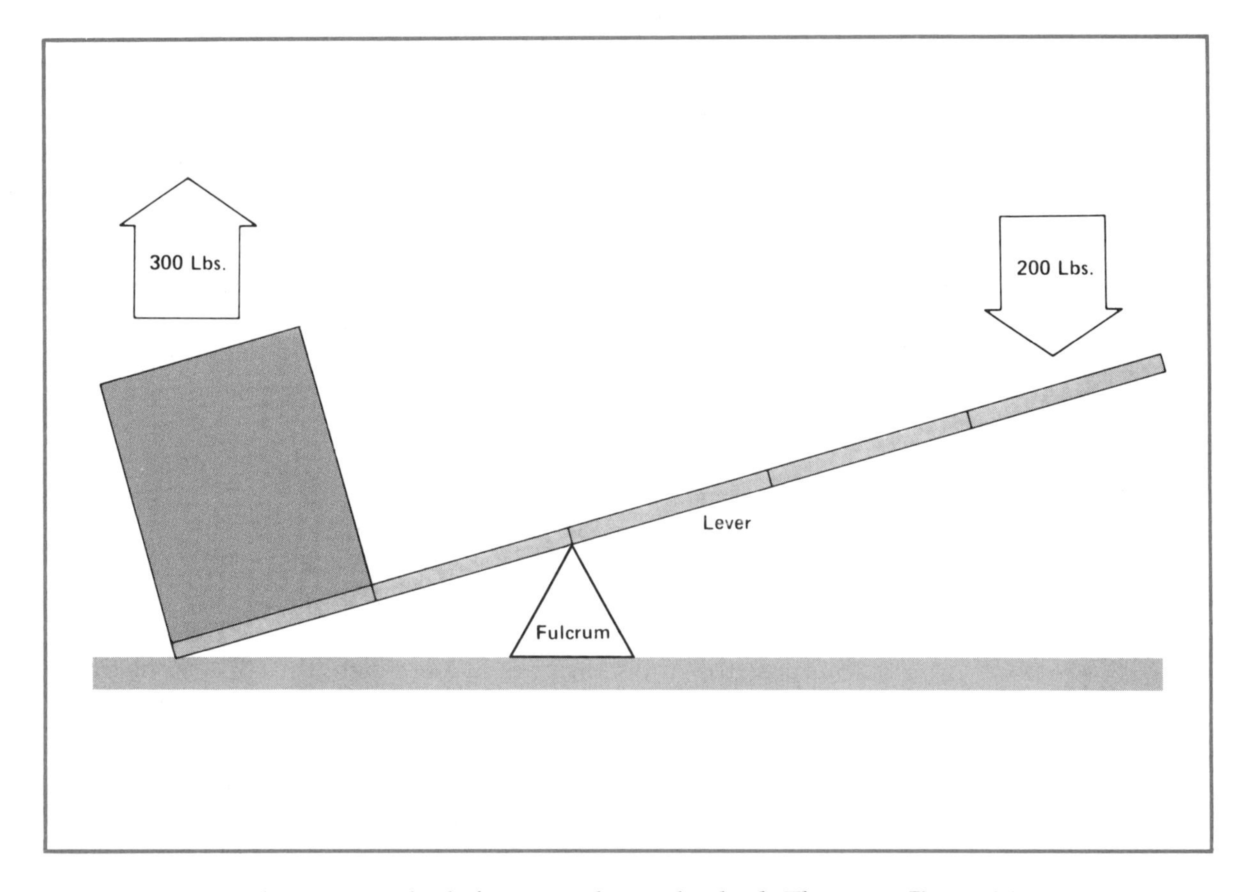

Figure 10-3. The lever and fulcrum increase the amount of work that can be performed through the application of force.

There are several ways in which force can be multiplied. Three common methods are:

- Levers
- Gears
- Hydraulics.

Levers

Did you ever play on a see-saw when you were younger? Suppose two children sit at opposite ends of a see-saw. One child weighs 40 pounds, and the other weighs 50 pounds. The end of the see-saw holding the heavier child will descend to the ground. The lighter end will rise.

This occurs because there is greater force, or weight, being exerted on one end of the see-saw. The see-saw actually is a *lever,* balanced on a *fulcrum.* A lever is a rigid (stiff) object, usually long and narrow, that *pivots* at some point along its length. Pivot means to turn, or rotate. A fulcrum is the point (or object) at which the lever pivots. Force applied to one end of the lever is transferred to the other end. The lever, in this case the see-saw, will tilt (pivot) on its fulcrum, in the direction of greater force.

However, *leverage* depends upon distance as well as force. Leverage is the mechanical advantage gained by the multiplication of force. If the see-saw were balanced, or positioned horizontally, the mathematical principle would be this: The product of the force and the distance from the fulcrum are the same on both sides of the fulcrum. Thus, if the heavier child moves closer to the fulcrum, the see-saw will be balanced. Also, balance will occur if the lighter child moves farther from the fulcrum. Or, each one could move a shorter distance to achieve the same effect. The mathematical equation is written this way:

$$\text{Force} \times \text{Distance} = \text{force} \times \text{distance}$$

Suppose that the see-saw is 12 feet in length. This means that it extends out from the fulcrum 6 feet on either side. The 40-pound child is sitting 5 feet from the fulcrum. Where does the 50-pound child have to sit to balance the see-saw? Use the equation, letting D equal the unknown distance:

$$\begin{aligned} 50 \times D &= 40 \times 5 \\ 50D &= 200 \\ D &= \frac{200}{50} \\ D &= 4 \end{aligned}$$

The solution is for the heavier child to sit 4 feet from the fulcrum.

Leverage enables you to lift a heavy object using force that is considerably less than the weight of the object. (See Figure 10-3.) For example, suppose you wanted to lift an object weighing 300 pounds, but you could exert only 100 pounds of force. If you had an 8-foot bar and a fulcrum, you could lift the 300-pound object. Simply position the bar, or lever, on the fulcrum so that the object is 2 feet from the fulcrum. Apply your force 6 feet from the fulcrum, and the object will rise. The principle is the same: The point of application of force is three times as far from the fulcrum as the object to be lifted. Use the same equation:

$$\begin{aligned} 300 \times 2 &= 100 \times 6 \\ 600 &= 600 \end{aligned}$$

EXERCISE 10-9

Read each question carefully before calculating the answer. Round calculations to two decimal places.

1. Allen and Alana are playing on a see-saw. Allen weighs 100 pounds and Alana weighs 75 pounds. If Alana is sitting 5 feet from the fulcrum, how far from the fulcrum should Allen sit to balance the see-saw?

2. Travis is attempting to move a 90-pound rock with an 8-foot lever. He has positioned the fulcrum 2 feet from the rock. How much force will he need to apply to move the rock?

3. Cathy wants to lift a 200-pound object. She has positioned a lever with the fulcrum 3 feet from the object. The lever is 15 feet long. How much force will she have to apply to lift the object?

4. Steve is using a lever that is 12 feet long. The fulcrum is 3 feet from one end. If 50 pounds of force are applied to the end that is 9 feet from the fulcrum, how much force will be delivered at the other end?

5. An adult and a child are playing on a see-saw. When the fulcrum is 8 feet from the child, the adult and the child balance. The child weighs 40 pounds. How far from the fulcrum must the 160-pound adult be sitting?

6. Emi has built a 20-foot lever and has positioned the fulcrum 10 feet from one end. If 125 pounds of force are applied to one end, how much force will be delivered at the other?

7. Sarah and Liz are playing on a see-saw. Sarah weighs 115 pounds and Liz weighs 86.25 pounds. If Liz is sitting 8 feet from the fulcrum, how far from the fulcrum must Sarah sit to create an equal balance?

8. Ramon wants to lift a 1,000-pound object with a 12-foot lever. He has placed the fulcrum 2 feet from the object. If he applies 150 pounds of force, how much force will be delivered to the object? What is the least amount of force he must apply to lift the object?

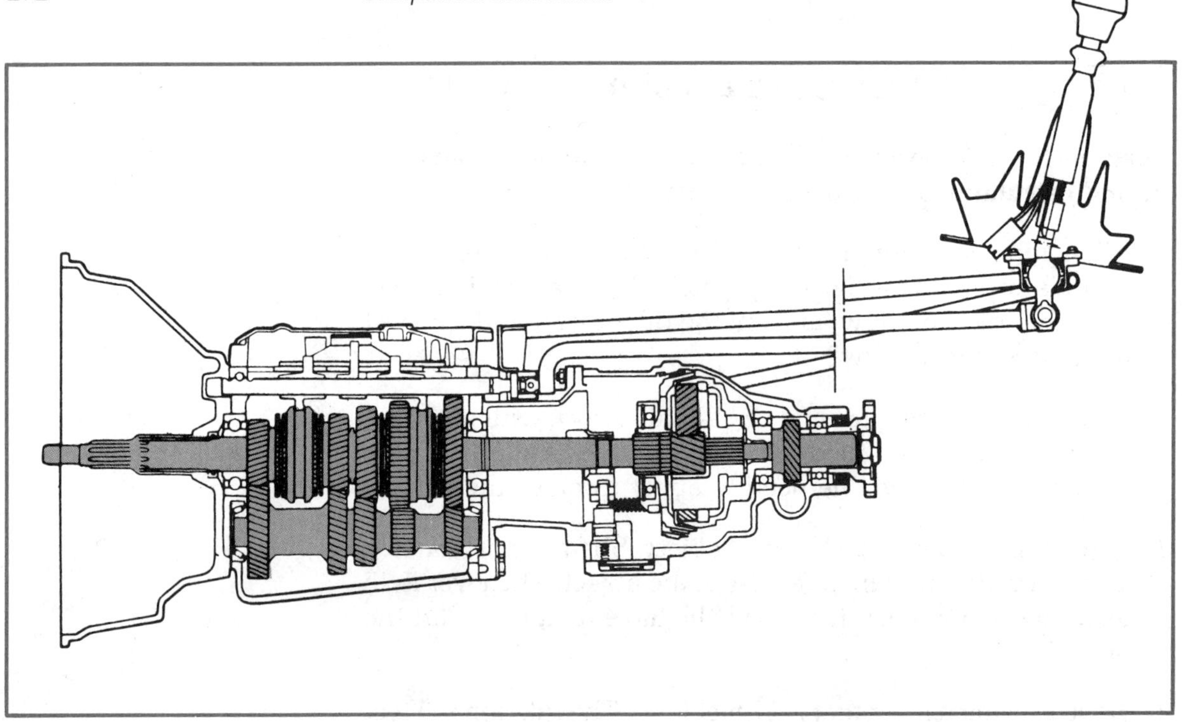

Torque produced by an automotive engine is multiplied by the lower gears of a transmission. Torque multiplication helps to get a vehicle moving from rest.

Gears

Gears of different sizes can be used to increase or reduce a force. Gears are used with turning or twisting forces. These kinds of forces are called *torque.* For example, an automotive engine produces power by turning a shaft, called a driveshaft, in a circular motion. The formula for torque is:

$$\text{Torque} = \text{Force} \times \text{Radius}$$

The *radius* of a circle is the distance from the center of the circle to its circumference. A gear radius is the distance from the center of the driving shaft to the outer edge of the gear. Torque is measured in *pounds-feet (lb.-ft.).* The radius is measured in feet. If a radius is measured in inches, it is necessary to divide the radius by 12 to obtain torque in pounds-feet.

Think about loosening a nut with a wrench. If you apply 10 pounds of force on the end of a wrench 1 foot long, you will apply 10 pounds-feet of torque to the nut. The length of the wrench is the same as the radius of a gear. If the nut is extremely tight, you may not be able to budge it. The next step is to try using a longer wrench. Applying 10 pounds of force at the end of a wrench 2 feet long produces 20 pounds-feet of torque. By doubling the length of the shaft, you have doubled the turning force (torque) applied to the nut.

EXERCISE 10-10

Read each question carefully before calculating the answer. Round calculations to two decimal places. Remember that measurements used to calculate torque must be in pounds and feet.

1. If 40 pounds of force are applied to a 2-foot wrench to loosen a nut, how much torque is delivered?

2. If 55 pounds of force are applied to a 6-inch wrench to loosen a nut, how much torque is delivered?

3. If 25 pounds of force are applied to the end of a wrench and 75 pounds-feet of force results, how long is the wrench?

4. Amelia delivered 295 pounds-feet of torque to a nut with a 2.5-foot wrench. How much force did she apply?

5. Yukio delivered 175 pounds-feet of torque with an 18-inch wrench. How much force did he apply?

6. If a 2.5-foot wrench delivers 250 pounds-feet of torque, how much force is applied to the wrench?

Torque Multiplication and Reduction

Gears, which rotate on shafts, produce turning or twisting forces in a similar manner. Switching from a small wrench to a larger wrench increased torque in the preceding example. Similarly, torque is increased, or multiplied, when a smaller gear drives a larger gear, as shown in Figure 10-4.

Think about a gear with a radius of 2 feet driving another gear with a radius of 3 feet. The radius is the distance from the center of the gear to its outside edge, or the teeth. Suppose the teeth of the smaller gear are transferring 20 pounds of force to

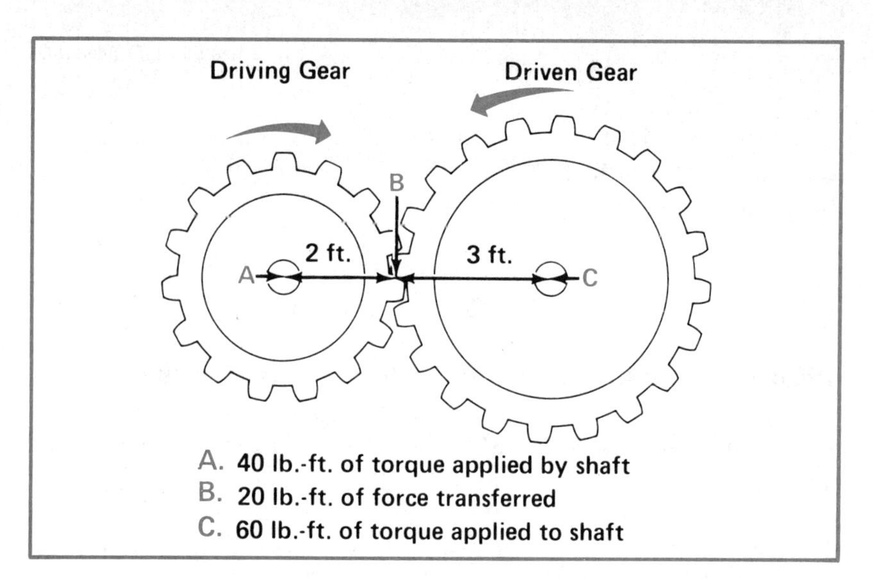

Figure 10-4. Torque multiplication occurs when one gear drives a larger gear. As torque is multiplied, speed of rotation is decreased.

the teeth of the larger gear. To calculate the amount of torque applied to the shaft being turned by the larger gear, use the same formula: Torque = Force × Radius.

20 lb. × 3 ft. = 60 lb.-ft.

How can you calculate how much torque is being applied by the shaft of the driving gear? For this, you just use the same formula: Torque = Force × Radius.

Torque = 20 lb. × 2 ft.
Torque = 40 lb.-ft.

Thus, the 40 pounds-feet of torque applied by the shaft of the driving gear is increased to 60 pounds-feet of torque at the shaft of the driven gear.

In addition, when torque is increased, or multiplied, the speed of revolution is decreased. In the case of the gears above, torque was increased by 50%, from 40 pounds-feet to 60 pounds-feet. However, the speed at which the larger gear revolves will be 50% slower than the revolving speed of the smaller gear.

Gears also can be used to reduce torque and increase speed. This occurs when a larger gear drives a smaller gear.

This is the principle used in automotive transmissions. Lower gears are larger. They multiply the torque of the engine to get the vehicle started. As the vehicle gains speed, less power, or torque, is needed to keep it moving. The driver can change to smaller (higher) gears to increase speed.

EXERCISE 10-11

Read each question carefully before calculating the answer. Round calculations to two decimal places.

1. Gear A is applying 45 pounds of force to Gear B. If the radius of Gear B is 2 feet, how much torque is being applied to its shaft?

2. Gear A has a radius of 1.5 feet. Gear B has a radius of 4 feet. If Gear A applies a 120-pound force to Gear B, how much torque will be applied to the the shaft of Gear B? How much force is being applied by the shaft of Gear A? By what percentage has torque been increased?

3. Torque at the shaft of a driving gear is 80 pounds-feet. Torque at the shaft of the driven gear is 320 pounds-feet. If the radius of the driving gear is $\frac{1}{2}$ foot, what is the radius of the driven gear? Which of the gears is rotating at a faster speed?

4. A driven gear has a radius of 1.25 feet. The driving gear has a radius of 0.25 feet and there are 40.5 pounds-feet of torque being applied by the shaft. What amount of torque is being applied to the shaft of the driven gear?

5. A driven gear has a radius of 3 feet and there are 150 pounds-feet of torque being applied to its shaft. The driving gear has 12.5 pounds-feet of torque at its shaft. What is the radius of the driving gear?

6. A driving gear has a radius of 1.75 feet. The driven gear has a radius of 0.25 feet. Fifty pounds of force are being applied by the driving gear to the driven gear. How much torque is being applied to the shaft of the driven gear?

7. If Gear A is rotating at a faster speed than Gear B, which gear shaft has the higher amount of torque?

8. The shaft of a driving gear is applying 225 pounds-feet of torque to the gear. The driven gear is transmitting 75 pounds-feet of torque to its shaft. If the radius of the driven gear is 1.25 feet, what is the radius of the driving gear?

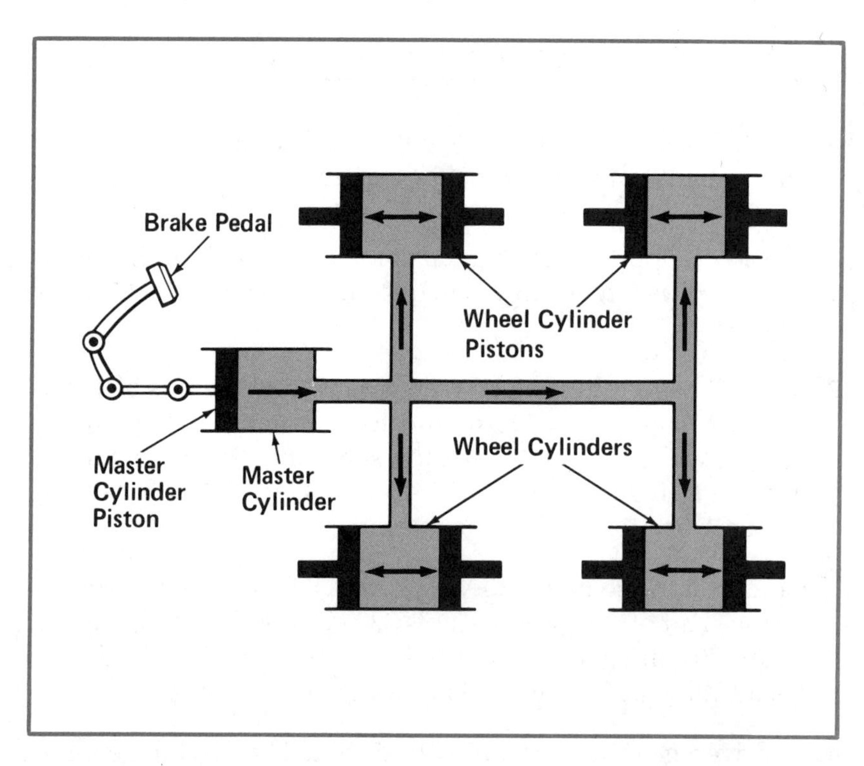

Figure 10-5. Pressure applied at one end of a hydraulic system to operate automotive brakes is transferred instantly to the other end.

Hydraulics

Another effective method of multiplying force is through *hydraulics.* Hydraulics is a science that deals with the effects of fluids, or liquids, in motion. Hydraulic devices are operated by pressure transmitted when a fluid is forced through a relatively small *orifice.* An orifice is a hole or entrance to a passageway, such as a hose or a tube.

A simple hydraulic system, such as those used to operate automotive brakes, is shown in Figure 10-5. When the brake pedal is depressed, a *piston* in the master cylinder, or reservoir, pushes against fluid in that cylinder. A piston is a *cylinder* that slides back and forth inside a cylinder. A cylinder, as shown in Figure 10-6, has the form of a geometric figure created by the circumference of a circle as it moves along a straight line. The ends of a cylinder are equal parallel circles.

As the piston moves, it exerts pressure on the fluid. Part of the fluid is forced through tubes and hoses to the wheel cylinders. As fluid enters the wheel cylinders, the pistons in those cylinders are forced to move. The movement of the wheel cylinder pistons applies the brakes. (See Figure 10-7.)

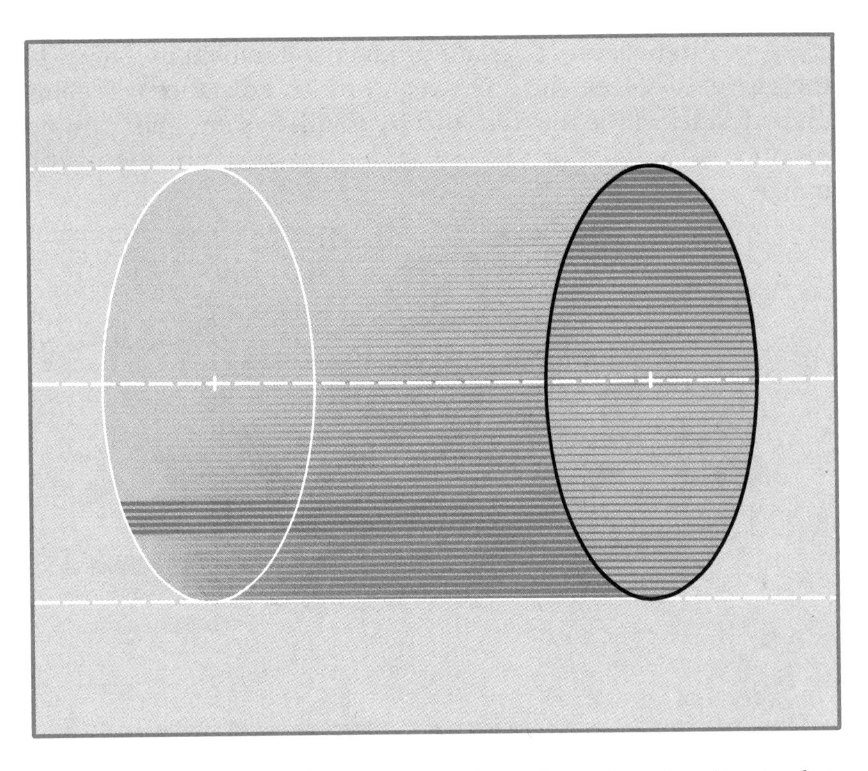

Figure 10-6. A cylinder is a shape formed by a circle moving along a straight line. The circles at the ends of the cylinder are equal in circumference.

One of the most important principles of hydraulics is that fluids cannot be compressed. As soon as pressure is applied at one end of a hydraulic system, that same pressure acts at the other end. If the cylinders and pistons are equal in size, no mechanical advantage is achieved.

However, if the driven piston is larger than the driving piston, a mechanical advantage occurs when pressure is applied. If the area of the larger piston is twice the area of the smaller piston, the advantage is 2:1.

Once again, distance is involved. The larger piston will move only half as far as the smaller piston. However, pressure is measured in *pounds per square inch (psi).* Since the pressure is constant, twice as much pressure is exerted against the larger piston. Think of it this way: The pounds of force being applied to each square inch are the same. However, there are twice as many square inches on the face of the larger piston.

For example, suppose a force of 100 pounds is placed against a driving piston. If the driven piston is eight times as large, the force against the larger piston will be 800 pounds.

As with the lever, a certain product will remain the same. In the case of the lever, the product of force and distance was equal on both sides of the fulcrum. In a hydraulic system, the constant product is the area of the piston and the distance the piston moves.

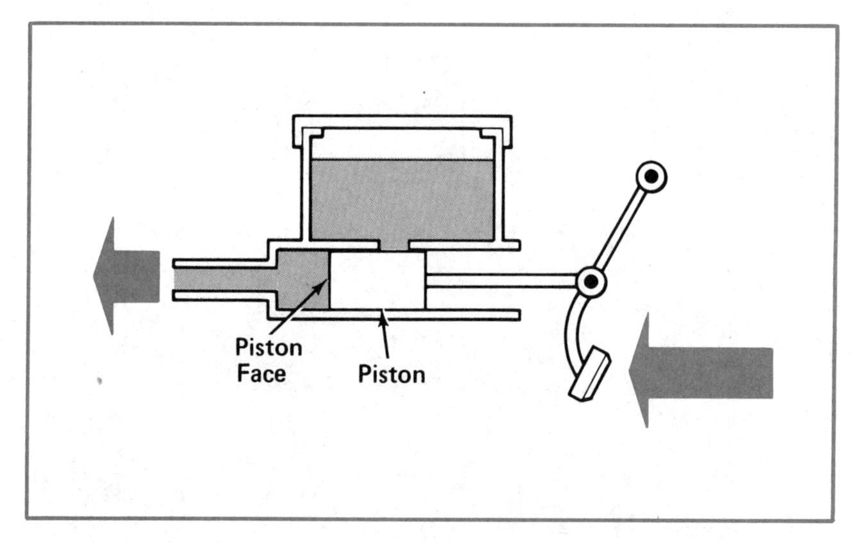

Figure 10-7. The face of a driving piston exerts pressure against a fluid. The face of a driven piston receives the pressure of the fluid.

Say that piston A has an area of 20 square inches and piston B has an area of 80 square inches. This gives a mechanical advantage of 4:1. If piston A is pushed 8 inches with a force of 100 pounds, how far would piston B move? Let D equal the distance piston B moves, and the equation would be:

$$80 \times D = 20 \times 8$$
$$80D = 160$$
$$D = \frac{160}{80}$$
$$D = 2$$

Thus, piston B would move 2 inches, one-fourth the movement of piston A.

This is the principle used in hydraulic equipment that can lift and support heavy weights. An automobile hoist is a common example. Another familiar example is the raising of the box of a dump truck.

Electrical energy and compressed air are common sources of power for moving hydraulic pistons. Your foot is another, when you push on the brake pedal of an automobile.

EXERCISE 10-12

Read each question carefully before calculating the answer. Round calculations to two decimal places.

1. The driven piston in a hydraulic system has an area of 120 square inches. When the driving piston moves 4 inches, the driven piston moves 1 inch. What is the surface area of the driving piston?

2. The driving piston has an area of 15 square inches. When the driving piston moves 4 inches, the driven piston moves 10 inches. What is the area of the driven piston?

3. When a driving piston moves 2.5 inches, the driven piston moves 6.25 inches. If the driving piston is 7.5 square inches, what is the area of the driven piston?

4. If a driven piston moves 4 inches when the driving piston moves 6 inches, by what percentage does the hydraulic system increase force?

5. The mechanical advantage of a hydraulic system is 3:1. That is, when the driving piston moves 3 inches, the driven piston moves 1 inch. If the area of the driving piston is 15 square inches, what is the area of the driven piston?

6. A driven piston moves 4 inches when the driving piston moves 14 inches. If the area of the driven piston is 6 square inches, what is the area of the driving piston?

7. When a driving piston moves 3.5 inches, the driven piston moves 5.25 inches. If the area of the driven piston is 13 square inches, what is the area of the driving piston?

8. A driving piston is 5 square feet and the driven piston is 2 square feet. If the driving piston in this system moves 5 inches, how far will the driven piston move?

SUMMING UP

- Work is the transfer of energy and the movement or change that results. Energy is the ability to do work. Power is the ability to apply energy to a job. Work can be measured in foot-pounds. The ability to perform mechanical work is measured in horsepower. One horsepower equals the energy required to lift 550 pounds one foot in one second.
- Horsepower requirements can be graded if the amount of work to be done is known.
- Two common methods of transmitting power are belts and gears.
- Idler gears transmit power between a driving gear and a driven gear without affecting the relationship between the gears. Idler gears can be used to change the direction of rotation of a driven gear.
- The energy applied to a task can be increased through mechanical advantage. Three common methods of increasing mechanical force are levers, gears, and hydraulic systems.
- The principle of leverage is that the product of force multiplied by distance is the same on both sides of the fulcrum, or pivot point.
- Torque is a turning or twisting force. Torque is measured in pounds-feet (lb.-ft.). The formula for determining torque is: Torque = Force × Radius. Different-size gears can be used to increase torque (decrease speed of revolution) or to increase speed (decrease torque).
- The basic principle of hydraulics is that fluids cannot be compressed. Therefore, force exerted on a liquid is transferred throughout the containers holding that liquid. A simple hydraulic system consists of two cylinders connected by a narrow tube or hose, in which a fluid is stored. A piston in the smaller cylinder exerts force against the fluid. As this pressure is transferred to the piston in the larger cylinder, the force is increased. The increase in force is in proportion to the difference in size between the pistons. Pressure is measured in pounds per square inch (psi).

TESTING YOUR WORKING KNOWLEDGE

Read each question carefully before calculating the answer. Round calculations to two decimal places.

1. How much work is accomplished when a 135-pound person carries a 35-pound dog up a 12-foot-high staircase?

2. How long would it take a $\frac{3}{4}$-horsepower engine to lift a 550-pound machine 1 foot?

3. A construction company is building an elevator that can lift 8,000 pounds 50 feet in 48.5 seconds. Determine the horsepower requirements for the elevator.

4. The belt that drives an assembly line is 18 inches wide and has a pull of 50 pounds per inch. The pulley that drives the belt rotates at 100 rpm and has a circumference of 2.5 feet. Determine the horsepower transmitted by the belt.

5. A 35-tooth gear rotating at 75 rpm must drive another gear at 125 rpm. How many teeth should the driven gear have?

6. A 40-tooth gear that rotates at 175 rpm drives an idler gear that, in turn, rotates a driven gear at 50 rpm. Does the driven gear turn in the same or opposite direction as the driving gear? How many teeth are on the driven gear?

7. The fulcrum of a 10-foot lever is positioned 3 feet from one end. If 150 pounds of force are applied to the end that is 7 feet from the fulcrum, how much force will be delivered at the other end?

8. A driven gear of a large clock has a radius of 3.5 feet. There are 10 pounds-feet of torque being applied to its shaft. The driving gear has 2 pounds-feet of torque at its shaft. What is the radius of the driving gear?

TERMS THAT COUNT

work
energy
power
machine
mechanical energy
foot-pound
horsepower
engine
motor
combustion
pulley
groove
circumference
revolutions per minute (rpm)
gear
mesh
gear train
idler gear
lever
fulcrum
pivot
leverage
torque
radius
pounds-feet (lb.-ft.)
hydraulics
orifice
piston
cylinder
pounds per square inch (psi)

PRESENTING INFORMATION

Read each question carefully. Then present the information in the format required. Round calculations to two decimal places.

1. As discussed in this unit, automobile transmissions are composed of gears of different sizes. This allows the driver to control the applications of torque produced by the engine to change speeds. Different sized driven gears are meshed with gears of the input shaft. For example, a driving gear may have a radius of 0.08 feet. First gear has a radius of 0.21 feet. Second gear has a radius of 0.16 feet; third gear, 0.13 feet; and fourth gear, 0.08 feet. On the average, the driving gear will apply 100 pounds of force to the driven gears.

   From these figures, determine the amount of torque applied to the shaft of each driven gear—first gear, second gear, and so on. Then, draw a line graph that shows how the amount of torque applied changes as a driver shifts from first through fourth gear. As you answer this question, recall that, as torque decreases, the speed at which the driven gears rotate increases.

2. Calculate the amount of horsepower necessary to lift each of the following objects 1 foot in 1 second: a 100-pound weight, a 200-pound weight, a 300-pound weight, a 400-pound weight, and a 500-pound weight. Then, draw a bar chart that plots how the horsepower requirements change as the amount of weight increases. Does the amount of horsepower required increase at a constant rate?

SKILL DRILLS

A. Find the sum in each problem.

1. 5,633.1 + 2,234 + 3.463 + 5,324.9 =
2. 23,112 + 153,535 + 64,437 + 523,436 + 2,659 =
3. 29,321 + 8,324.02 + 231,276.9 + 87,443.66 =
4. 4,311,132 + 8,546.655 + 980,755 + 8,837.015 =
5. 6.21 + 76.4309 + 1.4358 + 48.2132 + 32.75784 =

B. Find the difference in each problem.

1. 7,212.1 − 4,213.999 =
2. 15,323 − 87.67 =
3. 762.665 − 324.52 =
4. 3,744.7 − 545.655 =
5. 3,422.32 − 1,487.879 =

C. Find the product in each problem. Round calculations to two decimal places.

1. 12.87 × 34.87 =
2. 632.4 × 87.49 =
3. 231.226 × 8.25 =
4. 91.39 × 395.6 =
5. 3,547 × 2.967 =

D. Find the quotient in each problem. Round calculations to two decimal places.

1. 98.98 ÷ 9.98 =
2. 315 ÷ 53.53 =
3. 2,769 ÷ 56.43 =
4. 7,432.7 ÷ 425 =
5. 23,534.51 ÷ 9,322.5 =

11

Mathematics for Making Things

YOUR LEARNING JOB

When you have completed the exercises and assignments for this unit, you should be able to:

- ☐ Compare the costs of operating a manufacturing business.
- ☐ Use an equation to compute profit.
- ☐ Evaluate the costs of manufacturing a product.
- ☐ Assign costs to categories, such as materials, labor, overhead, and distribution.
- ☐ Present information in chart and graph forms to measure the results of business operations and decisions.

WARM-UP

A. Find the sum in each problem.

1. 3,456 + 5,763 + 8,439 + 4,992 =
2. 11,324 + 56,837 + 43,938 + 87,912 + 65,324 =
3. 34,546 + 456,392 + 34,293 + 3,487,584 + 789 =
4. 2,342,765 + 4,536,845 + 34,328 + 87,938,231 =
5. 32,273,362 + 5,327,473 + 8,214,325 + 345,276,987 =

B. Find the difference in each problem.

1. 386,908 − 289,879 =
2. 12,352,541 − 9,745,653 =
3. 4,835,376 − 3,879,797 =
4. 54,631,213 − 37,312,526 =
5. 243,842,453 − 211,469,879 =

C. Find the product in each problem. Round calculations to two decimal places.

1. 32,142 × 15 =
2. 3.18 × 3.6 =
3. 45.47 × 0.68 =
4. 267 × 4.346 =
5. 53.61 × 36.82 =

D. Find the quotient in each problem. Round calculations to two decimal places.

1. 45,376 ÷ 325 =
2. 3,918 ÷ 2,376 =
3. 45,407 ÷ 768 =
4. 12,326,872 ÷ 4,761 =
5. 3,453,618 ÷ 836,745 =

PRODUCT MATHEMATICS

The business of manufacturing can range from a one-person operation to giant corporations, such as General Motors and Westinghouse. The mathematics involved in operating any manufacturing business is basically similar, no matter what size the business is.

There are some basic costs involved in making products for sale. A simple manufacturing operation, for example, would include the following cost factors:

- Materials
- Labor
- Overhead
- Sales (and advertising, if applicable)
- Distribution.

The costs listed above would be paid from the total revenues, or income, of the business. When all costs of manufacturing have been paid, the amount of income remaining, if any, is profit.

The Business Pie

Think of business costs as pieces of a pie. The whole pie, shown in Figure 11-1, represents the *gross income* of a business. Gross income is the total revenue from all sales of the product or products manufactured by the business. Each slice of the pie represents a percentage of that gross income. All but one slice represent the costs of doing business. The final slice represents profit.

THE PROFIT EQUATION

If you are the owner of a business, your goal is to make the profit slice as large as possible. To do this, you must keep the other slices as small as possible. This idea is expressed as a simple equation:

$$I - C = P$$

The letters in the equation represent income (I), costs (C), and profit (P). The profit of a business is its income minus its costs.

To illustrate this equation, refer to the pie chart in Figure 11-1. *Materials,* also called *raw materials,* are the items from which products are manufactured. Suppose that, in your business, materials represent $\frac{1}{4}$ of the pie, or 25%. Also, suppose that you found a new source for those materials, and this cost could be reduced by $\frac{1}{5}$, or 20%. This would reduce your materials cost by 5% of the pie. Materials then would be 20% of gross income. If all other factors remained the same, your costs of operating the business would be reduced by 5%. As a result, your profit would be increased by 5% of gross income. Thus, if profit had been 5% of gross income, the 5% increase would double the profit to 10%.

The same applies to reducing any of the other slices, or business costs. The amount of profit earned by a business depends largely upon how efficiently the business is operated.

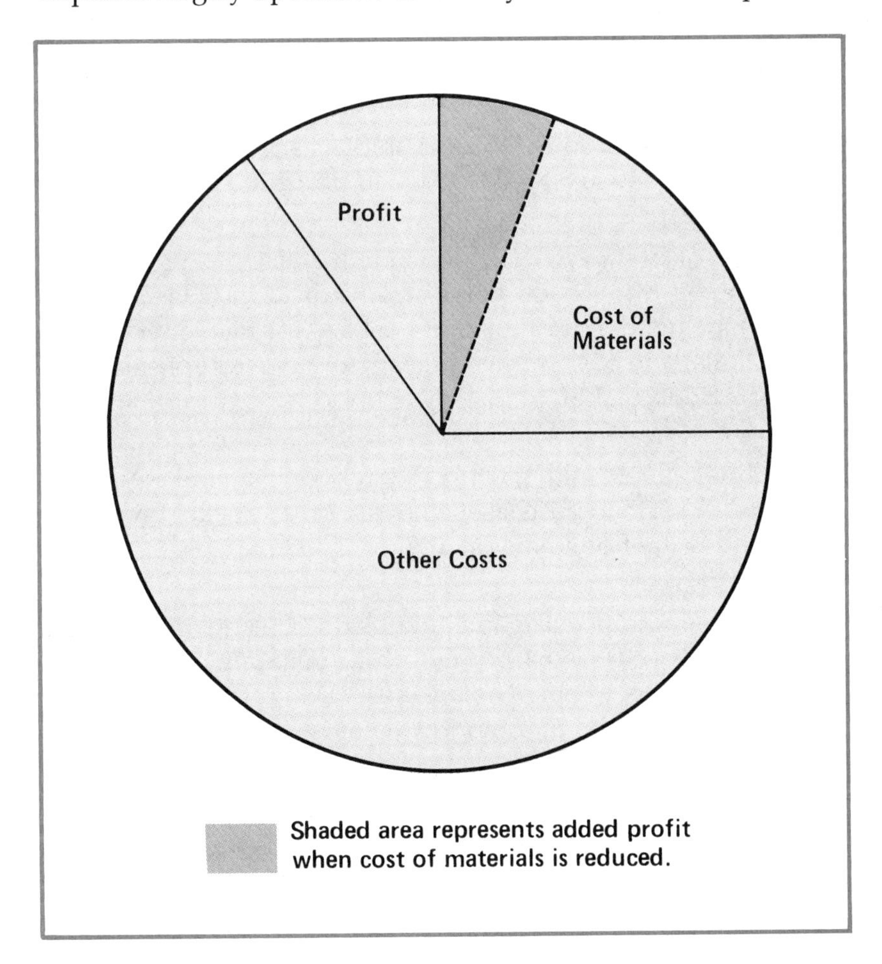

Figure 11-1. The sum of the slices of a business pie is the gross income of the business.

EXERCISE 11-1

Read each question carefully. Then use the profit equation, I – C = P, to calculate the answer. Round answers to the nearest cent.

1. Last year, gross income at Parallel Print Shop was $112,145.97. Costs totaled $73,452.68. How much profit did Parallel Print Shop earn?

2. Gross income at Barry's Bakery was $35,670.87 in February. Costs totaled $31,873.19. How much profit did Barry's Bakery earn for February?

3. The Auto House sells automobiles for $6,780. Automobiles cost $4,812 to purchase, take care of, and sell. How much profit does the Auto House make on 14 automobiles?

4. Gross income at Tomato Computers was $54,718 last month. Costs were $45,326.89. How much profit did Tomato Computers earn?

5. The Repair Shop had a total income of $23,876 last month. Costs included $4,892 for parts, $3,523 for machinery, rent, and tools, and $11,213 for wages. How much profit did The Repair Shop make?

6. The Book Nook sold 1,178 books for $3.75 a book. A single book cost $2.15 to purchase, take care of, and sell. How much profit did the Book Nook earn?

7. The Video Room sells game tokens for 25 cents. It costs the Video Room $1,856.64 per month to maintain video games and to operate, or run, the business. When the Video Room sells 15,472 tokens in a month, what is the profit?

8. Teresa is a driver and earns $5 for each delivery made. It costs her $2.73 to make one delivery. When Teresa makes 389 deliveries, how much profit does she earn?

THE COST EQUATION

Many different costs can be involved in manufacturing a product. However, costs usually can be broken down into a few general areas. Thus, for most businesses, the following equation would summarize costs:

$$C = M + L + O + D$$

Spelled out, the equation reads: Costs (C) equal materials (M) plus labor (L) plus overhead (O) plus distribution, including

Manufacturing companies use mathematical equations in every area of their operations, from production processes to measurements of business performance.
COURTESY OF CINCINNATI MILACRON INC.

sales, (D). All costs involved in offering a product for sale can be assigned to one of these four categories.

Materials. As discussed above, materials are the items, or substances, from which a product is made. In a machine shop, materials might be steel bars. In a fast food outlet, materials might include ground beef, cheese, tomatoes, lettuce, and rolls. In a clothing factory, materials might be cloth and the thread used in sewing the cloth into garments.

EXERCISE 11-2

Read each question carefully before calculating the answer. Round answers to two decimal places.

1. Last year, materials for Bio Laboratories represented 15% of total costs. If total costs were $123,000, how much was spent on materials?

2. Last month, materials for The Mirror Company represented 19% of total costs. The Mirror Company's total costs were $47,845. If, next month, the price of materials goes up 15% and other costs remain the same, how much will the company's total costs be?

3. Last month, materials at Julie's Machine Shop represented 18% of total costs. Total costs were $56,780. This month, Julie found a new source that will cut her material expense in half. How much lower will total costs be?

4. In December, total costs for Harry's Hamburgers were $13,456. Material costs represented 28% of this amount. If Harry can cut his material costs by one-quarter, how much will total costs be reduced?

5. On the average, total monthly costs at Kazumori's Luggage are $45,000 and material costs are $15,000. What percentage of total costs do material costs represent? If Kazumori can reduce the cost of materials 15%, how much will total monthly costs be?

Labor. *Labor* costs include wages and other costs of employing workers to make the product and perform other jobs within the business. Labor costs are discussed later in this unit.

EXERCISE 11-3

Read each question carefully before calculating the answer. Round answers to two decimal places.

1. Labor at Jose's Body Shop represents 45% of total costs. If total costs for last month equaled $23,781, what amount was spent on labor?

2. Last year, Carmela's Remodeling spent $15,700 on labor. Total costs were $25,400. What percent of total costs did labor costs represent?

3. Last year, Yasunari's labor costs represented 65% of total costs. Total costs last year were $65,000. How much was spent on labor? If labor costs are cut 25% this year and other expenses are unchanged, what percent of total costs will labor represent?

4. The Shoe Factory spent $25,670 on labor last month. Labor represents 25% of total costs. How much were total costs?

5. Currently, labor at Huck Finn's Restaurant represents 45% of total costs. If labor costs rise 25%, what percent of total costs will labor represent?

Overhead. *Overhead* is the cost of operating, or running, the *plant* where the business is operated. Plant includes the land, building(s), machinery, other equipment, furnishings, and fixtures used in conducting a business. Rent or *lease* payments are a major overhead factor. A lease is a contract for the use of property; the agreement usually runs for a year or more. Other overhead costs can include light and power, and the purchase, maintenance, and repair of machinery and equipment. A burglar alarm system, for example, is an item of overhead. Some overhead costs, such as leases, involve long-range decisions.

Office equipment is one type of business overhead cost. COURTESY OF VECTOR GRAPHIC

EXERCISE 11-4

Read each question carefully before calculating the answer. Round answers to two decimal places.

1. On the average, overhead for Pete's Printing represents 15% of total costs. Total costs per month average $15,060. On the average, then, how much is spent each month on overhead?

2. Last month, The Donut Stop spent $1,230 on overhead. This figure represented 23% of total costs. What were total costs?

3. On the average, overhead at Tuxedo Junction represents 25% of total costs. Management wants to move to a new location that will attract more business. However, rent on the new location will increase overhead 45%. If Tuxedo Junction moves, what percent of average total costs will overhead represent?

4. Mike Row Circuits spent $13,450 on overhead last month. Overhead represented 13% of total costs. If, next month, overhead increases 20% and other costs remain the same, what will total costs be?

5. Last month, overhead at the Gas Mart represented 10% of total costs, which were $15,000. This month, other costs remained the same while overhead increased by $750. What percent increase in overhead does this represent?

Distribution. *Distribution* costs include all expenses involved in getting the product to the customer. Sales commissions, discussed in Unit 4, usually are based on a percentage of the dollar value of sales. Other distribution costs might include travel and entertainment expenses and automobile mileage for salespersons. Expenses for a business telephone may be included in the cost of distribution. Also included are the packaging of products and the costs involved in their shipping.

EXERCISE 11-5

Read each question carefully before calculating the answer. Round answers to two decimal places.

1. The following figures represent distribution costs during January for Angie's Used Cars: commissions, $4,061; salespeople's phone bills, $890; travel and entertainment expenses, $545. Total costs were $23,700. What percent of total costs did distribution represent?

2. In March, total costs for O.J.'s Malt Shop were $4,820. Distribution costs represented 24% of this amount. If distribution costs during April dropped 6% from March figures, how much was spent on distribution during April?

3. Last month, distribution costs represented 15% of total costs for Fred's Market. If distribution costs were $3,547, how much were Fred's total costs? If, this next month, distribution costs increase 15% and other costs remain the same, how much will total costs be?

4. Gregorio's Pizza spent $587 on distribution in August. Total costs were $2,670. If distribution costs are 30% less in September and other costs remain the same, how much will total costs be?

5. Last month, The Lamp Lagoon's distribution costs represented 23% of total costs, which were $3,670. This month, other costs remained the same, while distribution costs increased $675. By what percent did distribution costs increase?

Controlling Costs

As a business owner or manager, you would want to have as much control as possible over costs in the areas listed above. Your level of control would depend largely upon several factors. One is the nature of the company's business and the types of products it manufactures. Another is the size of the company. A third is its location. These and other factors may determine your range of choices in controlling costs.

Material costs. Because materials usually are a major cost in manufacturing, this area may involve a major portion of cost-saving efforts. One way to reduce the cost of raw materials may be to order in large quantities. Several questions arise, including these:

- How much are you able and willing to spend to increase order quantities?
- If it is necessary to borrow to buy in greater quantity, what is the cost of borrowing the money?
- Is storage space available for additional raw materials? If so, could that space be used more profitably for another function?
- Do sales projections justify an added investment in materials at this time?

Mathematics is involved in every business decision of this type. In some cases, it may be possible to shop around for the best prices on materials. Still another method of reducing materials costs may be to reduce or eliminate waste in the manufacturing process.

EXERCISE 11-6

Read each question carefully before calculating the answer. Round calculations to two decimal places.

1. High-Low Plastics uses an average of 10,000 square feet of Material A per year. The material is purchased at $9.80 per square foot. Suppliers have offered to sell 10,000 square feet at $7.50 per square foot. If management borrows the money at 18% simple interest and pays the loan off in one year, how much money will the company save?

2. At Harry's Machine Shop, steel represents 85% of material costs. On the average, material costs are $3,560 per month. Managers have estimated that 35% of steel is wasted during the manufacturing process. If waste amounts can be reduced to 5%, how much will the machine shop save?

3. Currently, AAA Incorporated purchases 12,000 square feet of plastic every three months at $9.65 per square foot. Management has discovered that the cost of plastic can be reduced $\frac{1}{5}$ by ordering a full year's supply, or 48,000 square feet. However, to order such a large quantity, the company also must rent a 1,000-square-foot warehouse space for materials storage. The space would cost $9 per square foot, per year. Over a year's period, would the company save any money by ordering the larger quantity?

4. JDI Industries manufactures leather shoes. Currently, leather is purchased at $10 per square foot. However, by ordering 25,000 square feet, the company can obtain a 25% discount. If JDI borrows the money at 16% simple interest for one year, how much will the leather cost? What is the difference between the discount rate and the standard rate?

5. Your job is to find the lowest price on 500 reams of 20-pound paper. A ream equals 500 pieces of paper. You have received the following bids: Supplier A sells the paper for $4.50 per ream and gives a 10% discount for orders over 400 reams. Supplier B sells the paper for $4 per ream and gives no discount. Supplier C sells the paper for $4.75 a ream and gives a 13% discount on orders of 500 reams or more. Which supplier offers the best price for your company? How much will the order cost?

Overhead costs. Overhead expenses range from long-term decisions to minor adjustments that may be made several times a day. Signing a lease for building space may involve projecting your company's needs for several years. Many factors can enter into such a major decision. On the other hand, you might adjust a thermostat several times a day to control your heating or cooling costs.

Major overhead decisions also include purchasing machinery and office equipment and furnishings. Cost of such items is one decision factor. Others include efficiency, reliability, and ease of maintenance and repair. Saving money on the price of equipment may be false economy if the equipment frequently is out of service.

Some overhead costs, such as paper towels and restroom supplies, can be adjusted frequently. Several sources may be available for such items, and their costs can be minimized through competition.

Careful shopping is required for decisions on insurance and on security devices or systems. Again, lower initial cost is but one consideration.

EXERCISE 11-7

Read each question carefully before calculating the answer. Round calculations to two decimal places.

1. M. O. Manufacturing is moving to a larger building. Three possible locations have been chosen. Location A has 5,000 square feet and rents for $0.75 per square foot, per month. Location B is 6,000 square feet and goes for $0.69 per square foot, per month. Location C has 5,500 square feet and costs $0.73 per square foot, per month. Management has decided not to pay more than $3,800 per month. Which location will management choose? How much will yearly rent costs be?

2. Magneto's Laundry Service has spent $2,600 to purchase a computer system. Currently, Magneto pays two part-time workers a total of $240 per month to perform the same work the computer will do. Management expects to pay one employee $40 per month to operate the computer. Will the computer system save the company any money during the first year? If so, how much? If not, how much will the company lose? Once the cost of the computer is made up, how much will the company save per year?

3. You are about to sign a five-year lease on a new building. Therefore, you must project how much space your company will need during the next five years. For the last 10 years, your

company has doubled its space requirement every two years. Currently, you are in a 15,000-square-foot building. What is the smallest size building you should move into?

Overhead of a business includes the rental or purchase cost of the building or space in which the business operates.

4. The Furniture Doctor wants to move to a new location. Due to higher rent, overhead costs will increase 50%, but remaining costs will stay the same. If overhead represented 20% of total costs before the move, what percent will it represent after the move?

5. You have been assigned to purchase 20 tables and have received the following bids: Supplier A sells tables for $45 each and gives a 5% discount for every five tables purchased. Supplier B sells tables for $48 each and provides a 10% discount for 15 or more tables purchased. Supplier C sells tables for $42.50 each and gives no discount. Which supplier is offering the best price for 20 tables? What is the difference between the highest price and the lowest price?

Distribution costs. The cost of distribution includes everything involved in getting a company's products to its customers. Cost factors attributable to distribution can include:

- Advertising
- Direct sales (sales made by sales representatives)
- Product packaging
- Product handling
- Product distribution (shipping).

The kind of advertising done by a business depends upon the audience it wants to reach. If products are offered to the general public, the *mass media* may be the most efficient way to advertise. The mass media includes newspapers, magazines, radio, television, and direct mail advertising. An important consideration in designing an advertising program is *cost per thousand (CPM).* (The Roman Numeral ''M'' means 1,000.) CPM measures the cost effectiveness of advertising by determining how much money is spent in reaching each 1,000 readers, listeners, or viewers.

Some products are aimed at smaller markets. In such cases, other forms of advertising might be preferred. There is no point in paying to reach thousands, when the audience for your product may number only in the hundreds.

A sales force can have a major effect on the cost of distribution. A thriving business generally has salespersons who work efficiently and avoid running up unnecessary expenses. A sales staff that is irresponsible can cut deeply into a company's profit.

Packaging and handling of products can be approached with one major concept in mind: The less a product is handled, the better. Efficiency is the key in these areas. Remember, each person who handles a product on its way to the customer must be paid.

Efficient storage, or *warehousing,* of merchandise is another cost factor. It is important to have sufficient *inventory* to meet demand for a product. Inventory is the stock on hand. Too large an inventory, however, leads to extra costs. Inventory problems are discussed in Unit 12.

Location of a business may limit control over distribution. A plant in a relatively isolated community may have only one method of shipping available. Elsewhere, distribution decisions may involve evaluating many options, or choices.

One example of distribution decisions would be the makeup of a fleet of delivery vehicles. Should you buy or lease the vehicles? Should they be equipped with manual or automatic transmissions? Which manufacturer or dealer offers the best maintenance and service programs? The list can go on almost endlessly.

EXERCISE 11-8

Read each question carefully before calculating the answer. Round answers to the nearest whole number.

1. At Silo Industries, money spent shipping products to customers represented 25% of distribution costs. If purchasing new vehicles reduced shipping costs 50%, by what percent would distribution costs be reduced?
2. Last year, The Cookie Company spent $2,000 on advertising, $3,500 on product packaging, $2,700 on product handling, and $5,000 on product shipping. What amount did the Cookie Company spend on distribution? What percent of distribution costs did advertising represent? Product packaging? Product handling? Product shipping?
3. The Biscuit Factory spent $23,000 on distribution last year. Direct sales represented 15% of these costs. Advertising represented 10%. Product packaging was 25%. Product handling equaled 13%. Product shipment represented 37%. How much did the Biscuit Factory spend on each cost factor?
4. The sales and distribution department of Weber's Clothing has been given a $40,000 budget. Product packaging and handling costs will require 45% of this budget. Product shipping costs will use 25%. How much will be left for advertising and direct sales costs?
5. Last year, sales costs for Gerald's Publishing were $35,000. Of this amount, 65% was paid to salespeople. If salespeople earn a 5% commission on gross sales, what were gross sales for Gerald's Publishing last year? What percentage of gross sales did sales costs represent?

THE LABOR EQUATION

Labor costs can be critical in some businesses. A *labor-intensive* business is one in which the cost of labor is a major portion of overall costs. Many *service industries* are labor-intensive. Service industries are those in which services, rather than goods, or products, are bought and sold. Restaurants, hotels, and hospitals are examples of service businesses. Certain government services also are labor-intensive. Examples include education and public safety agencies, such as police and fire departments. Labor-intensive industries also include construction and clothing manufacturing.

Regardless of the type of business or industry, it is important to understand the real costs of labor. These costs can be expressed by the equation:

$$L = W + F + U$$

This equation states: Labor (L) equals wages (W) plus fringe benefits (F) plus unproductive labor (U). Thus, there is more to labor costs than the wages or salaries paid to employees.

One major consideration is employer contributions to social security and to state-operated programs, such as unemployment funds. These costs come under the heading of wages.

Unproductive labor costs include time spent in training sessions.

Fringe benefits can include a number of costs. Typical fringe benefits that might involve employer contributions include:

- Health insurance
- Life insurance
- Pension or other retirement plans
- Paid vacations
- Paid holidays
- Paid sick leave
- Safety equipment
- Work clothes or uniforms
- Educational programs
- Employee assistance programs
- Recreational programs.

In some industries, the cost of fringe benefits approaches the cost of wages. For instance, some union contracts call for substantial employee benefits, such as generous insurance and retirement programs. In many cases, employers offer fringe benefit packages to attract and keep employees.

In recent years, society has become increasingly oriented toward recreation and leisure activities. Therefore, paid time off from work has become an important fringe benefit. Such free time includes vacations and holidays. Sick leave also has become a major bargaining point between labor and management in many industries.

One area of labor costs that can be significant in some businesses is *unproductive labor.* Unproductive labor is paid time that is devoted to activities other than production of goods or providing services.

For example, a new employee may require a period of training to build job skills. In highly technical industries, training can occupy as much as half of an employee's first year on the job. Good training programs pay off in the long run. However, unproductive labor can cause this area of costs to grow out of proportion to normal productivity. This is true especially during periods of rapid growth in a company's work force or high employee turnover.

EXERCISE 11-9

Read each question carefully. Then use the labor equation, L = W + F + U, to calculate the answer. Round answers to two decimal places.

1. In August, The Hamburger Hut spent $1,769 on wages, $350 on fringe benefits, and $276 on unproductive labor. What amount was spent on labor?

2. Last year, Fred's Repairs spent $21,780 on wages, $3,560 on fringe benefits, and $2,100 on unproductive labor. What amount was spent on labor?

3. Labor costs for Genie's Car Rental were $15,790 in May. Wages represented 85% and fringe benefits represented 15%. What amount was spent on wages? On fringe benefits?

4. Gasoline Alley employs two people at $5 an hour to pump gas. Each employee worked 160 hours last month. In addition, the company paid out $80 in insurance premiums, $40 in sick leave, and $25 for uniforms and cleaning. What amount was spent on fringe benefits? What percent of labor costs did fringe benefits represent?

5. The Cab Company spent $34,000 on labor last month. Wages represented 65% of labor costs. If wages increase 28% next month, how much will labor costs be?

6. At A. K. Engineering, wages represent 60% of labor costs. If wages can be reduced 25%, what percent will labor costs be reduced?

7. On the average, Barry's Big Top spends $75,000 a year on wages. In addition, $5,000 is spent on paid vacations, holidays, and sick leave. Uniforms and outfits cost $25,800 and safety equipment costs $19,800. On the average, how much does the circus spend on labor for one year?

8. Labor costs at Benny's Warehouse were $35,890 last year. Wages represented 76% and fringe benefits represented 24%. If, next year, management expects wages to increase 15% and fringe benefits to increase 12%, how much will labor costs be?

Decision making is based on accurate meaningful information. Computer printouts are helpful in supplying this information.
COURTESY OF IBM CORPORATION INFORMATION SYSTEMS GROUP

MEASURING RESULTS

Accurate, meaningful information is the basis for most effective decision making. This is true particularly in business. The presentation of information:

- Aids management in making wise decisions
- Enables management to evaluate past decisions.

Measuring results of business decisions is the key to making product mathematics work. Such measurements can be made best when information is presented in clear, easily understood forms. Many business decision makers rely heavily on the information presentation methods discussed in Unit 2. Chief among these methods are charts and graphs.

Bar charts can be used to compare a number of values. Monthly sales totals, for example, can be shown in a bar chart. Such a chart, shown in Figure 11-2, can be useful in planning seasonal adjustments in production schedules.

Line graphs can illustrate trends, as shown in Figure 11-3.

Charts and graphs are especially useful for presenting information to groups of people. Pictorial representation of comparative figures and trends is an effective method of summarizing important ideas. Business is one area in which the language of mathematics truly improves communication.

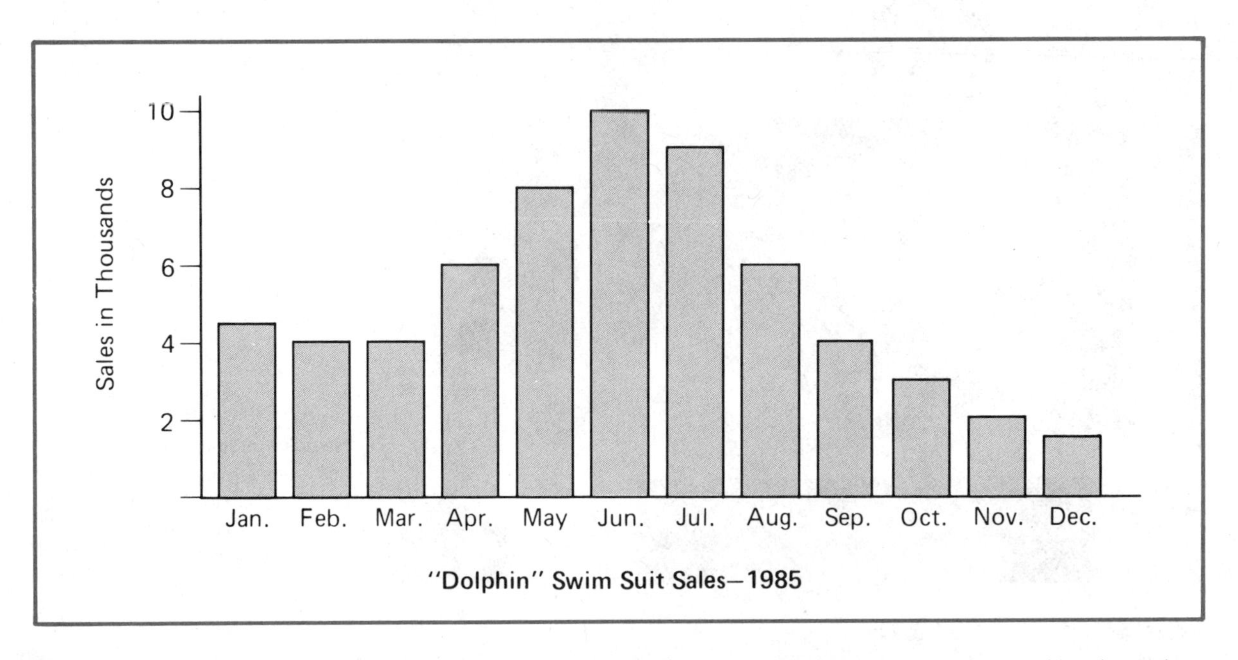

Figure 11-2. Bar charts can be used to evaluate many types of business activities. This chart compares monthly sales totals.

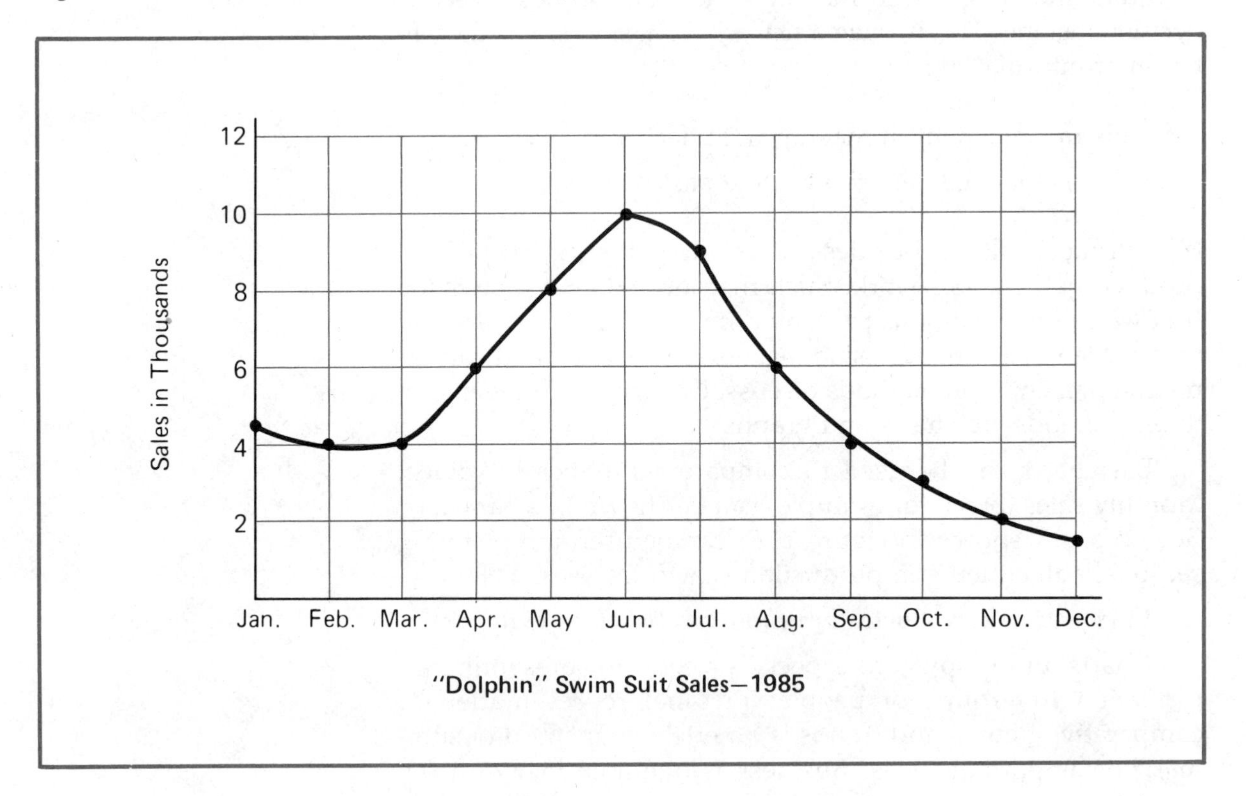

Figure 11-3. Line graphs are pictorial representations of trends and comparative values.

EXERCISE 11-10

Read each question carefully. Then present the information in the format required.

1. In 1984, Ted's Auto Body earned a profit of $15,000. In 1985, Ted's Auto Body made a $35,000 profit. In 1986, profits decreased to $20,000. But, in 1987, profits jumped to $45,000. Draw a bar chart that compares profits for the four years.

2. Ketch Machining manufactures metal bearings. In the company's first year of business, $45,000 in wages were paid out to make 15,000 bearings. It cost $60,000 in wages to produce 25,000 bearings in the second year. In the third, it took $75,000 in wages to produce 30,000 bearings. In the fourth, it took wages of $90,000 to produce 35,000 bearings. Draw a graph that relates the cost of wages to the number of bearings produced.

3. Four years ago, Wagner Print Shop employed 15 people and paid out $1,400 in fringe benefits. Three years ago, the print shop employed 20 people and paid out $1,700 in fringe benefits. Two years ago, the company employed 25 people and paid out $1,800 in fringe benefits. Last year, the print shop paid out $1,850 in fringe benefits for 35 employees. Draw a graph that relates the number of people employed to the cost of fringe benefits. From your graph, make some conclusion about how the cost of fringe benefits is affected by the number of employees.

4. Four years ago, Zero Industries employed 20 people and paid out $1,000 in unproductive labor costs. Three years ago, the company employed 30 people and paid out $1,500 in unproductive labor costs. Two years ago, the company employed 40 people and paid out $3,000 in unproductive labor costs. Last year, Zero Industries employed 50 people and paid out $6,000 in unproductive labor costs. Draw a graph that relates the number of people employed to unproductive labor costs. From your graph, make some conclusion about how unproductive labor costs are affected by the number of people employed at Zero Industries.

5. During the last four years, income and costs for Matilija Manufacturing were: Four years ago, income equaled $45,000 and costs totaled $20,000. Three years ago, income was $55,000 and costs were $25,000. Two years ago, income totaled $65,000 and costs equaled $30,000. Last year, income was $75,000 and costs were $35,000. Draw a graph with two lines, one showing income and the other showing costs. Then, darken in the area of the graph that shows profit. From your graph, determine if profits are increasing or decreasing.

SUMMING UP

- Product mathematics is a system of identifying, comparing, and evaluating costs of manufacturing.
- A basic business calculation is the profit equation: $I - C = P$ (income minus costs equals profit).
- Manufacturing costs can be separated into four general categories: materials, labor, overhead, and distribution. Within these categories, individual cost decisions can be made.
- Labor costs involve more than wages. Fringe benefits also are a major cost factor for many businesses. General fringe benefits may include health insurance and pension or profit-sharing retirement plans. Special benefits may include free uniforms, meal and transportation expenses, and others.
- Another labor factor that must be computed is the cost of unproductive labor. This includes training and other nonproducing activities during which employees receive regular wages.
- Mathematics provides efficient tools for measuring the results of business decisions. Charts and graphs are used to present information on important factors, such as sales volumes and plant costs. Charts and graphs are helpful for making comparisons and for evaluating trends.

TERMS THAT COUNT

gross income
materials
raw materials
labor
overhead
plant
lease
distribution
mass media
cost per thousand (CPM)
warehousing
inventory
labor-intensive
service industry
unproductive labor

TESTING YOUR WORKING KNOWLEDGE

Read each question carefully before calculating the answer. Round answers to two decimal places.

1. The Body Shop earned $28,876 last month. Costs were $4,892 for parts, $5,523 for materials, $1,200 for overhead, $2,400 for distribution, and $11,213 for labor. How much profit did The Body Shop make?

2. Last month, materials for The Mining Company represented 21% of total costs. Total costs amounted to $47,845. If, next month, the price of materials goes up 15% and other costs remain the same, how much will total costs be?

3. Before the move, overhead at The Blouse Bonanza represented 17% of total costs. After the move, however, rent on the new location increased overhead 33%. By what percentage did total costs increase?

4. Backcountry Incorporated manufactures bed frames from plastic. Currently, plastic is sold at $0.80 per pound of pellets when 1,000 pounds or more are purchased. Pellets then are melted and molded into frames. Suppliers have offered to sell 25,000 pounds of pellets at a 25% discount. Management is considering borrowing the capital at 16% simple annual interest. If the company pays the loan off in one year, how much will the plastic cost? (Be sure to include the finance charge.) How much will the company save on materials by purchasing the larger quantity?

5. Your company is about to sign a five-year lease on a new building. Your job is to project how much space your company will need during the next five years. For the last 20 years, the company has doubled its space requirement every 10 years. Currently, you are in a 50,000 square foot building. What is the smallest size building you should move into? If rent for the new building costs $0.80 per square foot, per month, how much will rent cost per year?

6. Last year, distribution costs for The Potato Chip Factory were $34,000. Direct sales represented 13% of these costs. Advertising represented 12%. Product packaging was 28%. Product handling was 23%. Product shipping used up 24%. How much did The Potato Chip Factory spend on each cost factor?

7. At Quasi Construction, wages represent 60% of labor costs. If wages can be reduced 25%, what percent of labor costs will wages represent?

8. Last year, wages represented 75% of labor costs, fringe benefits used up 15%, and unproductive labor was responsible for 10%. Labor costs equaled $175,900. If management expects wages to decrease 11%, fringe benefits to increase 5%, and unproductive labor to decrease 3%, how much will labor costs be next year?

9. Last month, The Chair Palace's sales costs represented 23% of total costs, which were $8,500. This month, other costs remained the same while sales costs increased by $1,875. What percent increase in sales costs does this represent?

10. Currently, The Doll Company purchases 1,000 square feet of plastic per year at $8.95 per square foot. The plastic is heated and formed into dolls. Management has discovered that the cost of plastic can be reduced one-fourth by ordering 5,000 square feet, or a five-year supply. However, if they do this, the company must rent an additional 500 square-foot warehouse space in which to store the materials. This additional space will cost $0.70 per square foot per month. Would the company save any money by ordering the larger quantity of plastic? If so, how much? If not, how much would be lost?

PRESENTING INFORMATION

Read each question carefully. Then present the information in the format required. Round answers to two decimal places.

1. Gross income at Precision Injection Molding was $135,680 last year. Material costs equaled $24,000, labor costs were $65,780, overhead costs were $12,900, and distribution costs equaled $11,640. Determine the profit. Then, draw a pie chart that relates profit and material, labor, overhead, and distribution costs to gross income.

2. During the last four years, income and costs for Kentucky Krafts were: Four years ago, income equaled $356,800 and costs totaled $220,600. Three years ago, income was $355,400 and costs were $245,000. Two years ago, income totaled $365,000 and costs equaled $250,000. Last year, income was $290,000 and costs were $250,000. Draw a two-line graph showing income and costs. Then, darken in the area of the graph that shows profit. From your graph, determine if profits are increasing, decreasing, or staying about the same.

SKILL DRILLS

A. Convert the following fractions to their percentage equivalents. Round answers to one decimal place.

1. $\frac{35{,}600}{45{,}891}$
2. $\frac{123{,}560}{99{,}870}$
3. $\frac{34{,}519}{35{,}518}$
4. $\frac{102{,}801}{218{,}000}$
5. $\frac{54{,}873}{49{,}549}$
6. $\frac{5{,}432{,}000}{7{,}658{,}019}$
7. $\frac{872{,}000}{943{,}000}$
8. $\frac{716{,}390}{4{,}849{,}980}$

B. Compute the following percentages. Round calculations to two decimal places.

1. 35% of 34,650
2. 19% of 4,390
3. 56% of 67,980
4. 23% of 902,198
5. 76% of 2,341,928
6. 91% of 453,782
7. 83% of 812,437
8. 69% of 712,456

C. Compute the following percentages of monetary values. Round calculations to two decimal places.

1. 34% of $321,546
2. 17% of $345.82
3. 45% of $5,673.09
4. 67% of $12,314.34
5. 23% of $7,654.08
6. 97% of $55,891.01
7. 73% of $7,123.56
8. 52% of $23,451,800.01

Mathematics of Supply

YOUR LEARNING JOB

When you have completed the exercises and assignments for this unit, you should be able to:

- ☐ Explain why it is important for a business to control inventory levels.
- ☐ Use the inventory equation to solve problems and make decisions involving restocking of inventory.
- ☐ Use graphs to present information on sales performance and inventory levels that will assist in inventory control decision making.
- ☐ Explain why certain distribution systems, such as hub-and-spoke operations, can work better than traditional, point-to-point systems.

WARM-UP

A. Find the sum in each problem.

1. 2,390.18 + 3,490.03 + 231.89 + 5,009.71 =
2. 1.0982 + 34.802 + 325.017 + 54.02876 + 80 =
3. 345.912 + 473.849 + 526.098 + 4,382.828 =
4. 413.0981 + 98,382 + 345.092 + 3,424.98 + 3,283.003 =
5. 9,387 + 5,493 + 0.1829 + 34.182 + 43,291.49243 =

B. Find the difference in each problem.

1. 45,394 − 32,328.91 =
2. 3,293 − 2,192.901 =
3. 467,281.091 − 121,390.927 =
4. 546,938.0826 − 325,423.3982 =
5. 435,432.382 − 342,328.8403 =

C. Find the product in each problem. Round calculations to three decimal places.

1. 456.04 × 93.001 =
2. 0.2983 × 0.4583 =
3. 34,928.298 × 0.721 =
4. 44.982 × 1.727 =
5. 287.03 × 512.017 =

D. Find the quotient in each problem. Round calculations to two decimal places.

1. 176.09 ÷ 29 =
2. 547.734 ÷ 45.11 =
3. 4,583 ÷ 82.38 =
4. 32.093 ÷ 0.45 =
5. 0.489 ÷ 13.901 =

THE INVENTORY PROBLEM

One of the most important challenges in many businesses lies in control of *inventory.* Inventory is the stock of merchandise, parts, and other items manufactured, sold, or used by a company.

Inventory level is an important business consideration. Businesses want neither too much nor too little stock on hand.

The word inventory also refers to a detailed list or catalog of items in stock. The activity of counting, or cataloging, stock on hand usually is described as *taking inventory.*

The inventory problem is this: Too much inventory is bad, and too little inventory is equally bad. The goal of any business is to have the right amount of inventory on hand.

Too Much Inventory

An excess of inventory represents extra costs. There are several areas in which too much inventory can hurt a business. They include:

- Overhead
- Labor
- Cost of inventory
- Cost of money borrowed to purchase inventory
- Decreased value of outdated merchandise.

Overhead. Inventory must be stored. Storage requires space, for which rent must be paid. Storage space may require heating and/or cooling, thus increasing the cost of utilities.

Labor. Each time an item is handled, labor costs are involved. Some types of merchandise require constant handling. Cars displayed on a lot, for instance, must be washed and kept in running order. They also may have to be moved frequently. This is true especially if the cars are used as demonstrators for potential buyers to test drive.

Cost of inventory. Each item of inventory represents an investment of money. Paying for excess merchandise or supplies ties up money that could be used for other purposes.

Cost of money. Businesses frequently borrow money to maintain or increase their stocks of merchandise. Interest on such loans can become too costly if the business has an oversupply of merchandise.

Decreased value of outdated merchandise. Some products, such as cars and major appliances, tend to lose their sales value when new models are introduced. Thus, items held in inventory for too long can lose value. If a business must reduce prices to sell items, money is lost.

EXERCISE 12-1

Read each question carefully before calculating the answer. Round answers to the nearest cent.

1. Martha's Furniture is moving inventory from the showroom to a 1,000-square-foot storage area. Rent on the new storage space will cost $0.75 per square foot, per month. Two employees earning $8 an hour and working 156 hours per month will operate the storage area. Utilities will cost $25 per month. How much will Martha's inventory storage space costs be, per month?

2. The Stereo Hut wants to purchase larger quantities of inventory. Management has two alternatives: Alternative A is to build a second level in the current store at a cost of $15,000. The bank will loan construction money at 8% simple annual interest for one year. Alternative B is to lease a 500-square-foot storage space for 65 cents per square foot, per month. Utilities would average $30 per month. Extra employee costs would be $250 per month. Over a five-year period, which alternative will cost the Stereo Hut less money? What is the cost difference between the two alternatives?

3. Dandy Distributors operates three warehouses and spends $75,000 per month on inventory among them. Overhead for the three warehouses represents 35% of inventory costs. Cost of goods is 55%. Labor is responsible for 10%. To lower inventory costs, management wants to move inventory to one large warehouse. It is believed that overhead costs would reduce by 50%, cost of goods would reduce by 25%, and labor costs would increase by 100%. According to management's predictions, how much will inventory costs be per month for one large warehouse? By what percentage will inventory costs decrease per month?

4. The Video Mart had 32 television sets left in stock when the manufacturer announced distribution of a new, less expensive model. The new model was to sell to the Video Mart for $50 per set and to the public for $100 per set. The Video Mart purchased the old model from the manufacturer for $125 per set. However, the old models had to be sold for less than the

new model. The price was $95. How much did the Video Mart lose on overstocked television sets?

5. O. R. Enterprises manufactures plastic parts for automobiles. The company purchases 1,000 pounds of plastic at 75 cents per pound, each year. Suppliers have offered to sell O. R. Enterprises 5,000 pounds, a five-year supply, for 50 cents per pound. In addition, they have offered to store the materials for $25 per month, for five years. How much would it cost O. R. Enterprises to purchase and store the five-year quantity? Would it cost less to purchase and store the five-year quantity or to purchase plastic by the year?

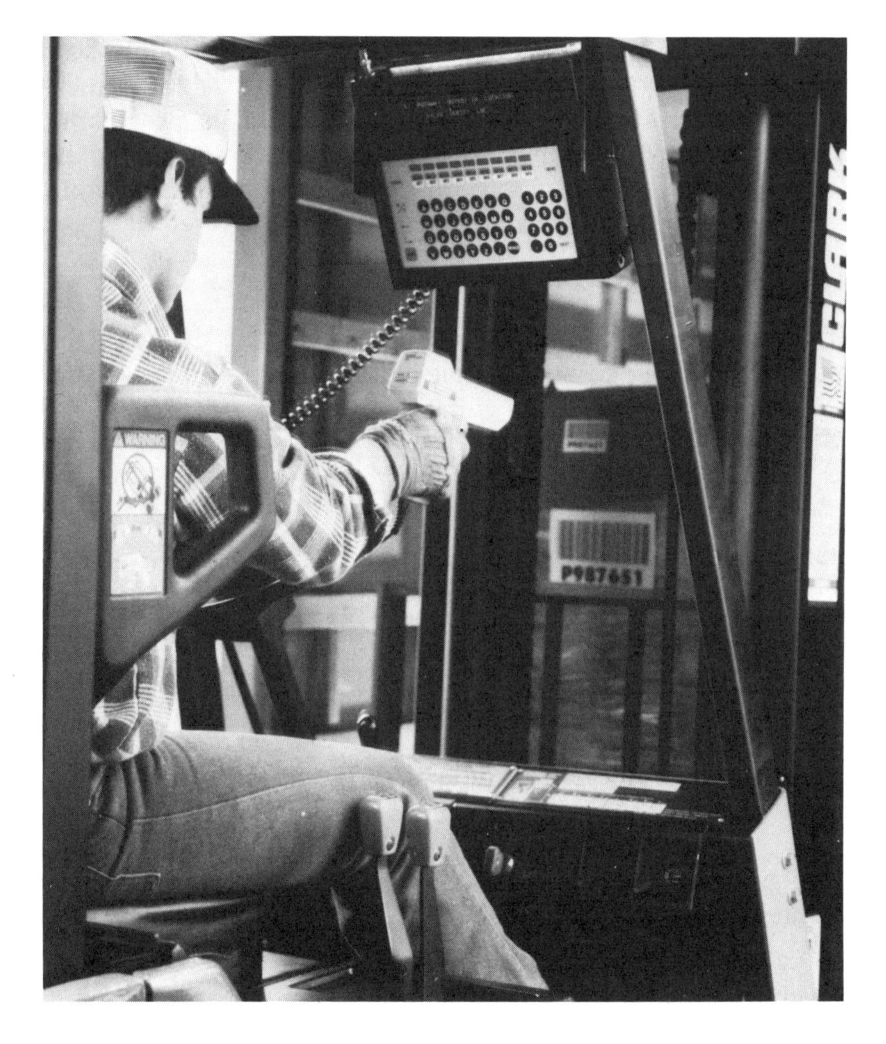

Laser gun used by the operator of this electronic forklift scans bar codes on items to be picked up in warehouse. The laser gun, linked to a computer via radio, increases both operator and vehicle productivity.
COURTESY OF L.X.E., INC., A DIVISION OF ELECTROMAGNETIC SCIENCES

Too Little Inventory

Many businesses cannot operate without inventory. There is an old saying: ''Out of stock means out of business.''

Think about what would happen if a gasoline filling station ran out of gasoline. If you were the operator, you probably would close the doors until more fuel could be delivered. Running out of merchandise is an obvious problem. However, other kinds of inventory also must be maintained at adequate levels.

For example, consider a factory that makes portable cassette players. Each unit is packaged in a special box. If the factory ran out of boxes, it would be unable to ship any cassette players.

There is a true story that illustrates the need to maintain proper inventories. A small group was touring a computer manufacturing plant when production work suddenly was halted. Asked what had happened, an embarrassed official explained that the factory had run out of *flux*. Flux is a substance used to promote bonding between surfaces being joined by soldering. Until that moment, flux had seemed a relatively insignificant inventory item.

THE INVENTORY EQUATION

The cost of inventory can be expressed mathematically. The statement of inventory cost is a simple equation that includes several cost factors. The inventory equation can be stated as follows:

$$I_C = O + L + C_I + C_M$$

Stated in English, the equation reads: Inventory cost equals overhead, plus labor, plus cost of inventory, plus cost of money. A good example of inventory cost involves the new-car dealer. Each shiny new car you see in a showroom represents several cost factors for the dealer. Placing a car on display for sale is called *flooring*.

The costs of flooring include all of the factors in the inventory equation. Think about each cost factor individually:

- Overhead: The dealer must buy or lease the space for storing and displaying the cars.
- Labor: Employees must be paid to move the cars when necessary and to keep them attractive and in good running

condition. Regular tasks include washing the cars and keeping their batteries charged.

- Cost of inventory: The dealer buys the cars from the factory. Each car on a lot represents a sizable investment.
- Cost of money: Money must be borrowed to purchase a stock of cars. The dealer must pay interest to a lending institution while each car remains on the lot. The longer a car sits before being sold, the more expensive it is as an item of inventory.

Flooring automobiles requires a large outlay of money. Costs include keeping cars clean and in good running condition.

For the car dealer, and other retail merchants, inventory is a two-sided problem. On the one hand, the car dealer must have a variety of models on hand to meet customer demands. On the other hand, inventory levels must be kept within reason. Mathematics provides a method of making inventory decisions based on the inventory equation.

EXERCISE 12-2

Read each question carefully. Then use the inventory equation, $I_C = O + L + C_I + C_M$, to calculate inventory cost. Round answers to the nearest cent.

1. You operate a warehouse that supplies cassettes and albums to music stores. Last month, you had 2,500 albums and 750 cassettes in stock. Each album cost $2.50; each cassette cost $3.25. You paid $850 to employees to handle inventory. Storage space cost $550. How much were inventory costs last month?

2. The Motorcycle Market purchased 125 motorcycles last year. Each motorcycle cost $1,200. Money to purchase inventory was borrowed at 16% simple annual interest. Storage space cost $9,000. Employees were paid $13,540 to handle inventory. How much did The Motorcycle Market spend on inventory last year?

3. Carlos borrowed $15,000 to purchase furniture to open a store. The simple annual interest rate was 15%. The loan was to be repayed in 12 monthly payments. He paid one employee $1,300 per month to handle inventory. In addition, he rented a 1,000-square-foot warehouse space in which to store furniture. The warehouse space cost 70 cents per square foot, per month. What were Carlos' average monthly inventory costs for his first year in business?

4. In September, inventory costs for Harry's Auto Parts were $37,735. Labor represented 40% of this amount. If inventory people worked 2,349 hours moving, preparing, counting, and ordering merchandise, what hourly wage did Harry pay?

5. Last year, Data Computers spent $35,700 on inventory. Of this amount, 23% represented labor, 14% was spent on overhead, and the rest was used to purchase inventory items. How much did management spend on labor? On cost of the inventory items? On overhead?

Inventory and Business Expansion

Think about a new-car dealer expanding business by contracting to sell a second brand of automobile. Assume that the sales volume of the original brand was expected to remain the same.

To expand to a second line of automobiles, each factor in the inventory equation would be increased. The dealer's overhead costs would grow. The expansion might require construction of a new building or an addition to an existing building. A second line of cars might require additional showroom space, service and repair space, and land for parking the additional cars. Labor costs also would rise. New employees would be needed to sell, maintain, service, and repair the second line of cars. Inventory purchase costs would be increased, possibly doubling. The same increase would apply to the cost of borrowing money to floor the additional cars. Using the inventory equation, the dealer could make realistic inventory cost decisions.

EXERCISE 12-3

Read each question carefully before calculating the answer. Round answers to the nearest cent.

1. Last year, Harry's Stereo spent $13,000 on inventory overhead. This year, management at Harry's wanted to increase inventory storage space. The company borrowed $10,000 at a simple annual interest rate of 10% per year to build a storage facility. The duration of the loan was 12 months. Utilities (another part of overhead costs) on the new storage space were $35 per month. By what percentage rate did inventory cost increase this year?

2. A year ago, Carrol's Linoleum and Tile doubled inventory storage space. As a result, inventory costs increased to $46,700, a 25% increase over previous costs. How much did previous inventory costs go up?

3. Last year, the Bicycle Barn spent $54,000 on inventory. Cost of bicycles represented 58% of this amount. Cost of money was 12%. Overhead used up 14%. Labor used up 16%. This year, The Bicycle Barn moved inventory storage to a larger warehouse. Operation of the new warehouse increased

overhead 25% and labor 8%. In addition, the costs of borrowing money and of merchandise doubled. Determine what inventory costs are for the new warehouse. By what percentage rate did inventory costs increase?

4. Applied Machining wants to build a new warehouse so the company can purchase larger amounts of inventory. Currently, the company spends $76,500 per year on inventory. Cost of raw materials represents 46% of this amount. Cost of borrowing money is 12% of the total inventory cost. Labor uses up 24%. Overhead represents 18%. Management believes that the new warehouse will increase overhead 50%, labor 25%, and cost of raw materials 75%. In addition, borrowing costs of money will increase 35%. What amount does management predict inventory costs will be in the new warehouse?

5. Management at the Furniture Bonanza wanted to reduce inventory costs 25%. Inventory costs were averaging $45,000 per year. Of this amount, 35% was spent on overhead and labor, and 65% was used to purchase inventory. To achieve its goal, management cut inventory storage space in half. Labor and overhead costs were reduced 31%. Cost of inventory was reduced 22%. Did management achieve its goal? By what percentage were inventory costs actually reduced?

THE CHAIN OF SUPPLY

Every product you buy reaches you through a *chain of supply*. A chain of supply is a series of activities that bring goods or supplies to consumers. The question of inventory must be handled at each point along the chain.

For example, think about the chain of supply that brings the water you use at home. That chain might include the following steps:

1. Rainfall and/or melting snow create water runoff from higher elevations.
2. Runoff fills or replenishes a water reservoir.
3. Water from the reservoir is brought to a city through pipes or aqueducts, usually through gravity (downhill) flow. At

Water comes to you through a system that probably includes a water treatment plant. In this plant, water is being purified by back-wash filters. COURTESY OF COVINA, CA, IRRIGATING COMPANY

some locations, pumping may be required to maintain water flow.

4. Water for local service may be kept in large storage tanks, usually located on a hill or other high ground.
5. Water pressure moves the water from a storage tank to your residence.

Water pressure may be generated in different ways. Gravity provides water pressure in two ways. One is the weight of the water stored in the tank. This weight pushes the water out of the tank into pipes when a valve is opened.

Another is the downward path of water flowing from the tank to the user. To utilize the pull of gravity, storage tanks usually are located on high ground or are elevated above the ground.

Water pressure also may be generated by pumping. Pumps may be required to deliver water to customers at higher elevations or to reach upper floors of tall buildings.

The point is that a number of steps must occur for you to enjoy the water you use for drinking and bathing. Furthermore, this chain of supply depends on storage of supply, or inventories, along the way. The reservoir and water storage tank are critical inventory points in the chain.

EXERCISE 12-4

Read each question carefully before calculating the answer. Round answers to the nearest cent.

1. OK Soft Drinks produces canned drinks. The packager sells drinks to a distributor at $2.24 per case. The distributor raises, or marks up, the price 25% and sells the drinks to markets. The markets then raise the price 35% and sell drinks to consumers. How much does a case of soft drinks cost at the end of the supply chain?

2. Comfort Enterprises makes office chairs and sells them to distributors for $23.50. Distributors mark up the price 25% and sell chairs to office furniture stores. These stores, in turn, mark up the price 50% and sell chairs to consumers. How much do consumers pay for a chair?

3. Tomato Computers purchased a computer system from Benji's Computer Warehouse for $1,570. It cost Tomato Computers $275 in labor and $75 in overhead to store and sell one computer system. Tomato Computers marked up the price 75% to cover inventory costs. Benji's Computer Warehouse has increased the price of their computer systems 15%. By what amount will Tomato Computers raise the price of a computer system to earn the same profit per unit?

4. Ted manufactures glass cups. He sells one cup to distributors for 20 cents. From there, cups are sold twice more before reaching the consumer. That is, the consumer is the fourth party to purchase a cup. In each transaction in the supply chain, cups are marked up 50%. How much do consumers pay for one of Ted's cups?

5. Writing Industries manufactures ink pens. The company sells pens to a distributor. The distributor then sells pens to a supplier. The supplier, in turn, sells them to stationery stores, where consumers can buy them for $1. The pen was marked up 30% at each sales point in the supply chain. Did the manufacturer sell the pen to the distributor for more or less than 60 cents?

DOING SOMETHING ABOUT INVENTORY

Inventory decisions for most businesses involve time and quantity. When to order stock is equally as important as how much to order. Factors involved in when to order include:

- Quantity of stock on hand
- Anticipated rate of stock reduction based on sales rate
- *Lead time,* or time required for new stock to be delivered.

To illustrate, suppose you ran a video equipment business. A certain video cassette recorder (VCR) has been selling at the rate of four per week. You have 20 of these VCRs in stock. How many weeks' supply do you have? The solution is to divide the stock by the weekly sales rate:

$$\text{Number of Weeks' Supply} = \frac{\text{Stock}}{\text{Weekly Sales}} = \frac{20}{4} = 5$$

Thus, with 20 VCRs in stock, you would have a five-week supply. However, there is another, equally important, time factor: lead time. Suppose that your supplier requires four weeks to deliver new stock. Based on sales levels, you would have to reorder within the week to avoid running out of stock. That is, you have enough VCRs to last five weeks, while it will take four weeks to receive new units. So, you have to order within a week or risk running out.

However, a retail business would not operate on such a narrow margin. An increase in sales volume could wipe out the inventory, leaving the store with no merchandise to sell.

Instead, you would want to have additional inventory on hand to cover any increased demand. This extra inventory is called *safety stock.* Say that your policy is to maintain a safety stock of no less than a five-week supply of VCRs. In addition, you would not want to reorder each time a delivery was made. Therefore, you might order a 12-week supply. This supply factor would be three times the number of units that might be sold during the four-week lead time for delivery of a new order. In addition, you need enough units to maintain a five-week safety stock. Your restocking level, therefore, would be:

$$\text{Restocking Level} = (\text{Lead Time} \times \text{Weekly Sales} \times \text{Supply Factor}) + (\text{Weekly Sales} \times \text{Safety Factor})$$

$$\text{Restocking Level} = (4 \times 4 \times 3) + (4 \times 5) = 48 + 20 = 68$$

Now, assume that the sales average of four units a week continues. At what point do you reorder? Again, a simple equation provides the solution. You want to maintain enough inventory to supply normal sales through the lead time, plus your five-week safety stock (safety factor). The equation is:

$$\text{Reorder Level} = (\text{Lead Time} \times \text{Weekly Sales}) + (\text{Weekly Sales} \times \text{Safety Factor})$$

$$\text{Reorder Level} = (4 \times 4) + (4 \times 5) = 16 + 20 = 36$$

The trigger point for reordering, therefore, would be when your inventory is reduced to 36 units. If the weekly sales rate is constant, your inventory would be right at 20 units when your new order arrives.

EXERCISE 12-5

Read each question carefully before calculating the answer. Round calculations to two decimal places.

1. King Hardware sells five vacuum cleaners per week. It takes the hardware store five weeks to receive vacuum cleaners after orders are placed. Management keeps a three-week safety supply on hand. Determine the reorder level of vacuum cleaners for King Hardware.

2. Medina Stationery sells 15 reams of paper per week and keeps a four-week safety stock on hand. Once a ream of paper is ordered, it takes four weeks for delivery. Determine the reorder level of paper.

3. The Auto Market sells three trucks per month and keeps a one-month safety stock. It takes two months to receive an order. What is the reorder level?

4. The Poultry Market sells 20 pounds of chicken per week. Orders for chicken take two weeks for delivery and the butcher keeps a three-week supply of safety stock. What is the reorder level of chicken for the Poultry Market?

5. Kathy sells 25 pairs of tennis shoes per week. Orders take three weeks for delivery. What is the reorder level for tennis shoes?

6. If Kathy keeps a four-week safety stock, what is the reorder level?

7. Sun Industries manufactures sunglasses from plastic. The company uses 500 pounds of plastic per week. Orders take four weeks to be delivered. In addition, management keeps a 12-week safety stock on hand. Determine the reorder level of plastic for Sun Industries.

8. Olaf's Hardware sells 85 sheets of $\frac{1}{2}$-inch plywood per week. Orders for plywood take 14 weeks for delivery. In addition, plywood supplies often run out. Therefore, management at Olaf's keeps a 28-week supply of safety stock on hand. What is the reorder level of plywood? If Olaf's purchases plywood for $3.80 per sheet, how much money is invested in safety stock?

Unit Cost

Another factor considered in determining the most appropriate order quantity is unit cost. Generally, buying in large quantities means a lower price per unit than buying in small quantities. Volume discounts can help reduce inventory costs and boost profits.

Volume discounts are available because of lower handling costs. Remember, the more handling an item requires, the greater its cost. Handling includes packaging and transportation of merchandise. Other costs include processing of orders and billing. Thus, a single large shipment is more economical than several smaller shipments.

For instance, suppose you manage a fast food restaurant. Each week, you order a dozen boxes of paper napkins, each box containing 100 napkins. Each order costs $15.

Your supplier tells you that a volume discount is available if you order a larger quantity, such as a *gross.* A gross is 12 dozen or 144 units. The gross is cheaper because less handling is involved. Instead of 12 different orders being transported from the supplier to your business, only one trip is made. Also, the supplier has to process only one order instead of 12 and has to bill only once. Instead of paying $15 each for 12 orders, you get a price of $150 for the gross. You save $30 over the 12-week period. You now pay $12.50 per dozen packages instead of $15.

EXERCISE 12-6

Read each question carefully before calculating the answer. Round calculations to two decimal places.

1. Taylor purchases five 100-pound bags of plaster every three months at a price of $12 per bag. His supplier offered to sell a year's worth, or 20 bags, for $180. How much will Taylor save by taking the offer? What is the cost per 100-pound bag?

2. May's Two-Wheel Repair usually purchases bicycle tires for $48 per dozen. May sells an average of 11 dozen tires a year. The supplier has overstocked bicycle tires and has offered to sell May 144 tires for $3 per tire. How much will she save on 144 tires?

3. Precision Molding uses plastic to make motorcycle parts. The company orders 4,000 pounds of material at 75 cents per pound every six months. A new supplier has entered the market and is willing to sell 8,000 pounds for $5,600. How much can Precision Molding save by ordering the larger quantity? What is the price per pound?

4. Carlos sells four color television sets per week. Carlos' supplier called and offered to give him a 25% discount on any orders placed that day. Usually, the price is $145 per color television set. If Carlos orders 38 television sets, how much money will he save?

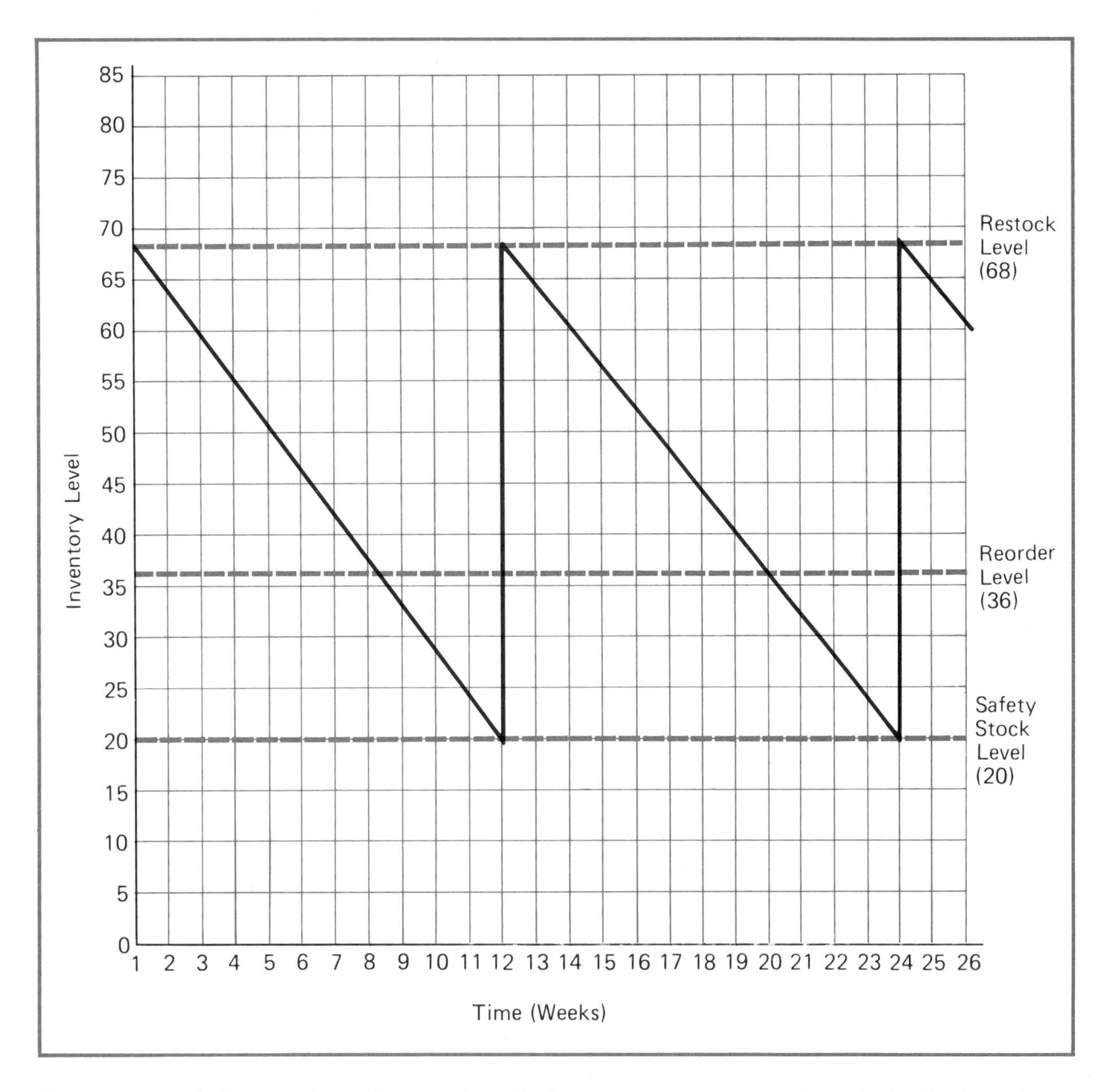

Figure 12-1. Graphs can be used to trigger orders for restocking of inventory.

Presenting Information: Triggering Orders

As discussed in Unit 11, charts and graphs are used frequently to illustrate sales and other business information. Tracking sales on a graph is one way to trigger orders of new stock. An example is shown in Figure 12-1.

A new order would be triggered when inventory reached 36 units. This figure represents the sum of a four-week sales supply (16) and a five-week safety stock (20). As shown in the graph, the optimum inventory would be the new-order quantity (48) plus safety stock (20), or 68 units when a new shipment arrives.

EXERCISE 12-7

For each question, your job is to maintain inventory levels. You are given the initial number of items on hand. You also are given safety-stock and lead-time levels, and a record of how many items were used or sold each day for a three-week period. Track the use of inventory items and determine when to reorder. Keep inventory between the restocking level and the safety-stock level. Present the information in the format required.

Example:
On the average, Hector's Auto Body uses five cans of gray paint daily. Lead time on orders is four days. In addition, Hector keeps a four-day safety stock and the restocking level is 60 cans. As of Feburary 28, Hector had 42 cans in inventory. On March 1, three cans were used; March 2, six were used; March 3, three were used; March 4, five were used; March 5, four were used; March 8, seven were used; March 9, six were used; March 10, seven were used; March 11, two were used; March 12, six were used; March 15, five were used; March 16, five were used; March 17, six were used; March 18, three were used; March 19, four were used.

Create a table that tracks changes in daily inventory level, quantity used, and quantity received. Also, use the table to calculate when to make orders. On days that you make orders, show the quantity you have ordered. Then draw a graph that shows how inventory levels fluctuated for the three-week period. Be sure to show the restocking and safety-stock levels.

Solution: See chart and graph on facing page.

1. M. O. Auto Parts sells an average of 50 spark plugs per day. Lead time on orders is two days. Management keeps a four-day safety stock and the restocking level is 500. As of June 30, the company had 450 in stock. On July 1, 56 spark plugs were sold; July 2, 49 were sold; July 3, 58 were sold; July 6, 43 were sold; July 7, 48 were sold; July 8, 39 were sold; July 9, 59 were sold; July 10, 62 were sold; July 13, 56 were sold; July 14, 60 were sold; July 15, 47 were sold; July 16, 57 were sold; July 17, 46 were sold.

   Create a table that tracks daily inventory levels, quantity used, and quantity received for spark plugs. Also, on days

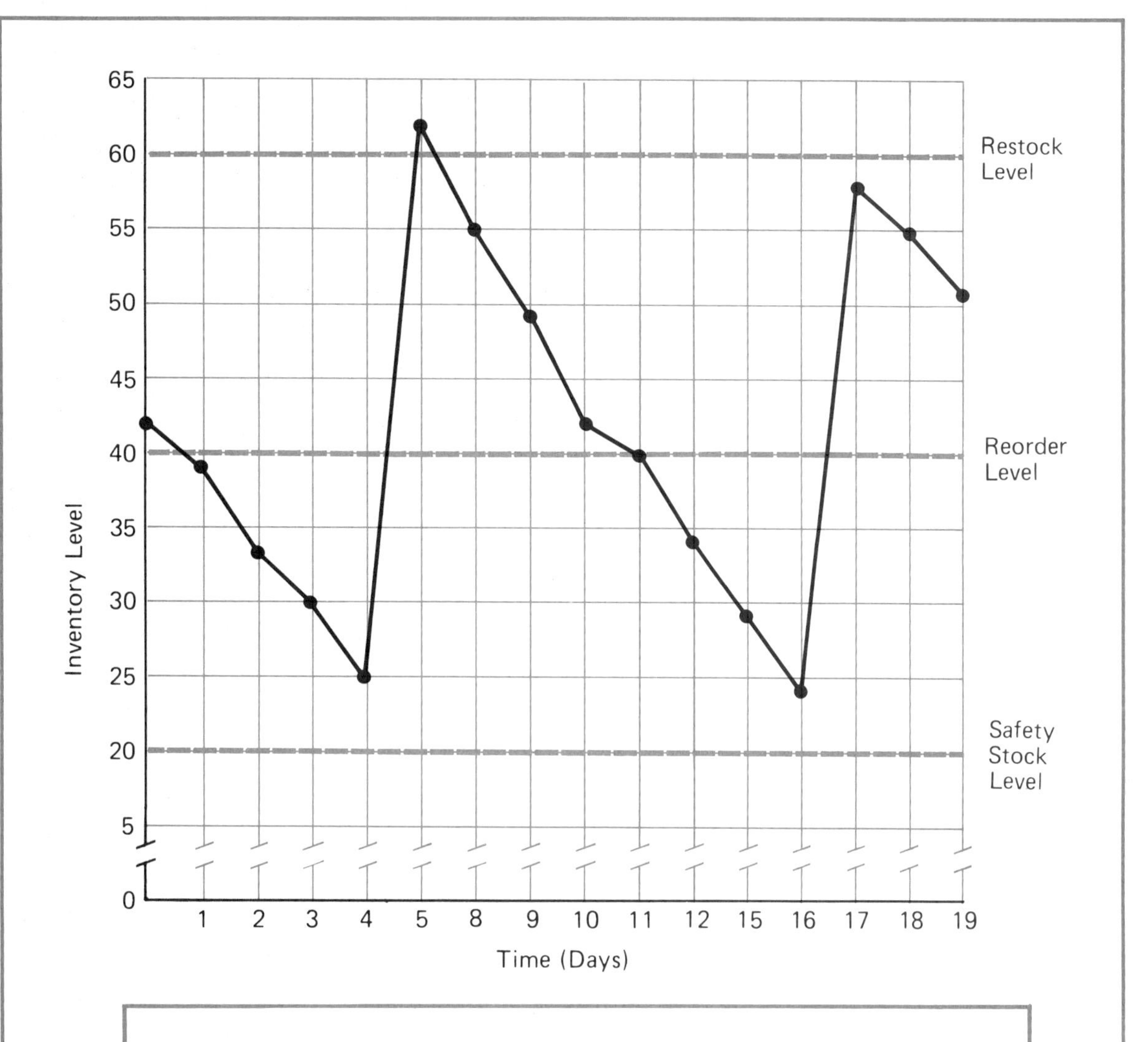

Date	Qty Used	Qty Recvd	On Hand	Reorder Point
March 1	3	0	39	March 1
March 2	6	0	33	
March 3	3	0	30	
March 4	5	0	25	
March 5	4	41	62	
March 8	7	0	55	
March 9	6	0	49	
March 10	7	0	42	
March 11	2	0	40	
March 12	6	0	34	March 11
March 15	5	0	29	
March 16	5	0	24	
March 17	6	40	58	
March 18	3	0	55	
March 19	4	0	51	

when orders are placed, show the reorder quantity ordered. Then, from your table, draw a graph that shows how inventory levels fluctuated for the three-week period described. Be sure to show the safety-stock and restocking levels.

2. Mary's Market sells approximately 12 gallons of milk per day. Lead time on orders is four days. Management holds a five-day safety stock. The restocking level is 145. As of October 4, the market had 145 gallons of milk on hand. On October 5, 11 were sold; October 6, 13 were sold; October 7, 15 were sold; October 8, 13 were sold; October 9, 10 were sold; October 12, 15 were sold; October 13, 13 were sold; October 14, 10 were sold; October 15, 12 were sold; October 16, 9 were sold; October 19, 24 were sold; October 20, 18 were sold; October 21, 11 were sold; October 22, 12 were sold; October 23, 12 were sold.

From the information above, determine on which days Mary will need to order more milk from suppliers. For each day you've chosen, determine how many gallons of milk Mary will need to order.

HOLDING AND MOVING INVENTORY

Mathematics and imagination are key ingredients in creative, efficient systems for holding and moving inventory. Your local supermarket provides some good examples. The next time you visit a market, make some observations based on the concepts that follow.

First, notice that the busiest section of the market is the front portion, where the checkout stands are located. This is where the heaviest shopper traffic is concentrated. The lightest traffic tends to be at the rear of the market.

Now, observe the contents of the shelves and sections in relation to location. The fastest-moving items usually are stocked near the checkout stands. These items include gum, candy, razor blades, photographic film, batteries, and certain magazines and newspapers. This type of merchandise is positioned for *impulse buying,* purchases made on the spur of the moment.

Major food items, such as meats, dairy products, and fresh produce tend to be included in shopping lists. These items are included in menu and meal planning. Shoppers will seek out such items, so they can be located at the rear or at the far sides of most markets.

Some products tend to be sold in low volume. Spices are a good example of products that usually involve purchasing decisions made before entering a market. Low-volume merchandise is situated in areas where shopper traffic is relatively light.

Seasonal Inventory Decisions

Merchandise at the front of a market also may be seasonal in nature. For example, items such as barbecue sauce and charcoal may be featured in special displays up front as summer begins. Shoppers look for such products when the outdoor cooking season arrives. However, those same items may be moved to the rear or less traveled sections of the store in January.

In January, the best display locations might be reserved for windshield ice scrapers, hot cereals, and cold medicines.

The point is that inventory decisions by retail merchants can be based on a variety of factors. Some of these factors are reflected in the names of familiar sale events, which include:

- Back to school
- Vacation specials
- Holiday specials
- Special occasions, such as graduations in June
- End of model year.

These kinds of marketing activities affect inventory decisions by retailers and their suppliers throughout the year. Furthermore, distributors and wholesalers must be prepared to move large quantities of merchandise to retailers at certain times. Efficient transportation and distribution systems are vital to all kinds of businesses.

Combining Efficiency and Economy

Two major factors in the inventory equation are overhead and labor. Close examination of these factors has led to some distribution decisions that may seem illogical at first glance. For example, consider the company that has one manufacturing plant on the East Coast but distributes its products nationwide.

In the past, this company had an efficient distribution system that included large warehouses at several key geographic locations. From these half-dozen locations, the company could ship its products quickly to any customer in the country.

In studying its operations, however, the company discovered some surprising facts. Each of its warehouse operations required large expenditures for overhead. Labor costs also were high, due to the necessity of handling the products on arrival and on dispatch.

The company concluded that the speed of air freight services made its warehouse system an unnecessary expense. Its products could be delivered as rapidly as necessary to any part of the country from a single location. The company closed its warehouses and now makes all shipments from its factory.

This type of system may not be best in all cases. Some products are transported more economically and efficiently by shipping in bulk to distribution points. Smaller quantities then are transported from the distribution points to customers.

A familiar example is your daily newspaper. Trucks carry large quantities of newspapers to dropoff points. Individual delivery workers pick up their inventory at those distribution locations.

Distribution systems vary, depending upon the type of product and the location of customers. Newspaper stands still operate in many inner cities, while home delivery of newspapers is more common in suburbs.

EXERCISE 12-8

Read each question carefully before calculating the answer. Round answers to the nearest cent.

1. Ludlum Enterprises operates three warehouses. Last year, the company spent $300,000 to distribute 148,000 tires. Management has performed a study to determine if one central warehouse would be less expensive to operate than three. The study has determined that it would cost approximately $2.10 per tire to distribute tires from one warehouse. Which method will cost less? What is the difference in cost between the two methods?

2. MOL Laboratories sends large amounts of finished goods directly to customers. Labor, shipping, and overhead costs average $0.50 per item. O.K. Distributors, a company with 14 warehouses, has offered to store and distribute MOL products. The cost would be $750 to handle and ship 400 items. Would MOL lose or save money by allowing O.K. Distributors to handle products? How much would MOL lose or save per item?

3. Marble Manufacturing currently operates three warehouses. Expenses at Warehouse A are $13,000 for labor, $3,500 for overhead, and $10,900 for shipping and handling. Warehouse B spends $12,000 on labor, $3,250 on overhead, and $13,400 on shipping and handling. Costs at Warehouse C are $1,500 for overhead, $6,450 for shipping and handling, and $4,500 for labor. Management is considering moving all distribution activities to one central warehouse. Expected costs for one warehouse would be $15,000 for overhead, $25,000 for labor, and $25,000 for shipping. Which method would cost Marble Manufacturing less money? What is the difference between labor costs of the two methods?

4. Nova Custom Clothes ships purchased items to customers through the mail. No customer is farther than 100 miles from the manufacturing facility. On the average, 2,000 items are shipped each month at a cost of $2.35 per item. The company wants to cut shipping costs by operating a delivery service. The cost of a vehicle would be $1,750 per month. Wages for a driver would cost $2,000 per month. Insurance and other

costs would average $150 per month. Would Nova save money by operating a delivery service? If so, how much? If not, how much would the company lose?

5. Management at Universal Rubber performed a study of distribution costs. The company operates four warehouses. To deliver one product to the final consumer, it costs Universal $5 in labor, $15 in shipping, and $7.40 for storage space. Management then estimated the cost of operating a central warehouse. The results showed that, on the average, storage space for one product would cost $8.50, labor would cost $9.50, and shipping would cost $9.50. If the company shipped 340,000 items per year, how much would it cost to distribute them through a central warehouse? Through the current four-warehouse system? What is the cost difference?

The Relay Teamwork System

In its brief but dramatic existence, the Pony Express provided future delivery systems with some useful concepts. One of these was the *relay system* of transportation. A Pony Express rider would run the horse as fast and as far as possible. This effort placed horse and rider at a relay station. The rider and his cargo would transfer to a fresh horse for the next leg of the delivery run. On longer runs, fresh riders would relieve their comrades at certain points. After a rest period, riders and horses would return the way they had come, now delivering in the opposite direction.

This general concept was adopted years ago by the nation's largest private parcel delivery company, United Parcel Service (UPS). Large trucks carry UPS parcels on intercity runs all over the United States. However, most UPS drivers can return home each night. A relay system makes this possible on long hauls.

Suppose that a truckload of parcels is being dispatched from Buffalo to Boston. This is a distance of approximately 465 miles. At the same time, a truckload of parcels is dispatched from Boston to Buffalo. Driving straight through, each driver could complete the run in slightly more than eight hours at 55 miles per hour. However, UPS would have to pay for overnight lodging for both of the drivers. Their meals away from home also would be at company expense.

Deliveries of new stock must be timed to ensure an adequate inventory level.

The company solved this problem by having the drivers meet at the approximate midway point between cities. There, they switch trucks. The driver who left Boston now takes the truck from Buffalo and returns home to Boston. The Buffalo driver does the same with the truck from Boston.

Mathematically, the concept is simple. The drivers spend the same amount of time on the road, without the expense of overnight lodging and meals. This type of arrangement is conducted all over the United States. Think about the savings for a company that employs hundreds of long-haul drivers.

For example, suppose that a parcel delivery company employed 500 drivers for intercity runs. Even with a special discount, motel charges alone might run $25 per driver. The cost for one night would be calculated this way:

$$500 \times \$25 = \$12{,}500$$

Thus, it would cost this company $12,500 a night to provide lodging for its 500 drivers. Now, suppose that, on average, each driver requires lodging three nights a week. The following equations show monthly and yearly costs:

$$\$12{,}500 \times 3 \times 4 = \$150{,}000$$
$$\$150{,}000 \times 12 = \$1{,}800{,}000$$

An annual motel bill of $1.8 million clearly illustrates the value of the UPS relay system.

EXERCISE 12-9

Read each question carefully before calculating the answer. Round calculations to two decimal places.

1. Dullus Trucking Company does not use a relay system. The company employs 25 drivers. Each driver spends 14 nights per month away from home. This costs the company $35 per driver, per day, for hotel bills and $15 per driver, per day, for meals. How much money would the company save, per year, by organizing a relay system for drivers? Assume that the system would eliminate hotel and meal expenses.

2. The Mississippi Parcel Company uses a relay system for deliveries within the state. The company was able to cut driver costs 35% by implementing the system. If the company currently spends $145,000 per year for driver costs, how much was spent before implementation of the relay method? How much did the company save?

3. Mobile Trucking Company is implementing a relay system for deliveries of less than 450 miles. Currently, the company spends $15,340 per month on hotel and food bills. This amount represents 35% of total shipping costs. What will shipping costs total after implementation?

4. Parallel Trucking uses a relay system and employs 156 drivers. Currently, total driver costs are $468,000 per month. Without the relay system, driver costs would increase $750 per driver, per month. Without the relay system, how much, including payment for meals and hotels, would each driver cost per month? By what percentage rate would total driver costs increase?

5. Golden Trucking employs 580 drivers. Driver costs average $2,175,000 per month. This includes hotel and food bills. When Golden Trucking implemented a relay system, driver costs were reduced 25%. On the average, how much, per month, was each driver spending on hotel and food bills?

Parcel delivery service has become highly competitive and efficient, due to the demand for quick delivery of business packages.
COURTESY OF UNITED STATES POSTAL SERVICE

Flying Freight in a Star Pattern

Competition has become keen among those parcel delivery companies that guarantee overnight delivery between major cities throughout the United States. To serve hundreds of possible destinations, these companies constantly strive to improve the efficiency of their delivery systems.

For example, Federal Express some years ago adopted a *star pattern* to move its customers' parcels throughout the country. The speed of jet transportation makes this star pattern an efficient system.

Think of the pattern as a hub from which spokes radiate in many directions. All shipments move along the spokes to the hub and then back along other spokes to their destinations.

Federal's system works this way: Every parcel is flown from its origination point to Federal's central distribution place in Memphis, Tennessee. As parcels arrive, they are processed and routed to flights that will take them to their destinations. Most of this activity takes place during the late night and early morning hours. Both company aircraft and commercial flights are used to move parcels to and from Memphis.

You might ask, "Why Memphis?" Geographic location is one major factor in the choice of Memphis. As the map in Figure 12-2 illustrates, Memphis is one of the most centrally located communities in the country. Consider the air distances from Memphis to some major cities:

City	Mileage
New York	964
Boston	1,139
Washington, D.C.	742
Miami	860
Chicago	491
Detroit	610
Houston	469
Denver	880
Los Angeles	1,619
San Francisco	1,807
Seattle	1,870

From Memphis, Seattle is the most distant major city in the United States. Thus, flying time between Memphis and any point in the continental 48 states is no more than about four hours. Since Memphis is located in the Central Time Zone, the greatest time differential, for cities on the West Coast, is only two hours. In theory, it should take no longer than six hours for a package to reach Memphis from any originating point.

To illustrate, suppose you wanted to send a package from Los Angeles to San Francisco. You have Federal pick up your package by 5 p.m. It is to be delivered no later than 10 a.m. the following day.

The two California cities are less than 340 miles apart by air. It would seem logical to send a parcel direct on a flight requiring less than an hour. However, this type of service would require costly processing and routing operations in each city. The inventory equation applies here in dramatic fashion. Think of the overhead and labor costs involved in having such space and operations duplicated in dozens, or even hundreds, of locations. It is far less costly, and more efficient, to route all shipments through a central location.

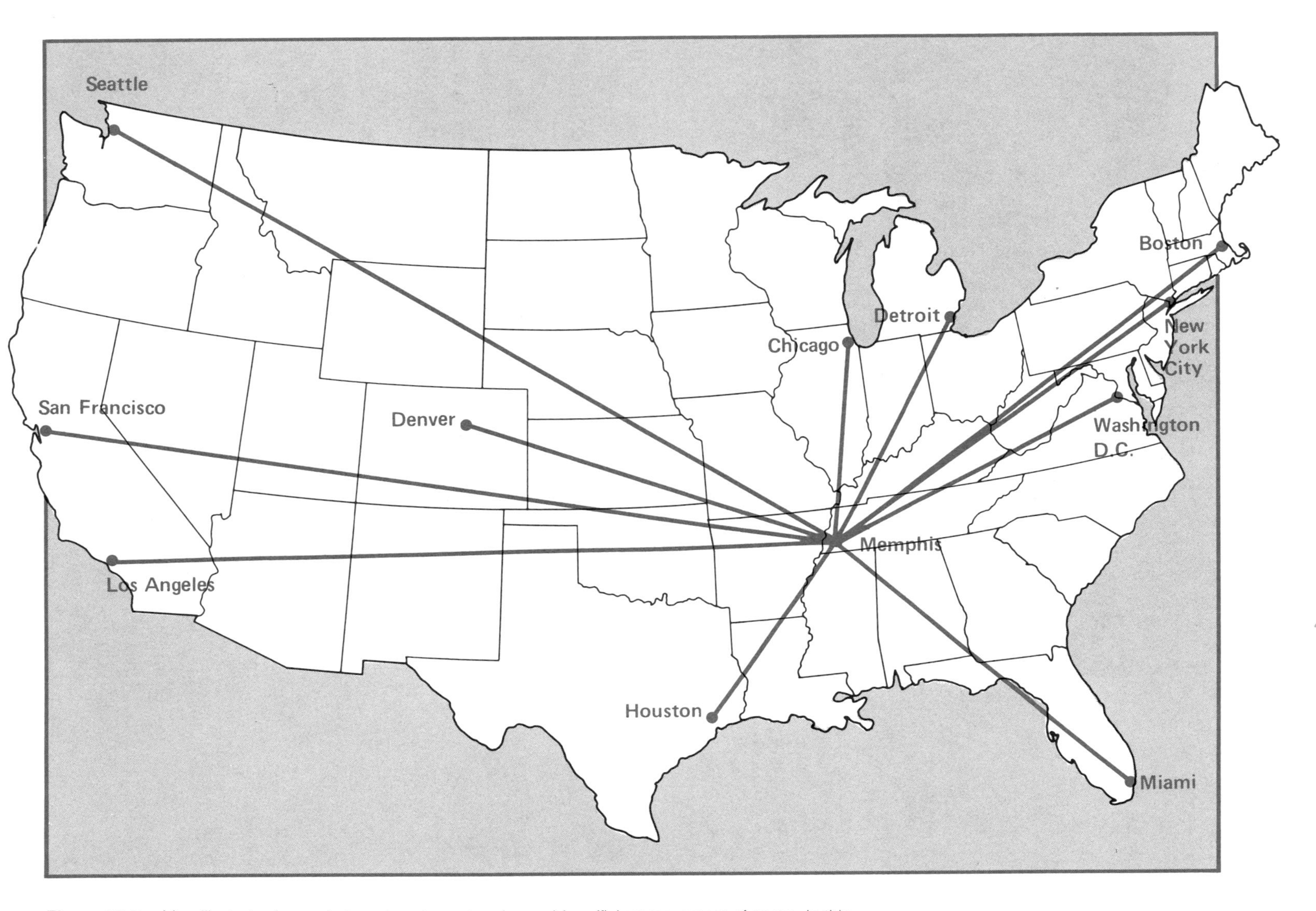

Figure 12-2. Map illustrates how a hub-and-spoke system is used for efficient movement of cargo. In this example, all shipments are routed to Memphis and then forwarded to their destinations.

Hub-and-spoke schedules enable airlines to operate with as full a passenger load as possible. Passengers having the same destination are brought together from many different originating points. Thus, passengers leaving from Pittsburgh may have to change planes instead of flying non-stop to their destinations. COURTESY OF CARNEGIE LIBRARY OF PITTSBURGH

'Have You Been to Atlanta Lately?'

Several major airlines also operate *hub-and-spoke schedules*. Examples are Eastern and Delta Airlines, which have hubs in Atlanta. Trans-World Airlines has its hub in St. Louis, and Continental Airlines has hubs in Denver and Houston.

Passenger airlines operate on a basis of passenger miles. Profitable operation requires that as many seats as possible be filled for each flight. More than a few unoccupied seats can mean that a flight loses money. The hub-and-spoke concept helps to fill airplanes.

Suppose you wanted to fly from Houston to Miami. If you flew on Eastern or Delta, your flight would go from Houston to Atlanta. You would change planes in Atlanta to fly to Miami. That may seem inconvenient. However, with the speed of jet travel, the time lost is minimal.

Think about it from the airlines' standpoint. There probably are dozens of other travelers who want to reach Miami. These other people may be traveling from New York, Boston, Chicago, St. Louis, Baltimore, and several other locations. They will all join you in Atlanta for that flight to Miami. The Atlanta-to-Miami flight is not likely to have many empty seats. The same is true for flights leaving Atlanta for other major terminals.

SUMMING UP

- Inventory is the stock of merchandise sold and supplies used by a business.
- Costs are involved in acquiring, handling, and storing merchandise. These costs must be held to a minimum.
- On the other hand, it is equally important to maintain safe levels of inventory. ''Out of stock, out of business'' is an old saying.
- Inventory problems can be solved with the aid of an inventory equation: Inventory costs equal the sum of overhead, labor, cost of the inventory, and cost of money borrowed to purchase inventory. This equation also helps in making other business decisions, such as whether to expand.
- Overhead includes cost of land and buildings, utilities, insurance, and other expenses of storing quantities of merchandise. Labor costs are involved each time someone has to handle inventory.
- Inventory is a factor all along the chain of supply that delivers goods and services to consumers.
- Lead time for merchandise deliveries is a vital factor in restocking inventory. Weekly or monthly sales rates are another important factor. Also important is a provision for safety stock, which helps avoid running out of inventory.
- Another consideration in restocking is unit price. Volume discounts, supplier price changes, and new-model introductions can affect order quantities.
- Mathematics can be used to present information on sales performance and inventory volume. Plotting sales and inventory data on a graph provides a method to trigger restocking activity. An order is placed when inventory reaches a certain level.
- Handling inventory includes decisions on how to display merchandise. Seasonal items may be featured at appropriate times. Items that promote impulse buying usually are displayed in the highest traffic areas of markets.
- Transportation and distribution systems are designed for cost savings as well as efficiency. An example of a cost-effective distribution system is a driver relay system such as that used by United Parcel Service. Another example is the hub-and-spoke systems used by some airlines and by overnight parcel delivery companies.

TERMS THAT COUNT

inventory
taking inventory
flux
flooring
chain of supply
lead time
safety stock
gross
impulse buying
relay system
star pattern
hub-and-spoke schedule

TESTING YOUR WORKING KNOWLEDGE

Read each question carefully before calculating the answer. Round answers to the nearest cent or to two decimal places.

1. Auto Publishing operates three warehouses. Total inventory costs for all three are $150,000. Of this amount, labor represents 35%, overhead accounts for 10%, and cost of items is 55%. Management wants to move inventory to one, large warehouse. Studies have predicted that, in one location, overhead costs would increase 50%, but labor would decrease 25%. Cost of inventory also would decrease 10% because larger quantities could be ordered. If the company moves inventory to one central location, will inventory costs be less or more? By how much?

2. Last year, Cloud Industries spent $35,000 on overhead, $15,000 on labor, $155,000 on inventory items, and $30,000 to borrow money. What percent of inventory costs did overhead represent? Labor? Inventory items? Cost of borrowing money?

3. Current yearly inventory costs at Hercules Manufacturing are $25,000 on inventory items, $3,500 on overhead, and $24,000 on labor. Management is determined to reduce inventory costs by 35% and cut storage space in half. As a result, labor costs were reduced 25%, overhead costs were reduced 25%, and inventory costs were reduced 40%. Did management achieve its goal? By what percentage were inventory costs actually reduced?

4. The Shoe Factory sells a pair of tennis shoes to shoe distributors for $8.65. Shoe distributors raise, or mark up, the price 25% and sell shoes to local warehouses. Local warehouses mark up the price 25% and sell shoes to shoe stores. Stores mark up prices 50% and sell shoes to consumers. How much do consumers pay for shoes made by the Shoe Factory?

5. Peter's Plastics performed a study to examine the costs of operating four warehouses. The result was this: To deliver one product to the final consumer, it cost Universal $8 in labor, $13 in distribution, and $2 for storage space. The company then performed another study to find out the cost of

operating one central warehouse. The results showed that storage space for one product would cost $1.50. Labor would cost $9.50. Distribution would cost $10.50. If the company ships 170,000 items per year, how much would it cost to distribute them through a central warehouse? Through the current four-warehouse system? What is the cost difference?

6. Wendy sells an average of 25 cowboy hats per week. Currently, Wendy has 175 cowboy hats in stock. Her supplier has overstocked the hats and has offered to sell them for $3.80 each. Usually, the price is $4.20. How much money per hat will Wendy save by ordering her restocking level at the lower price?

7. Manny's Motorcycle Repair keeps a safety stock of 25 rebuilt motors. Rebuilt motors take four weeks for delivery. Manny sells five rebuilt motors per week. What is the reorder level for rebuilt motors?

8. The Corner Grocery sells an average of 27 soft drinks per day. The store keeps a 14-day safety stock on hand. Orders take 4 days for delivery. What is the reorder level for soft drinks?

PRESENTING INFORMATION

Read the following question carefully. Then present the information in the format required.

Mountain Market sells approximately 25 gallons of milk per day. Lead time on orders is four days. Management holds a five-day safety stock. The restocking level is 250. As of August 4, the market had 245 gallons of milk on hand. On August 5, 23 were sold; August 6, 26 were sold; August 7, 28 were sold; August 8, 21 were sold; August 9, 25 were sold; August 12, 27 were sold; August 13, 20 were sold; August 14, 21 were sold; August 15, 23 were sold; August 16, 29 were sold; August 19, 19 were sold; August 20, 18 were sold; August 21, 28 were sold; August 22, 26 were sold; August 23, 25 were sold.

Create a table that shows daily inventory levels, quantities used, and quantities received for milk. Also, on days when orders are placed, show the quantity ordered. Then, from your table, draw a graph that shows how inventory levels fluctuated for the three-week period described. Be sure to show restocking level and safety-stock level.

SKILL DRILLS

A. Find the sum in each problem.

1. $4{,}456.58 + 3{,}533.32 + 657.87 + 7{,}812.21 =$
2. $1.4356 + 33.531 + 436.097 + 97.63226 + 99 =$
3. $125.934 + 748.456 + 652.784 + 4{,}533.339 =$
4. $113.5647 + 65{,}434 + 343.562 + 3{,}564.768 + 3.283003 =$
5. $0.387 + 0.493 + 0.3215 + 32.143 + 23{,}321.00074 =$

B. Find the difference in each problem.

1. $23{,}432 - 12.2431 =$
2. $3.243 - 2.0000901 =$
3. $2.98225 - 1.4300534 =$
4. $5.43676 - 3.300769 =$
5. $6{,}435{,}445.748 - 622{,}462.0087 =$

C. Find the product in each problem. Round calculations to three decimal places.

1. $0.04 \times 3.001 =$
2. $0.83 \times 5.4987 =$
3. $34.3468 \times 0.21 =$
4. $87.002 \times 1.332 =$
5. $87.93 \times 12.9709 =$

D. Find the quotient in each problem. Round calculations to two decimal places.

1. $32.59 \div 2.9 =$
2. $54.74 \div 23.65 =$
3. $484 \div 78.92 =$
4. $43.001 \div 5.61 =$
5. $0.198 \div 3.609 =$

P A R T
IV

Mathematics and Your Environment

IN THIS PART

Look around your home. Mathematics was involved in all areas of its design and construction. Carpenters, plumbers, and electricians all use mathematics in their work. Laws that set minimum building standards are based on mathematical calculations.

The role of mathematics in structures does not end with the completion of construction. For example, you might decide to have new carpeting installed. Mathematics would be used to determine the amount of carpeting needed.

Mathematics is equally important outdoors. Unit 14 discusses how mathematics is related to the life-support systems of an area. Many of the numbers are quite large, matching the size of the projects being described. For instance, did you ever wonder how many truckloads of earth were removed from a large excavation?

Acres of land, acre-feet of water, and large volumes of materials are all covered in this part. You also will be introduced to the advantages of containerized cargo, a fast-growing trend in freight transportation.

13 Mathematics Indoors

YOUR LEARNING JOB

When you have completed the exercises and assignments for this unit, you should be able to:

- ☐ Calculate shapes, sizes, and slopes for parts of structures.
- ☐ Estimate amounts of materials needed for structures.
- ☐ Calculate electrical usage or needs based on the relationships among voltage, amperage, and wattage.
- ☐ Use area calculations to estimate flooring needs and room capacities.
- ☐ Explain how degree day units are used to estimate heating fuel requirements.

WARM-UP

A. Find the sum in each problem.

1. 4,451.03 + 1,242.8 + 741.213 + 5.014 + 2,009.009 =
2. 34.029 + 451.035 + 3,019.54 + 7,293.002 + 43.801 =
3. 756.049 + 67.2202 + 1,293 + 5.0069 + 56.0231 =
4. 8,493.93 + 45.209 + 1,102.003 + 0.4932 + 34.928 =
5. 345.019 + 194.929 + 65.2927 + 27.092 + 5.0192 + 29.2038 =

B. Find the difference in each problem.

1. 486.45 – 32.0192 =
2. 4,536.098 – 4,236.1121 =
3. 65.4326 – 1.00293 =
4. 2.9473 – 2.039275 =
5. 341.425 – 54.79373 =

C. Find the product in each problem. Round calculations to three decimal places.

1. 53.012 × 34.019 =
2. 112.01 × 39.092 =
3. 93.039 × 2.934 =
4. 165.274 × 74.38 =
5. 578.0394 × 3.0293 =

D. Find the quotient in each problem. Round calculations to one decimal place.

1. 387.029 ÷ 45 =
2. 21.308 ÷ 4.34 =
3. 64.2932 ÷ 5.42 =
4. 54 ÷ 3.2983 =
5. .001 ÷ 5674.93 =

MATHEMATICS AND STRUCTURES

Mathematics is used extensively in every phase of the construction industry. Decisions involving shapes, sizes, and materials all require mathematical calculations. Think about planning your own house. You would use mathematics to make many decisions, including:

- Shape of the structure
- Number of rooms
- Square feet of floor space for each room
- Number and location of doors and windows
- Type of roof and pitch
- Heating and/or cooling requirements
- Number and location of electrical service outlets.

The list could go on and on. With few exceptions, such as color and materials, mathematics would be involved in every decision you would make.

For example, think how mathematics is used to determine the *pitch* of a roof. Pitch is the slope, or slant, for part of a structure. Roofs on most small structures, such as residences, are sloped to shed moisture. Several considerations are involved in determining the pitch of a roof. In cold climates, roof pitches tend to be steep to prevent excess buildup of snow in winter. Deeply packed snow can exert a dangerous crushing weight, or pressure, on a flat roof.

Other considerations include visual appearance, desired attic space, and building code requirements. A building code is a set of rules specifying construction standards in a community. County and city governments usually adopt and administer building codes.

Angles and Structures

Pitch, or slope, is determined by the *angle* the roof forms with the *horizontal.* An angle is the shape formed by two lines that extend in different directions from the same point. The term, horizontal, describes a flat line or level surface. For example, the classroom floor is a horizontal surface. The form of mathematics that deals with surfaces, lines, angles, shapes, and sizes of objects is called *geometry.*

To understand how angles are measured, think about a circle. The exact center of the circle also can be the beginning point of an angle. Start with a horizontal line drawn from the center of the circle, as shown in Figure 13-1. If that line were moved completely around the circle, it would travel 360 *degrees.* In geometry, a degree is $\frac{1}{360}$ of a circle.

Next, notice what happens in Figure 13-1 when the line is moved one-fourth of the way around the circle. The horizontal line now becomes *vertical.* Vertical means straight up and down. A simple calculation gives the number of degrees in this angle:

$$360 \times \frac{1}{4} = \frac{360}{4} = 90$$

Thus, the angle formed between a horizontal line and a vertical line contains 90 degrees. An angle of 90 degrees is called a *right angle.*

If the vertical line were moved another 90 degrees, or 180 degrees from its beginning, a straight, horizontal line would be formed. The line would be halfway around the circle, or 180 degrees. Therefore, an angle of 180 degrees is a straight line.

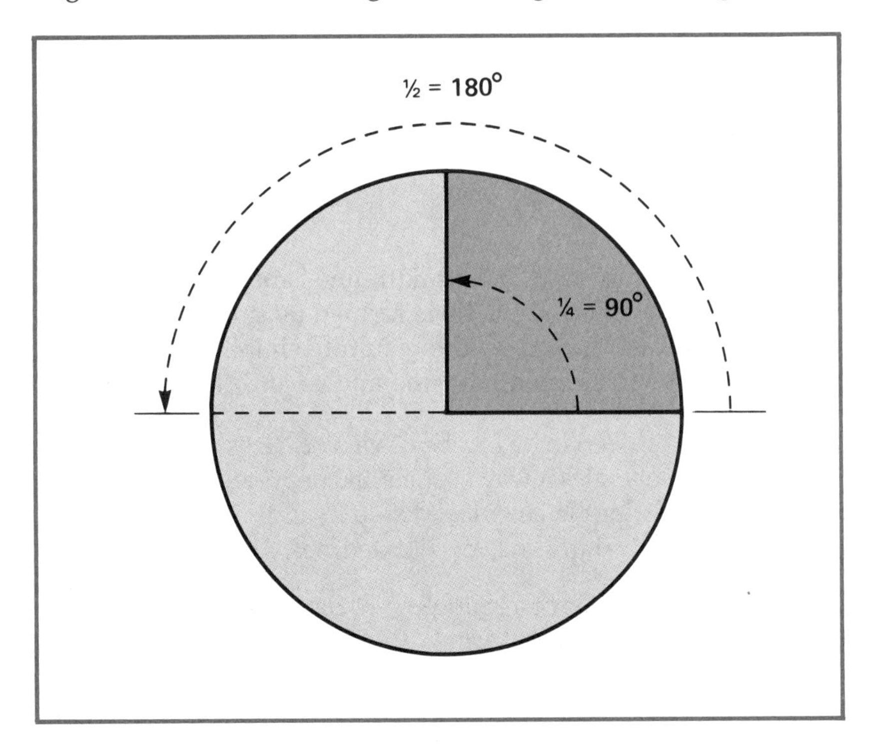

Figure 13-1. A right angle is formed by rotating a horizontal line one-fourth of the distance around a circle. A circle contains 360 degrees. Thus, a right angle is one-fourth of 360, or 90 degrees.

EXERCISE 13-1

Read each question carefully before calculating the answer.

1. When a vertical line crosses a horizontal line, what kind of angle is formed? How many degrees is it?

2. If a circle is divided into five equal angles, how many degrees will each angle be?

3. If a right angle is divided into two equal angles, how many degrees will each angle be?

4. What term describes a line or surface that runs straight up and down?

5. If a right angle is divided into three equal angles, how many degrees will each angle be?

6. If a circle is divided into two angles, one of which is a right angle, how many degrees will the other angle be?

Triangles

Now, look at Figure 13-2. If a straight line is bent at two places, two angles are formed. If the lines formed by the two angles intersect, or cross one another, the resulting shape is a *triangle.* A triangle is a shape containing three angles and three sides.

The angles in a triangle can be of various sizes. However, the sum of the three angles always equals 180 degrees. In Figure 13-3, the angles of a triangle are placed side by side. The sum of the angles forms a straight line, or 180 degrees.

If one of the angles of a triangle is 90 degrees, or a right angle, the triangle is called a *right triangle.* Right triangles are very useful shapes in mathematics, and especially in structures. A right angle, for example, is the angle between the floor and a vertical

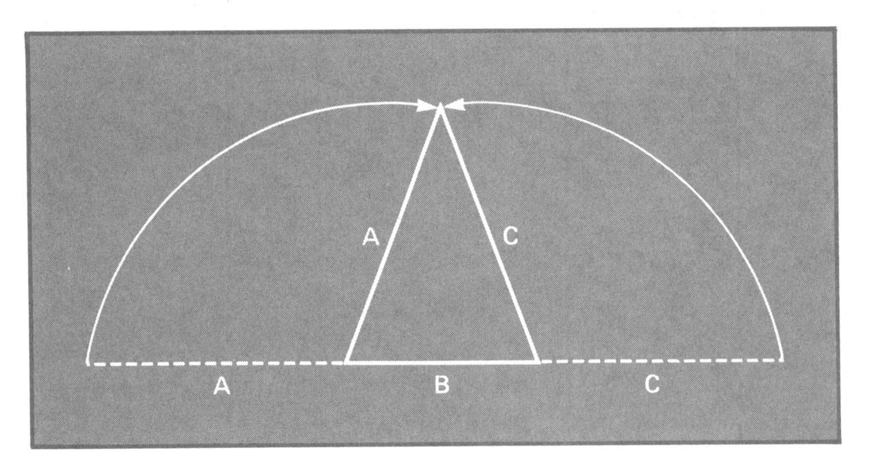

Figure 13-2. Bending a straight line at two places creates two angles. If the lines intersect, a third angle is formed. The resulting shape is a triangle, which is a three-sided figure.

wall. Another very useful fact about right triangles is the relationship among the sides. The *hypotenuse* is the side opposite the right angle. The hypotenuse always is the longest side in a right triangle. The *square* of the length of the hypotenuse is equal to the sum of the squared lengths of the other two sides. In geometry, this formula is called the *Pythagorean Theorem.* Its creator, Pythagoras, was a Greek philosopher and mathematician who lived during the sixth century B.C.

To square a number means to multiply the number by itself. For example, the square of 16 is: $16 \times 16 = 256$. In reverse, the *square root* of 256 equals 16. Further, the square of 4 is 16 ($4 \times 4 = 16$) and the square root of 16 equals 4.

The Greeks used Pythagoras' Theorem to design and construct such structures as arches and domes. Since then, the theorem has been used to create some of the most beautiful structures in the world.

Figure 13-3. Placing the three angles of a triangle side by side forms a straight line. This is because the sum of the three angles equals 180 degrees.

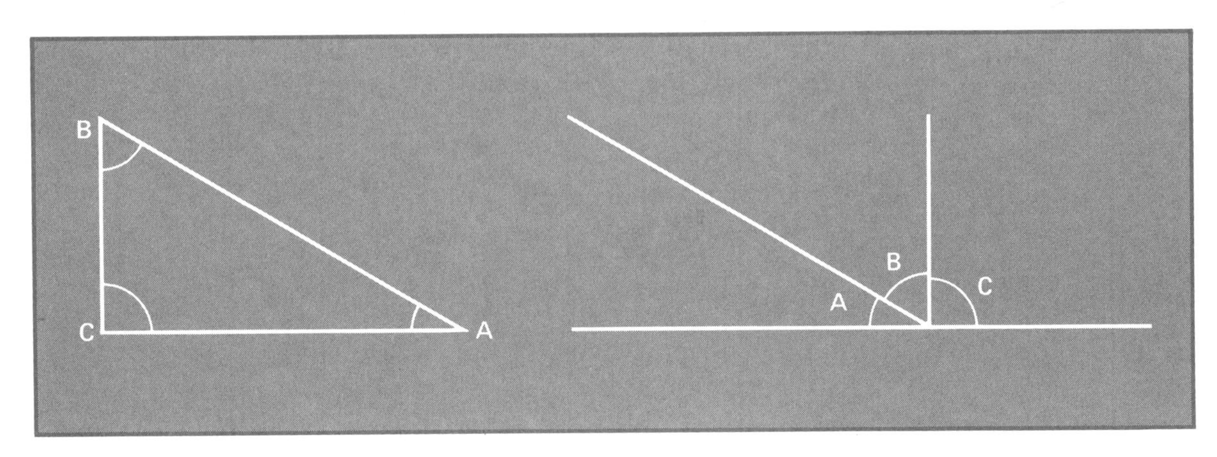

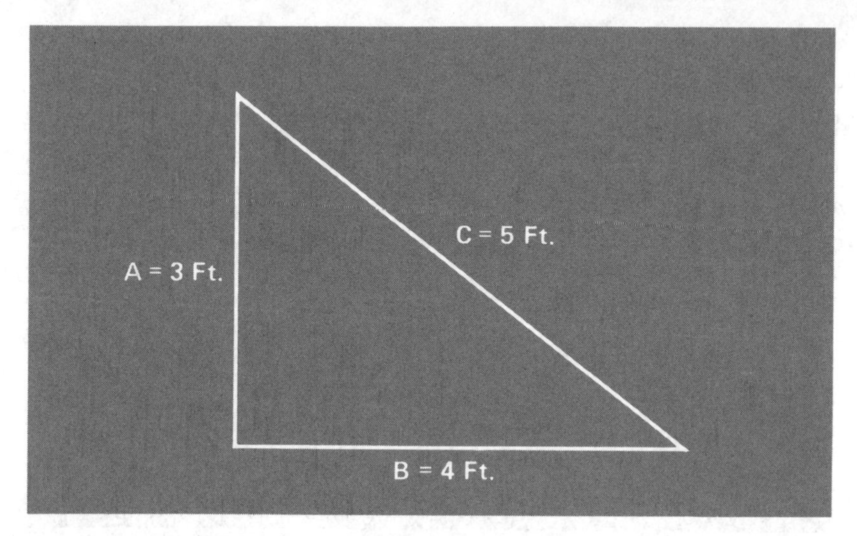

Figure 13-4. The hypotenuse of a right triangle is the side opposite the right angle. The square of the hypotenuse of a right triangle equals the sum of the squares of the other two sides. This rule is called the Pythagorean Theorem.

For an example of how the theorem is applied, look at the right triangle in Figure 13-4. The two shorter sides measure 3 feet and 4 feet, respectively. To find the hypotenuse, square the two shorter legs, add these values, and find the square root of their sum. For square root calculations, refer to the table in Appendix C, or use the square root function on a calculator. Square root is indicated by the symbol $\sqrt{\ }$. The equation for solving the problem illustrated in Figure 13-4 is:

$$3 \times 3 = 9$$
$$4 \times 4 = 16$$
$$9 + 16 = 25$$
$$\text{Square root of } 25 = \sqrt{25} = 5$$

Suppose that you knew the length of the hypotenuse and one other side of the triangle. Could you find the length of the third side? To do this, square the hypotenuse and the other known side. Subtract the square of the shorter side from the square of the hypotenuse. Using the square root key of a calculator, find the square root of the result:

Side A = 5 feet
Side C = 9 feet (hypotenuse)
Find Side B

$$9 \times 9 = 81$$
$$5 \times 5 = 25$$
$$81 - 25 = 56$$
$$\text{Square root of } 56 = \sqrt{56} = 7.48$$

Side B = 7.5 feet

EXERCISE 13-2

Use the Pythagorean Theorem to calculate the length of the unknown side. Round calculations to one decimal place.

1. Side A = 4. Side B = 5. Find Side C.
2. Side C = 12. Side B = 5. Find Side A.
3. Side A = 2. Side B = 13. Find Side C.
4. Side A = 25. Side C = 35. Find Side B.
5. Side A = 67. Side B = 11. Find Side C.
6. Side C = 13. Side B = 6. Find Side A.
7. Side C = 47. Side A = 23. Find Side B.
8. Side C = 112. Side B = 56. Find Side A.

Using Right Angles for Structures

To understand the value of right triangle calculations, think about building a ramp for access to a loading dock. Suppose that the dock is 4 feet higher than the pavement. Also, suppose that you want the ramp to extend 10 feet from the vertical front wall of the loading dock. How would you calculate the length of the sloping surface of the ramp? Such a structure is shown in Figure 13-5.

If you view the ramp structure from the side, you can see that it forms a right triangle. Since you know the lengths of two sides, it is simple to calculate the third. In this case, you want to find the length of the hypotenuse of the right triangle. Let C equal the length of the ramp surface, or hypotenuse. The problem can be solved by this simple equation:

$$
\begin{aligned}
4^2 + 10^2 &= C^2 \\
(4 \times 4) + (10 \times 10) &= C^2 \\
16 + 100 &= C^2 \\
116 &= C^2 \\
10.77 &= C
\end{aligned}
$$

This calculation of a hypotenuse tells you that you need lumber at least 11 feet long for the ramp surface. On the job, you would cut the boards to fit.

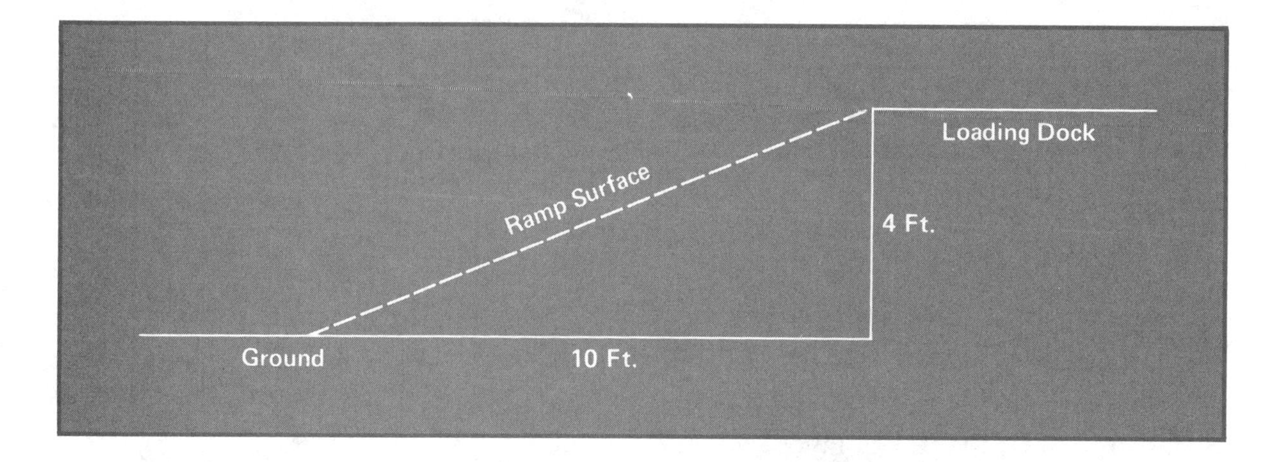

Figure 13-5. When any two sides of a right triangle are known, the third side can be calculated using the Pythagorean Theorem. This calculation is used in designing such structures as ramps.

EXERCISE 13-3

Read each question carefully before calculating the answer. Round answers to one decimal place.

1. Charlie is building a ramp to load and unload his motorcycle from his trailer. The ramp has to rise 2.5 feet, and the horizontal length must be 10 feet. Determine the length of the slanted ramp surface.

2. Mountain Junior College is building a ramp that will be used in carrying books to and from the student store. The horizontal length of the ramp will be 18 feet, and the slanted surface will be 24 feet. How high will the ramp rise?

3. Union Construction is building an on-ramp for a bridge. Over a horizontal distance of 150 feet, the ramp must rise 25 feet. How long will the top surface of the ramp be?

4. Janine wants to build a ramp to a 5-foot-high platform. She has a 14-foot board to use for the slanted surface. What will the horizontal length of her ramp be?

5. Faroah's Construction is building a walking ramp to connect the parking lot with the entrance to a concert hall. The parking lot is 30 feet below the level of the hall entrance. The ramp will have three sections. Each section will rise 10 feet and have a horizontal length of 75 feet. The ramps will zigzag to the hall

entrance. How far will concert goers have to walk to climb from the parking lot to the entrance?

Construction is a business in which almost all decisions are based upon mathematics.

6. You have been hired to build an exit ramp for a parking structure. The ramp must drop 20 feet over a horizontal distance of 200 feet. How long will the top surface of the ramp be?

7. Barrel Builders have been hired to construct a driveway from the street to the Smiths' garage. The horizontal distance from the street to the garage is 25 feet. The surface of the driveway will be 33 feet. How far above the level of the street is the entrance to the Smiths' garage?

8. The parking lot of Tallows Restaurant is 2 feet above the level of the street. The distance from the street to the parking lot is 30 feet. Determine the length of the driveway surface.

Calculating Pitch

Roofs have different pitches depending upon the design of the structure. As can be seen in the side view in Figure 13-6, a typical roof is formed by right triangles. Pitches can range from nearly flat to a 45-degree angle, which might be used on an A-frame house.

Instead of calculating angles, pitch is usually determined by using a simple formula. The dimensions used, shown in Figure 13-6, include:

- Span
- Run
- Rise.

The *span* is the width of the building being covered by the roof. The *run* is one-half the span, representing the center of the roof to the exterior wall. The run forms the bottom, or horizontal, leg of the right triangle. The *rise* is the distance from the tops of the walls to the top of the roof. The rise forms the vertical leg of the right triangle. The hypotenuse of the right triangle is then formed by the roof line.

There are four commonly used pitches for roofs. They are expressed as fractions: $\frac{1}{6}$, $\frac{1}{4}$, $\frac{1}{3}$, and $\frac{1}{2}$. A pitch of $\frac{1}{6}$ means that the rise is one-sixth the length of the span. A pitch of $\frac{1}{4}$ means that the rise is one-fourth the length of the span. These relationships are shown in Figure 13-7.

If you were building a relatively flat roof ($\frac{1}{6}$ pitch) over an 18-foot-wide house, the rise would be calculated this way:

$$\text{Rise} = \text{Span} \times \text{Pitch}$$

$$\text{Rise} = 18 \times \frac{1}{6} = 3$$

Thus, the rise would be 3 feet. The rise for a steeper roof ($\frac{1}{3}$ pitch) would be calculated in the same manner:

$$\text{Rise} = 18 \times \frac{1}{3} = 6$$

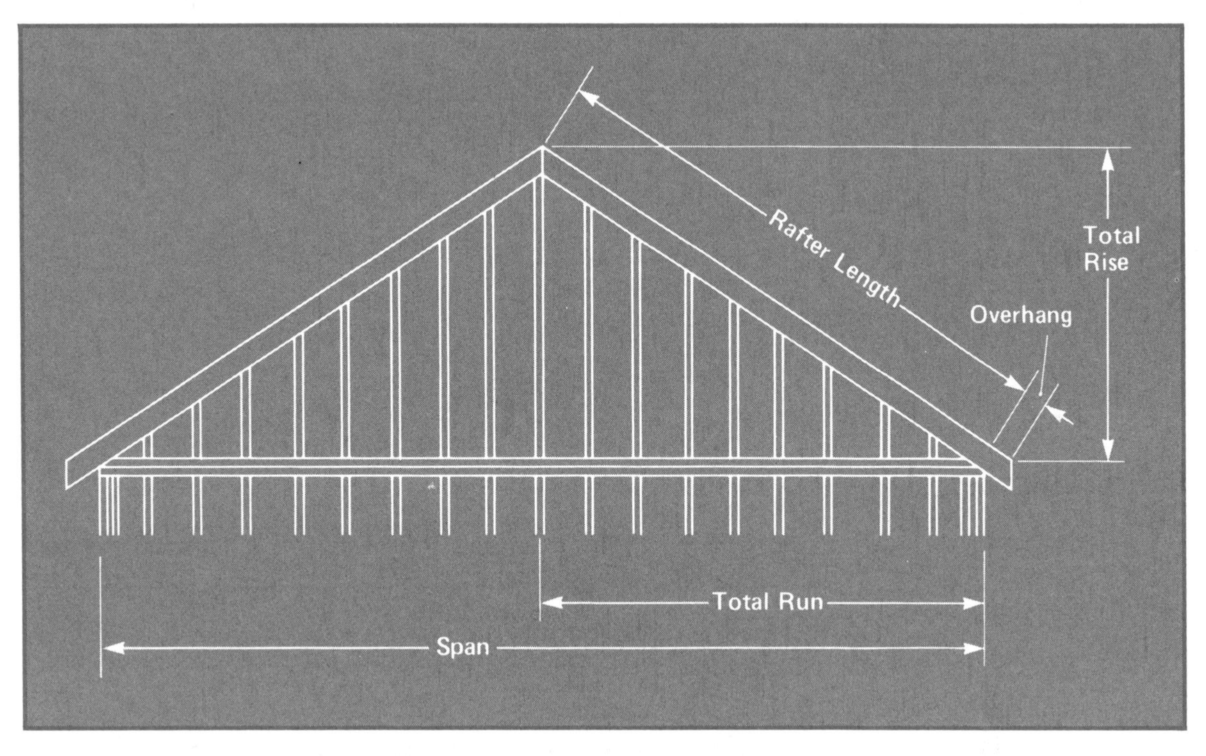

Figure 13-6. Each side of a sloped roof represents the hypotenuse of a right triangle.

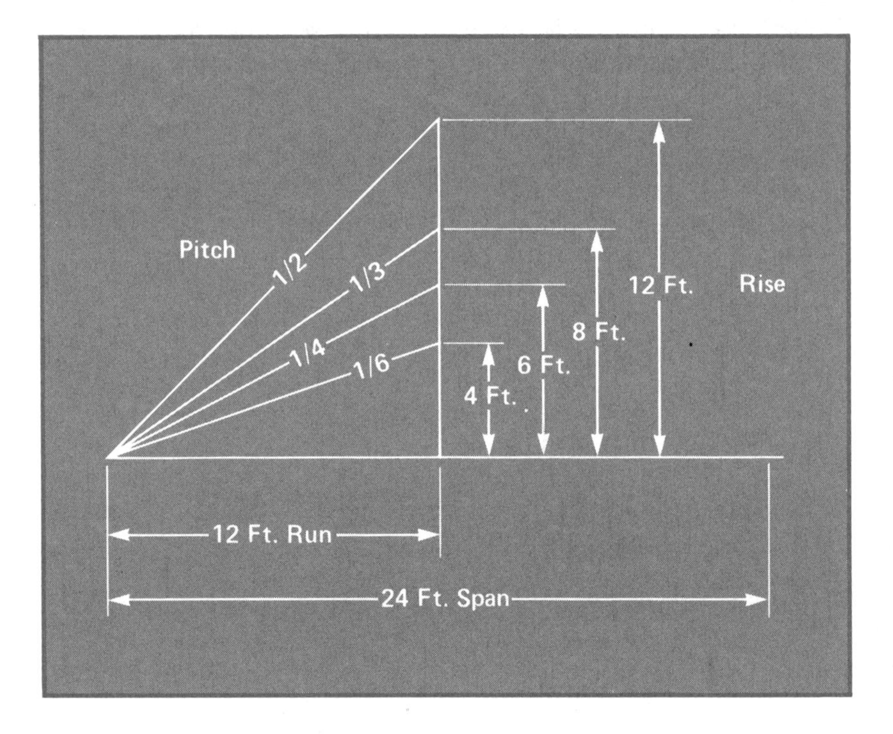

Figure 13-7. Relationships among the span, rise, and pitch of roofs are shown in this chart.

Using the same equation as a basis, the span or run can be determined:

$$\text{Rise} = \text{Span} \times \text{Pitch}$$
$$\text{Span} = \text{Rise} \div \text{Pitch}$$
$$\text{Run} = \frac{\text{Rise} \div \text{Pitch}}{2} = \frac{\text{Span}}{2}$$

To determine pitch:

$$\text{Rise} = \text{Span} \times \text{Pitch}$$
$$\text{Pitch} = \text{Rise} \div \text{Span} = \frac{\text{Rise}}{\text{Span}}$$

EXERCISE 13-4

Read each question carefully before calculating the answer. Round answers to one decimal place.

1. Otis is building a roof with a $\frac{1}{6}$ pitch on a house that is 24-feet wide. What will the rise of the roof be?
2. Barbara's house is 16 feet wide. The rise of her roof is 4 feet. What is the pitch?
3. The pitch of Chuck's roof is $\frac{1}{3}$. The rise is 6 feet. What is the span?
4. The span of Judith's roof is 36 feet. The pitch is $\frac{1}{2}$. What is the rise?
5. The pitch of Marcos' roof is $\frac{1}{4}$. The span is 28 feet. Determine the rise.
6. The rise of Oki's roof is 4 feet and the pitch is $\frac{1}{3}$. What is the span of Oki's house?
7. Gustavo is building a house with a 32-foot span. He wants the rise of the roof to be 8 feet. What pitch will be needed?
8. Ben is building a house with a 24-foot span. The pitch is $\frac{1}{4}$. Determine the rise of Ben's roof.

Using Carpenters' References

Carpenters also refer to pitch in terms such as ''4 in 12'' or ''6 in 12.'' These terms are based on the same principles described above. For example, a 4 in 12 pitch means that the roof rises 4 inches for each foot (12 inches) of run. On the 18-foot-wide house above, the run would be 9 feet. Multiply 9 × 4 to obtain 36 inches, or 3 feet, of rise. Now, divide the rise by the span:

$$\text{Run} = \frac{\text{Span}}{2} = \frac{18}{2} = 9$$

$$\text{Rise} = 9 \times 4 = 36 \text{ inches} = \frac{36}{12} = 3 \text{ feet}$$

$$\frac{\text{Rise}}{\text{Span}} = \text{Pitch}$$

$$\frac{3}{18} = \frac{1}{6}$$

The quotient is the fraction $\frac{1}{6}$. Thus, a 4 in 12 pitch is the same as a $\frac{1}{6}$ pitch. The same calculations can be made for other pitches. Examples include:

$$6 \text{ in } 12 = \frac{1}{4}$$

$$8 \text{ in } 12 = \frac{1}{3}$$

$$12 \text{ in } 12 = \frac{1}{2}$$

EXERCISE 13-5

Read each question carefully before calculating the answer. Round answers to one decimal place.

1. Using calculations, show that a pitch of $\frac{1}{4}$ is the same as a pitch of 6 in 12.

2. Using calculations, show that a pitch of $\frac{1}{3}$ is the same as a pitch of 8 in 12.

3. Using calculations, show that a pitch of $\frac{1}{2}$ is the same as a pitch of 12 in 12.

4. Jerry's house has a pitch of 4 in 12. The span of the house is 42 feet. What is the rise?

5. The rise of Julie's roof is 5 feet. The pitch is 6 in 12. What is the span?

6. The pitch of Leroy's house is 8 in 12. The span is 40 feet. Measured in feet, what is the rise?

7. Leonette owns a house with a 54 inch rise. The span is 27 feet. Determine the pitch. Express the answer in both carpenters' terms and as a fraction.

8. Jasmine's house is 28 feet wide. The pitch is 12 in 12. What is the rise when measured in inches? In feet? In yards?

Rafters: The Third Side of the Triangle

The pitch of a roof determines the length of the *rafters* in relation to the other sides of the right triangle. Rafters are the beams that support the roof (see Figure 13-8). Once you have determined the pitch of a roof, and its rise, you can calculate the required length of the rafters. The rafters are the hypotenuse in the right triangle. The solution involves squaring the rise and the run, adding their squares, and finding the square root of the sum.

For example, the 18-foot-wide house above has a run of 9 feet. If the pitch is 4 in 12 ($\frac{1}{6}$), the rise is 3 feet. Squaring the run and the rise gives 81 and 9, respectively. Their sum is 90, and the square root of 90 is approximately 9.5. Therefore, the length of the rafter will be 9 feet, 6 inches (9′6″) from the top of the roof to the top of the exterior wall.

The roof extends past the outer wall, of course, to permit proper moisture runoff, as shown in Figure 13-6. A standard overhang is 18 inches. Thus, you would add 18 inches, or $1\frac{1}{2}$ feet (1′6″) to the rafter length before cutting the lumber. As a result, the lumber for rafters would be cut into 11-foot lengths (9′6″ + 1′6″).

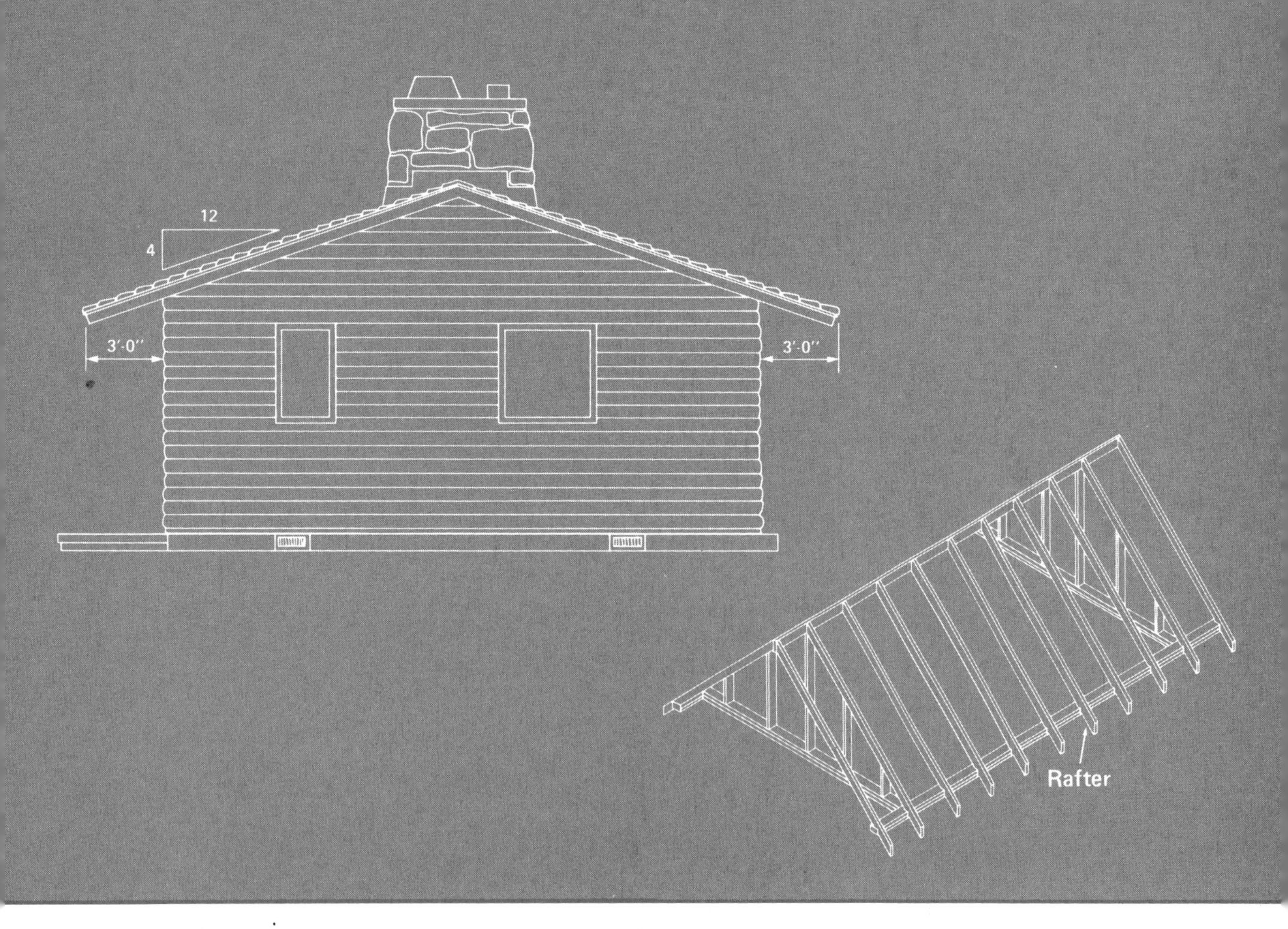

Sizing rafters and other framing lumber is not so precise as cutting wood for furniture and cabinets. In framing, small fractions of an inch can be ignored. In most cases, measurements can be rounded to the nearest inch. Square root calculations involving lengths measured in feet usually will produce a decimal remainder. Multiplying the decimal value by 12 will produce the number of inches.

Figure 13-8. A rafter becomes the third side of a triangle used to calculate the pitch and rise of a roof.

For example, you might want to calculate the length of rafters for a roof. Suppose the roof has a 4-foot rise and covers a house 30 feet wide. The rafter length from the roof peak to the side of the house is the hypotenuse of a right triangle. The square of the rise is 16, and the square of the run is 225. Adding the two numbers produces 241, which is the square of the hypotenuse. The square root of 241 is 15.524174. Rounding to two decimals leaves a rafter length of 15.52 feet, not including the overhang. Multiply 0.52 by 12 to convert the remainder to inches: 6.24. Round again to the nearest inch, and the rafter length is 15 feet 6 inches. With overhang the rafter comes to 17 feet.

EXERCISE 13-6

Read each question carefully. Use the standard 18-inch roof overhang in all calculations. Round decimal remainders to two decimal places and convert to inches. Round converted measurements to the nearest inch.

1. Jill is building a roof on a house with a 28-foot span. The pitch is $\frac{1}{4}$. How long will the rafters be?

2. Bernard cut rafters for a house with a span of 24 feet and a pitch of 6 in 12. How long are the rafters?

3. You have been hired to cut rafters for a new housing tract. The rise of the roofs is 72 inches. The span of each house is 36 feet. What length will you cut the rafters?

4. Rita owns a house with rafters that are 24 feet long. The rise of the roof is 11 feet. What is the span of the house?

5. Roxanne cut rafters that were 12.5 feet long for a 20-foot-wide house. What was the rise of the roof?

6. Moe built a house with a 30-foot span. The pitch of the roof was 8 in 12. How long were the rafters that Moe used?

STRUCTURES AND STANDARDS

The construction industry has many standards that help maintain uniformity. For example, lumber and other structural materials are sold in standard sizes. Standard sheets of plywood, hardboard, and particle board are 4 feet by 8 feet. Lumber is graded according to the type and quality of wood. When lumber is rough cut, meaning cut from the log, it measures the full dimensions of its *nominal size.* Nominal size refers to the stated dimensions that identify the lumber. Thus, a 2-by-4 (2×4) rough board measures the full 2 by 4 inches. Rough boards, once dried, are dressed, or surfaced, for smoothness and to make them conform to standard, or actual, dimensions. Dressing or surfacing is a process of removing material from the surfaces of lumber. Thus, a 2×4 actually measures $1\frac{1}{2}$ by $3\frac{1}{2}$ inches after dressing.

Lumber sold in volume usually is measured by the *board foot.* A board foot is equal to a board 1 inch thick, 12 inches wide, and 12 inches long. A 2-inch-thick board that is 6 inches wide and 12 inches long also is considered a board foot. The following formula is used to calculate board feet:

$$\frac{\text{Thickness (in.)} \times \text{Width (in.)} \times \text{Length (ft.)}}{12} = \text{Board feet}$$

or

$$\frac{T'' \times W'' \times L'}{12} = \text{Board feet}$$

Lumber sizes are written numerically in the same order: thickness (inches), width (inches), and length (feet). (Abbreviations are used for inches and feet: 2″ means 2 inches, and 6′ means 6 feet.) At the lumber yard, these three numbers offer the information you need to determine the sizes of wood to buy.

EXERCISE 13-7

Read each question carefully. Then calculate the number of board-feet. Round calculations to two decimal places.

1. How many board-feet are in a single 4 × 8 sheet of plywood that is $\frac{1}{2}$ inch thick?

2. How many board-feet are in a 6-inch-thick board that is 8 inches wide and 96 inches long?

3. If a 2 × 4 contains 1 board-foot of lumber, how long is it?

4. O.K. Lumber received 9,600 board-feet of 4 × 8 sheets of plywood. Each sheet was $\frac{1}{2}$ inch thick. How many sheets of plywood were received?

5. How many board-feet are contained in a 10-foot 4 × 4?

Standards and Codes

Building codes specify minimum standards for materials and design of structures. Parts of building codes vary from place to place. Geography, climate, and local tradition affect regulations dealing with carpentry, plumbing, electricity, insulation, and other areas of construction.

One example of construction standards is the placement of *studs,* the 2-by-4s that form the framing of vertical walls. Studs are positioned 16 inches apart, measured at their centers.

When a house is being framed, walls are built on the ground and raised to a vertical position when finished. The first wall to be erected usually has a double-stud corner post at each end. Such walls usually are referred to as "front walls" or "back walls."

The adjacent walls, which are called "side walls," do not have double-stud corner posts. When a side wall is erected, the end studs are nailed to the corner posts of the front and back walls. Thus, two different equations are required to calculate the numbers of studs required for these different walls.

Figure 13-9 shows a typical front wall. Starting from the left end, a stud is positioned every 16 inches. Suppose that the overall length of the wall, in inches, is divisible evenly by 16. An example would be a wall that measures 20 feet long. Twenty feet is 240 inches, and 240 divided by 16 is 15. In such a case, the equation for determining the number of studs needed is:

$$\text{No. of studs} = \frac{\text{Length of wall (feet)} \times 12 \text{ (inches in 1 foot)}}{16} + 3$$

$$\text{No. of studs} = \frac{20 \times 12}{16} + 3 = \frac{240}{16} + 3 = 15 + 3 = 18$$

Note that the number 3 was added to the total of 16-inch spaces. The 3 represents the two studs at the left corner post and an additional one at the right corner post.

Most walls don't come out to a length divisible by 16, however. Remember, 16 inches is the maximum space between studs. The final space between a stud and the right corner post may be less than 16 inches. In other words, when you divide the

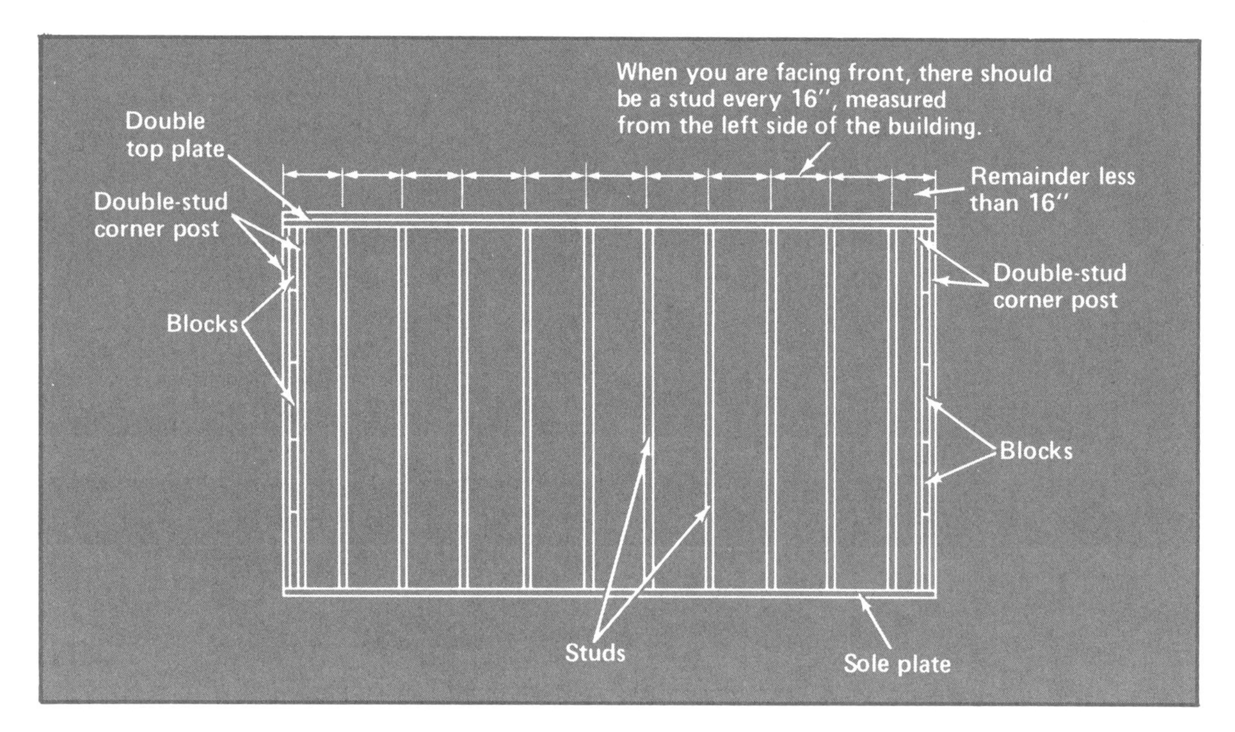

Figure 13-9. Studs are spaced 16 inches apart, center to center, along the length of a wall.

length by 16, there may be a remainder. Whenever there is a remainder, simply round up to the next whole number. Suppose you are building a wall 21 feet long. Multiply by 12 to obtain 252 inches. Divide by 16 to obtain 15.75 spaces. Round up to 16, add 3, and the requirement is 19 studs. The equation remains the same:

$$\text{No. of studs} = \frac{\text{Length of wall (feet)} \times 12}{16} + 3$$

$$\text{No. of studs} = \frac{21 \times 12}{16} + 3 = \frac{252}{16} + 3 = 15.75 \text{ or } 16 + 3 = 19$$

Figure 13-10 shows how a side wall is joined to a front or back wall. Since there are no corner posts, there are no extra studs to add to the equation. Simply divide the length of the wall in inches by 16. If there is a remainder, round up to the next whole number. This ensures that the final space between side wall stud and corner post will be 16 inches or less. The equation for the side wall is:

$$\text{No. of studs} = \frac{\text{Length of wall (feet)} \times 12 \text{ (inches in 1 foot)}}{16}$$

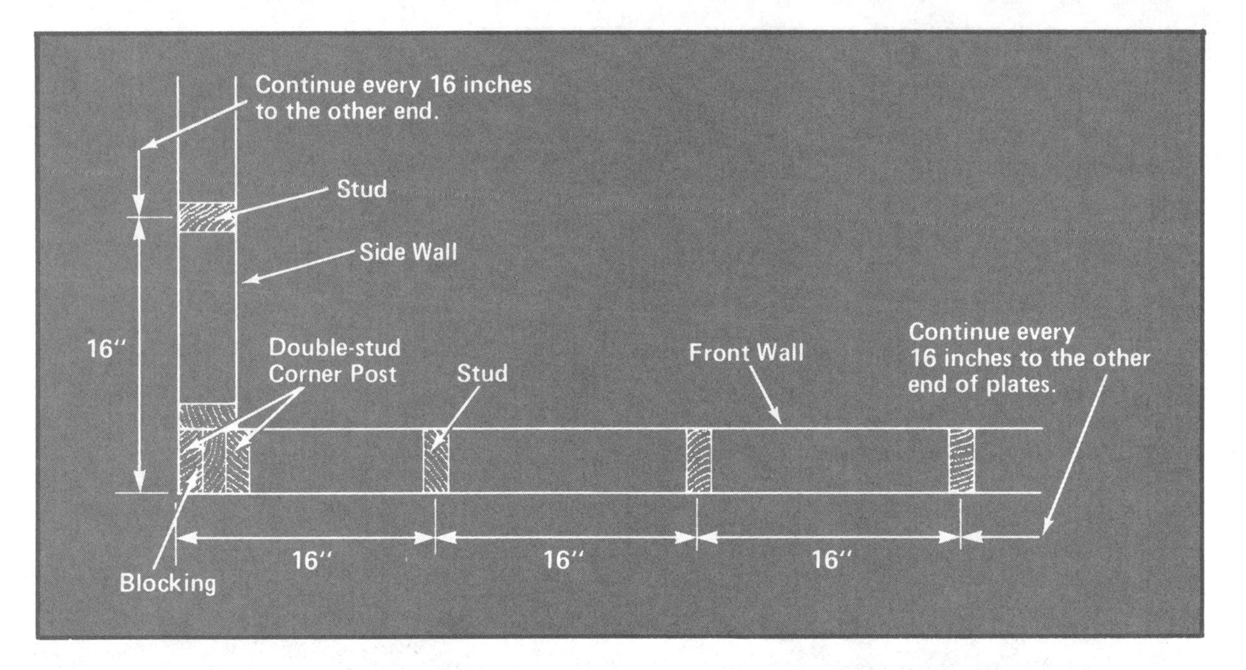

Figure 13-10. The first stud of a side wall typically is nailed to the corner post of a front wall.

EXERCISE 13-8

Read each question carefully before calculating the answer.

1. How many studs are required to construct a front wall 15 feet long?

2. Emilia is building a 12-foot-long side wall that is 8 feet high. How many studs will she need? She plans to cover the inside surface of the wall with 4-by-8 sheets of drywall. Drywall is material made of plaster that is similar to plywood in shape and function. How many sheets of drywall will Emilia need?

3. Craig is purchasing materials to frame a shed. The shed will be 10 feet square. How many studs will he need? The four walls will be 8 feet high. If he covers the insides of the walls with drywall, how many 4-by-8 sheets will he need?

4. Bonnie is putting new drywall on three walls in her bedroom. Each wall is 8 feet high. One wall is 12 feet long, and each of the other two walls is 10 feet long. How many 4-by-8 sheets of drywall will she need to purchase?

5. How many studs are required to build a front wall 36 feet long? If the wall is 8 feet high, how many 4-by-8 sheets of plywood will be needed to cover *both* sides of the wall?

Plumbing Standards

Standards also apply to plumbing equipment and fixtures. Water delivery pipes may be made of galvanized steel, brass, or plastic pipe. Copper tubing also may be used. Pipe and tubing are manufactured in sizes ranging from $\frac{1}{4}$ inch to $2\frac{1}{2}$ inches. However, the sizes commonly used in residences are $\frac{1}{2}$-, $\frac{3}{4}$-, and 1-inch diameters.

Cast iron and plastic pipe for drain, waste, and vent (DWV) systems come in diameters of 2 inches and larger. For home use, diameters of 2, 3, and 4 inches are most common. Cast iron pipe comes in 5- and 10-foot lengths.

Plumbing connectors and fixtures, such as valves and faucets, also come in standard sizes.

Electrical Standards

Standards for electrical installations usually are based on the National Electric Code (NEC). In many cases, however, local building codes are stricter to promote greater safety.

Electrical standards are exacting. Electric current travels in *circuits.* A circuit is a path from a *source* through a *conductor* to a *load* and back to the source. A source may be the electric line that feeds power into a house, or it may be a battery. A conductor is something through which electric current can flow, usually a wire. A load is an object, such as an electric light or an appliance, that uses electricity.

To understand how electricity flows in a circuit, think about water flowing through a hose. The comparison is not precise, but the basic concepts are similar. Three terms important in dealing with electricity are:

- Amperes
- Volts
- Watts.

An *ampere* is a unit used to measure the amount of electricity flowing past a given point in a circuit. Amperes commonly are called *amps.*

A *volt* is a measure of the pressure that is being used to push the electricity through a conductor.

Building codes list minimum standards for the design and construction of homes and other structures. COURTESY OF CINCINNATI DEPARTMENT OF BUILDINGS AND INSPECTION

A *watt* is a measurement of the amount of electrical power being used at any given moment.

The relationship among these three electrical terms can be expressed in simple equations:

$$\text{Watts} = \text{Amps} \times \text{Volts}$$

$$\text{Amps} = \frac{\text{Watts}}{\text{Volts}}$$

$$\text{Volts} = \frac{\text{Watts}}{\text{Amps}}$$

Thus, a 60-watt bulb on a 120-volt household circuit draws $\frac{1}{2}$ amp, based on the following calculation:

$$\text{Amps} = \frac{\text{Watts}}{\text{Volts}} = \frac{60}{120} = \frac{1}{2}$$

Knowing these basic facts can help you determine electrical power requirements and usage on most circuits.

EXERCISE 13-9

Read each question carefully before calculating the answer. Round calculations to two decimal places.

1. On the average, a refrigerator on a 115-volt circuit draws 4.75 amps. On the average, then, how many watts does the refrigerator use?
2. The microwave oven in Jill's house uses 600 watts on a 120-volt circuit. How many amps does her microwave draw?
3. If a 200-watt guitar amplifier draws 1.82 amps, how many volts does the circuit carry?
4. A computer draws 2 amps of electricity from a 120-volt circuit. How many watts does the computer use?
5. A photocopier in an office uses 200 watts on a 115-volt circuit. How many amps does it draw?
6. A 19-inch television set uses 79 watts on a 110-volt circuit. How many amps does the television set draw?
7. The light bulb in Bob's garage uses 150 watts of electricity on a 115-volt circuit. How many amps does the bulb draw?
8. The light bulb of a slide projector uses 500 watts on a 120-volt circuit. How many amps does it draw?

Electrical Outlet Requirements

According to the National Electric Code, any wall space wider than two feet must have at least one outlet. Outlets must be no more than 12 feet apart, measured horizontally along the walls. This rule is designed to minimize the use of extension cords. Knowing this, how many outlets would you install in a room that measures 20-by-15 feet? (This measurement means the two facing walls are 20 feet long, and the two walls connecting them are 15 feet long.) First, determine the total distance along the four walls:

$$(20 \times 2) + (15 \times 2) = 40 + 30 = 70$$

Then, divide the total wall measurement by the maximum distance between outlets:

$$70 \div 12 = 5.83$$

Round the solution to the next highest whole number. You would need a minimum of six outlets for a 20-by-15 room. These types of calculations are used in estimating jobs.

EXERCISE 13-10

Read each question carefully before calculating the answer. Round calculations to the next highest whole number.

1. Reggie is building a room that will measure 10 feet by 12 feet. What is the minimum number of electrical outlets Reggie should install in the room?

2. Jenny is building a room that will be 14 feet by 13 feet. How many outlets should she install?

3. Hercules Construction built a four-room house. Room A is 10 feet by 14 feet. Room B is 8 feet by 12 feet. Room C is 16 feet by 20 feet. Room D is 14 feet by 14 feet. How many outlets were required for each room?

4. Randy added two rooms to his house. One is 15 feet by 18 feet. The other is 12 feet by 16 feet. How many outlets was Randy required to place in each room?

5. Spectre Construction built a 15-unit apartment complex. Each unit, or apartment, was exactly the same. The kitchens were 8 feet by 8 feet. The living rooms were 20 feet by 12 feet. The bathrooms were 5 feet by 10 feet. Each bedroom was 10 feet by 12 feet. How many outlets were needed in each room? How many outlets were required for a single apartment?

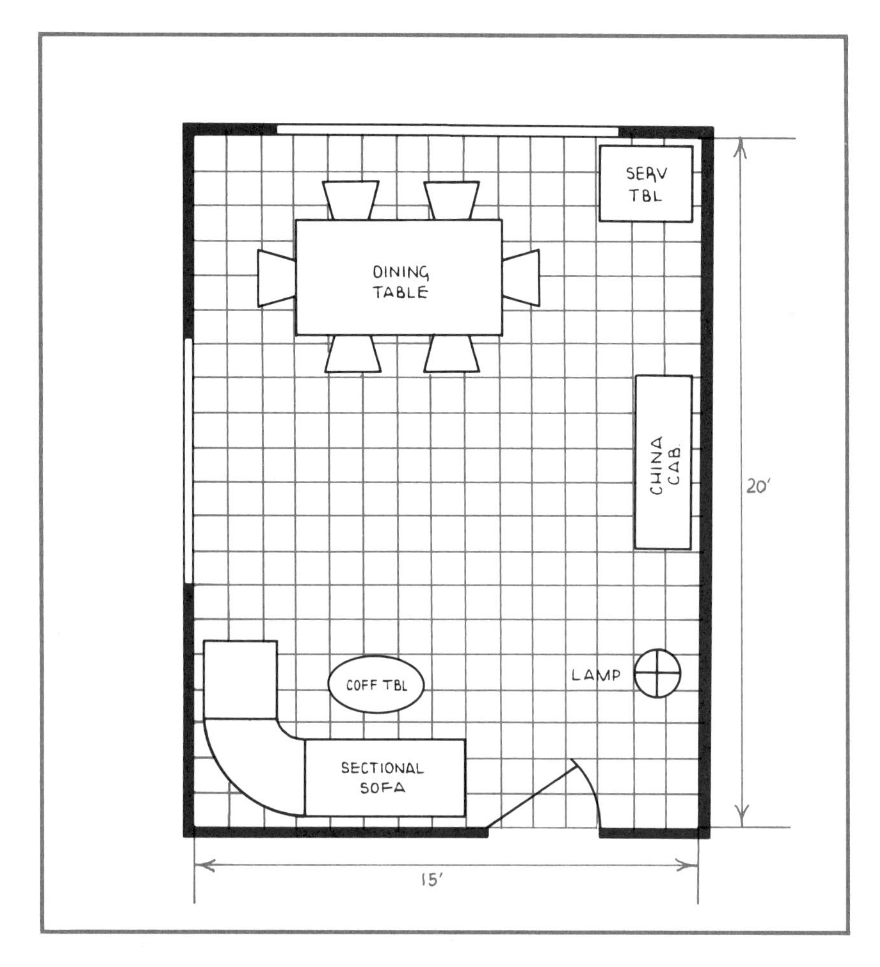

Area is determined by multiplying the length by the width. Most rooms, as shown in this plan, are four-sided areas.

MATHEMATICS AND AREAS

For structures, many decisions are made on the basis of area. Area is measured in square feet or square yards. Most rooms are in the form of *rectangles.* A rectangle is a four-sided area in which all four corners are right angles.

For example, consider a living room that measures 20 feet by 15 feet. The area of the room is determined by multiplying the length by the width. In this case, $20 \times 15 = 300$. The area of the room is 300 square feet.

Measuring Area

Suppose that you wanted to lay new carpet in a bedroom. Say that the bedroom measured 12 feet by 12 feet. Multiplying length by width produces an area of 144 square feet. However, floor coverings, such as carpeting and linoleum, generally are sold by the square yard. So, you need to determine the area in terms of square yards. There are two ways to do this.

One way is to divide each *linear* measurement by 3, since there are 3 feet in a yard. Linear means a straight line. The 12-by-12 room measures 4 yards by 4 yards. Multiply length by width, and the result is 16 square yards.

$$\frac{12}{3} = 4$$
$$4 \times 4 = 16$$

The other method is to change the square feet into square yards. If 3 feet equal 1 yard, how many square feet are there in a square yard? Just multiply 3×3: There are 9 square feet in a square yard. Therefore, to calculate square yards, simply divide square footage by 9:

$$12 \times 12 = 144$$
$$144 \div 9 = \frac{144}{9} = 16$$

Accurate measurements are important for figuring the number of tiles or the amount of carpeting needed for an area.

EXERCISE 13-11

Read each question carefully before calculating the answer. Round calculations to two decimal places.

1. You are placing new floor covering, in a kitchen. The kitchen is 12 feet by 8 feet. What is the area of the room? The linoleum costs $4.50 per square yard. How much will it cost to purchase the linoleum?

2. Hester wants to recarpet her living room, which is 12 feet by 16 feet. What is the area of the room? The carpet she wants costs $17.85 per square yard, installed. How much will the carpet cost?

3. Otis plans to put color tile on his kitchen floor. The tile he wants costs $0.35 per tile. Each tile measures 1 square foot. The kitchen is 10 feet by 6 feet. What is the area of the room? How much will it cost to purchase the tile?

4. Georgette owns a four-room house. Room A is 12 feet by 15 feet. Room B is 6 feet by 9 feet. Room C is 18 feet by 21 feet. Room D is 15 feet by 15 feet. She plans to recarpet the entire house with material that costs $22.50 per square yard, installed. What is the area of each room? How much will the job cost?

5. Ted has hired Fred's Floors to recarpet his five-room house. The living room is 15 feet by 21 feet. The kitchen is 6 feet by 9 feet. The master bedroom is 15 feet by 18 feet. The other two bedrooms are each 12 feet by 14 feet. The carpet costs $15.75 per square yard, and O.K. Carpet charges $5 per square yard for installation. How much will Ted pay to have the house recarpeted?

MATHEMATICS AND HEATING

Indoor heating requirements depend on a variety of factors. The volume of the space that is to be heated is one major consideration. Another consideration is the efficiency of the structure in retaining heat. Insulation and building materials affect the capacity for heat retention.

Efficiency and costs of different kinds of fuels are other factors to consider. Major sources of heat energy in the United States include fuel oil, natural gas, electricity, and firewood. Geographic location has a major effect on choosing the source of heat energy. In some parts of the country, firewood is in plentiful supply. In such areas, fireplaces are used to provide much, if not most, of the heat for homes.

The most obvious factor affecting heating requirements is climate. People who live in Maine and Montana need greater heating capacity than those who live in Alabama and Arizona.

Degree Days

In cold sections of the country, it is important that fuel suppliers be able to estimate the heating needs of their customers. Calculations of fuel requirements are based on a unit called a *degree day.* A degree day is the *mean* number of degrees the outside air temperature falls below 65 degrees F in a 24-hour period. Mean refers to the halfway point between two measurements or numbers.

Say that a community's average mean temperature in the month of January is 28 degrees F (Fahrenheit). Degree days would be calculated by subtracting 28 from 65 and multiplying the difference by the number of days. In this case, the community would have 1,147 degree days in January.

February might be still colder. Suppose the average mean temperature in February is 24 degrees. More fuel would be needed for heating on a daily basis. However, the monthly requirement for February would be almost the same as that for January. This is because there are fewer days in February. The method of calculating the degree days is the same:

$$\text{Degree days} = (65 - 24) \times 28 = 41 \times 28 = 1{,}148$$

In leap years, February has 29 days. Thus, the degree day total would increase by 41 under the same average temperature readings.

Making Projections

Areas of the country are divided into sections based on degree day averages. Average mean temperatures for a number of years are used to project probabilities. The coldest areas of the country are rated at 10,000 degree days or more. However, these are sparsely populated areas, such as the highest elevations of the Rocky Mountains in Colorado.

The coldest populated zone has a 9,000 rating. This zone includes parts of Maine, Michigan, Wisconsin, Minnesota, Montana, and North Dakota. Large populated portions of the Rocky Mountain states are situated at high elevations. These areas of Colorado, Idaho, Montana, Utah, and Wyoming also rank in the 9,000 category.

The warmest areas of the country are below 2,000 in degree day units. Parts of southern Florida rate as low as 100 degree days annually.

EXERCISE 13-12

Read each question carefully before calculating the answer.

1. In Juneau, Alaska, the average mean temperature during January was 8 degrees F. How many degree days did the city have in January?

2. During December, the average mean temperature in Helena, Montana, was 12 degrees F. How many degree days did Helena have during December?

3. During January, the average mean temperature in Portland, Oregon, is 38 degrees F. How many degree days does this city have in January?

4. There are approximately 97 days per year when the temperature in Denver, Colorado, drops below 65 degrees. The daily mean temperature during the winter averages 30 degrees. How many degree days does the city anticipate each winter?

5. Each year, in Portland, Maine, there are approximately 115 days when the temperature drops below 65 degrees. During the winter, the average daily mean temperature is 21 degrees. How many degree days does this city expect each winter?

SUMMING UP

- Mathematics is involved in almost every aspect of the design and construction of structures. Shapes and sizes are determined by using mathematics. Mathematics also is used in estimating materials, such as lumber for framing.
- Right angles and right triangles are important concepts in structures. For example, a vertical wall and a horizontal floor form a right angle at the point where they meet.
- Mathematical rules involving right triangles are helpful in calculating dimensions, whether horizontal, vertical, or sloping. Practical applications include calculating the pitch and rise of a roof.
- Geometry is a form of mathematics that deals with the measurement, properties, and relationships of shapes. An important formula in geometry is the Pythagorean Theorem: The square of the hypotenuse of a right triangle is equal to the sum of the squares of the other two sides. This formula can be used to calculate dimensions involving sloping surfaces of structures, such as the length of roof rafters.
- Uniform measurements are vital for proper construction. Therefore, measurements of materials and their placement are standardized. Materials usually come in standard sizes and shapes.
- Building codes adopted by local governments set minimum standards for construction. Local codes govern work done by carpenters, plumbers, and electricians.
- Electrical service needs are determined by considering three factors: volts, amperes, and watts. Voltage is the pressure required to move electrical current through a circuit. Amperage is the amount of current flowing past a given point at any time. Wattage is the amount of electrical power being used. A simple formula helps determine electrical service needs: Watts = Volts × Amperes.
- Area is another important consideration in structures. Room sizes are stated according to square feet of floor space. Square feet and square yards are useful in planning floor coverings.
- Climate is the major factor in determining heating requirements. One measurement of climatic differences is the degree day unit. A degree day is the number of degrees Fahrenheit the outside air temperature falls below 65 degrees F during a 24-hour period. Degree day calculations aid in projecting fuel requirements for structural heating.

TESTING YOUR WORKING KNOWLEDGE

Read each question carefully before calculating the answer. Round calculations to two decimal places.

1. If a right angle is divided into three equal angles, how many degrees will each angle be?

2. City Construction is building a ramp for a parking garage. Over a horizontal distance of 250 feet, the ramp must rise 45 feet. How long will the surface of the ramp be?

3. You are building a roof with a pitch of $\frac{1}{4}$. The house has a span of 20 feet. What is the rise?

4. Denise owns a four-room house. Room A is 9 feet by 15 feet. Room B is 9 feet by 9 feet. Room C is 15 feet by 21 feet. Room D is 12 feet by 18 feet. What is the area of each room? Denise plans to recarpet each room with carpeting that costs $19.50 per square yard, installed. How much will the job cost?

5. Charlie is building a roof for a 22-foot-wide house. The pitch of the roof is 4 in 12. The overhang is 18 inches. How long should he cut the rafters?

6. If a 4-by-4 contains $\frac{1}{2}$ board-foot, how long is it?

7. Barry is putting new drywall on two walls in his living room. Each wall is 8 feet high. One wall is 16 feet long, and the other wall is 20 feet long. How many 4-by-8 sheets of drywall will he need?

8. A stereo amplifier uses 100 watts on a 110-volt circuit. How many amps does the stereo draw?

TERMS THAT COUNT

pitch
angle
horizontal
geometry
degree
vertical
right angle
triangle
right triangle
hypotenuse
square
square root
Pythagorean Theorem
span
run
rise
rafter
nominal size
board foot
stud
circuit
source
conductor
load
ampere
amp
volt
watt
rectangle
linear
degree day
mean

PRESENTING INFORMATION

Read the following story carefully. Then present the information in the format required. Round calculations to two decimal places.

Mr. and Mrs. Thomas have received four bids to recarpet their five-room home. The living room is 21 feet by 18 feet. The kitchen is 15 feet by 6 feet. The bathroom is 6 feet by 9 feet. The master bedroom is 18 feet by 18 feet. The guest room is 12 feet by 12 feet. Bid A is $14.50 per square yard for carpet plus $6.50 per square yard for installation. Bid B is $22.50 per square yard for carpet, including installation. Bid C is $11.75 per square yard for carpet and $10 per square yard for installation. Bid D is $13.45 per square yard for carpet and $7.80 per square yard for installation.

1. Draw a horizontal bar chart that shows the total cost of each bid. Determine which bid offers the best price.

2. Draw a pie chart that shows how the area of each room relates to the total area to be carpeted.

SKILL DRILLS

A. Find the sum in each problem.

1. 2,543.67 + 7,532.52 + 112.425 + 19.742 + 256.322 =
2. 34,376 + 387.611 + 334.598 + 2,875.887 + 543.96 =
3. 327.498 + 6,375.2 + 156.5 + 5,421 + 531.643 =
4. 3,422.87 + 4,543 + 1,321.766 + 436.76 + 4,435 =
5. 263.438 + 574.322 + 34,767 + 23,215 + 554.768 + 2,690.6 =

B. Find the difference in each problem.

1. 647.45 − 45.7332 =
2. 326.218 − 128.6599 =
3. 13.5090 − 9.657 =
4. 2.43 − 2.000043 =
5. 41.114 − 34.786759 =

C. Find the product in each problem. Round calculations to two decimal places.

1. 13.325 × 65.769 =
2. 213.31 × 16.432 =
3. 23.435 × 4.874 =
4. 65.34 × 43.79 =
5. 78.145 × 5.057 =

D. Find the quotient in each problem. Round calculations to two decimal places.

1. 42.436 ÷ 13 =
2. 721.08 ÷ 9.57 =
3. 683.546 ÷ 97.2 =
4. 963 ÷ 46.6 =
5. 2.98 ÷ 12.505 =

14 Mathematics Outdoors

YOUR LEARNING JOB

When you have completed the exercises and assignments for this unit, you should be able to:

- ☐ Define infrastructure and explain the role of mathematics in providing necessary support services for an area.
- ☐ Calculate volumes of spaces and of materials to be stored.
- ☐ Measure the volume of material, in cubic yards, needed for a construction project.
- ☐ Given the dimensions of a reservoir, calculate its capacity or contents in acre-feet.
- ☐ Measure firewood in terms of cords.
- ☐ Calculate freight capacities based on weight and volume.
- ☐ Given appropriate information, determine profit or loss on moving a truckload of cargo.
- ☐ Explain potential savings possible through containerization of freight.

WARM-UP

A. Find the sum in each problem.

1. 7,531.76 + 2,532 + 74,352 + 5,918 + 7,543.437 =
2. 56,039 + 4,192.45 + 6,547.06 + 4,325.548 + 49.768 =
3. 481.142 + 54.3644 + 4,523.67 + 94.321 + 723.038 =
4. 5,434.03 + 25.436 + 456.6334 + 9.3253 + 52.367 =
5. 918.231 + 827.367 + 19.4364 + 43.6467 + 46.039 + 346.9378 =

B. Find the difference in each problem.

1. 142.534 – 123.98 =
2. 7,901.001 – 3,726.8764 =
3. 623.094 – 76.5129 =
4. 72.0165 – 54.63564 =
5. 9,478.09 – 23.41232 =

C. Find the product in each problem. Round calculations to three decimal places.

1. 312.312 × 53.545 =
2. 382.82 × 82.937 =
3. 54.9378 × 245.9 =
4. 15.435 × 235.8 =
5. 102.0092 × 43.862 =

D. Find the quotient in each problem. Round calculations to one decimal place.

1. 382.092 ÷ 76.45 =
2. 918.81 ÷ 34.92 =
3. 4.1039 ÷ 3.83 =
4. 12 ÷ 2.9989 =
5. 56.192 ÷ 46.192 =

MATHEMATICS AND LIFE SUPPORT

In the early 1980s, the term *infrastructure* became prominent in the news. An infrastructure is an underlying foundation, or basic framework, of a system or organization. There are many elements within the infrastructure of an area or country. Major examples include:

- Roads, highways, and parking areas or structures
- Railroads
- Highway and railway bridges
- Airports
- Navigable waterways
- Dams for flood control, water supply, irrigation, and power generation
- Water delivery systems
- Sewage and waste disposal systems
- Electric generating, transmission, and delivery systems
- Coal, oil, and natural gas production and delivery systems.

This infrastructure became newsworthy in the early 1980s because of the deterioration of many of its elements. In many parts of the country, highway repairs were needed. Some old bridges either collapsed or were declared unsafe. Some railroad lines were in need of repair. A massive effort was undertaken by the federal government to upgrade and modernize portions of the infrastructure that had become outdated.

Mathematics plays a major role in the design, construction, and operation of elements of the infrastructure. Mathematics also is used to plan for their maintenance, repair, and expansion.

Mathematics for Design

Airports are an example of the many ways mathematics is used in the design process. One of the first elements to be determined in airport design is the prevailing wind patterns at the location.

Aircraft are supported in flight by the lifting force of air flowing past the wings. The faster the airflow, the greater the lift. Thus, it is desirable for aircraft to be facing into the wind when they take off and land. Airflow moving against the front of an object is called *headwind*. As headwind speed increases, the ground speed required for takeoff decreases. During landing, headwinds provide a slowing, or braking, force as well as lift. Thus, aircraft should take off and land facing into the wind. Figure 14-1 illustrates how airflow acts on a wing to lift an airplane.

Airport runways can be positioned to take maximum advantage of prevailing wind patterns. Mathematics helps determine the best runway layout for a location. Calculations consider anticipated air traffic volume. An example of runway layout is shown in Figure 14-2.

The length of runways is critical. Designers must determine mathematically how many feet of runway are needed to accommodate the aircraft that will use the airport. A medium-size passenger airliner may be able to take off and land safely on a 5,000-foot runway. Larger, heavier, or faster aircraft may require up to 10,000 feet of runway.

Large aircraft place tremendous weight on runway pavement when they touch down. Part of this downward force is reduced by the shock-absorbing capacity of an aircraft's landing gear. Nonetheless, the pavement and underlying base material usually are thicker in the areas where touchdowns occur. In areas where wind patterns shift, both ends of runways may be designed to accommodate landing planes.

Figure 14-1. Air flow past an airfoil provides the lift that enables an aircraft to fly.

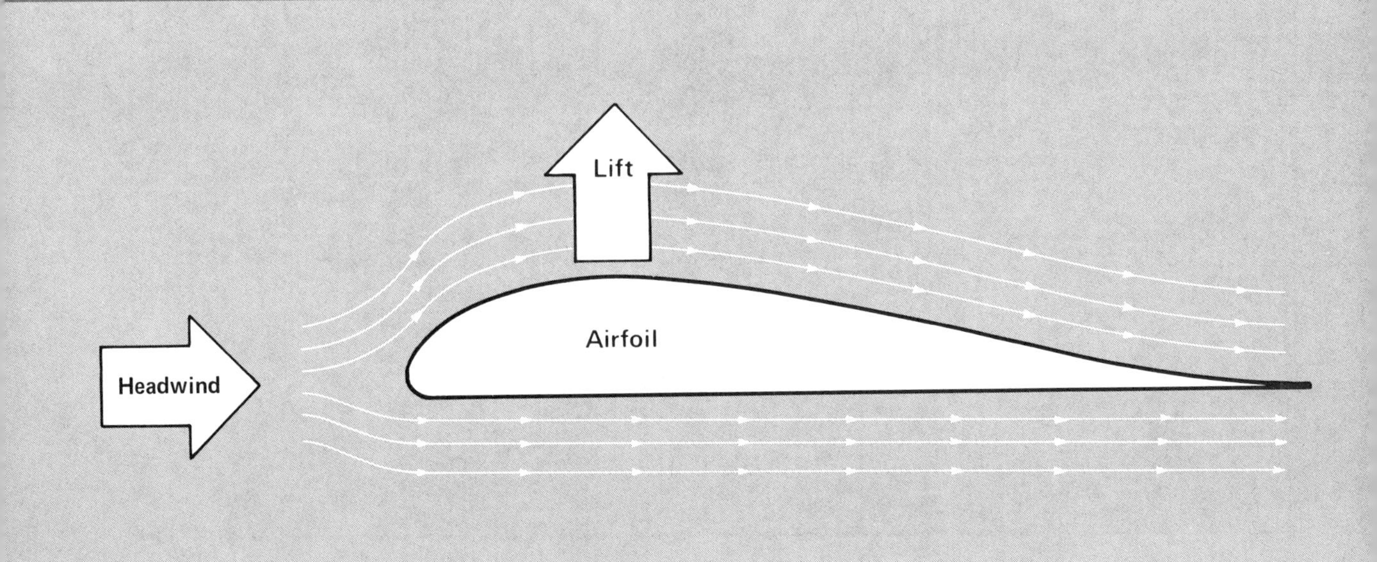

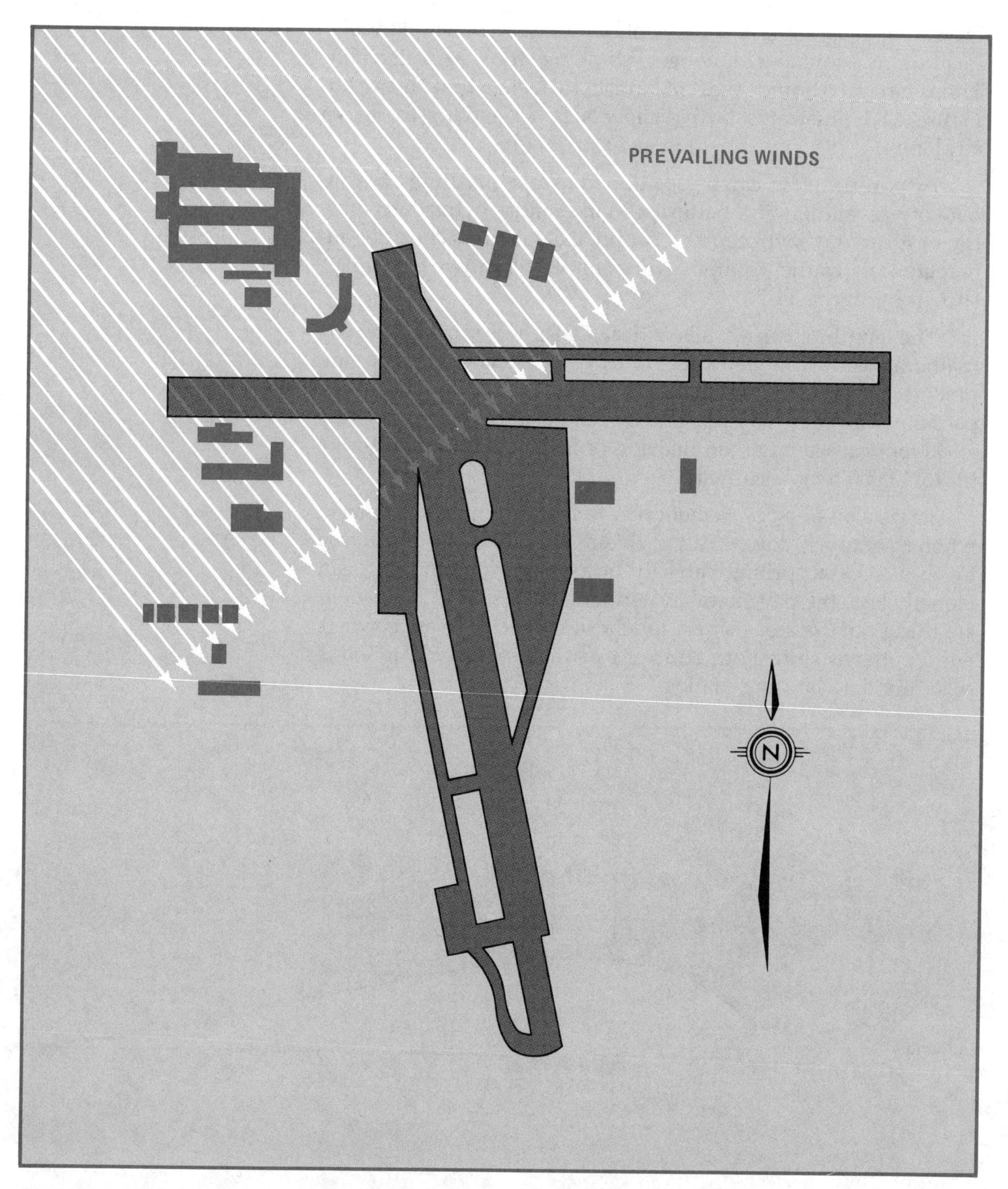

Figure 14-2. Airport runways are positioned to take advantage of prevailing winds. Headwinds are preferred for aircraft takeoffs and landings.

EXERCISE 14-1

Read each question carefully before calculating the answer. Round answers to two decimal places.

1. Carver City is designing an airport. Wind patterns in the city vary, depending on the season. In the spring and summer, winds blow from east to west. In the fall and winter, winds blow from west to east. Draw a chart that shows how runways should be aligned. Be sure to mark the directions of north, south, east, and west.

2. From the information in Question 1, determine the direction in which planes should take off and land during each season.

3. Engineers have determined that the ends of the runways must be three times as thick as the middle, to absorb shock. If the middle of the runway will be 5.85 feet thick, how thick should the ends be?

4. Engineers have determined that a plane will land or take off every 5 minutes from 7 A.M. to 7 P.M. and every 20 minutes from 7 P.M. to 7 A.M. How many planes will use the runways in a 24-hour period?

5. Designers have determined that a ground crew of 15 will be needed to prepare an airplane for flight once it has landed. This includes loading and unloading luggage and supplies, servicing the engines, and guiding the plane to loading and unloading areas. Designers also have determined that, between 8 A.M. and 6 P.M., there always will be five planes on the ground requiring service. How many flight-preparation employees will the airport need on hand between 8 A.M. and 6 P.M.?

MATHEMATICS AND MATERIALS

Many outdoor construction jobs involve calculations of *volume*. Volume is the space occupied by something. It also can be the material that occupies a space. Volume also is the measure of the capacity of containers.

An easy way to understand volume is to think about a container, such as a shoe box. Looking down at the top, or cover, of the box, you see two *dimensions:* length and width. A dimension is a measurement made in one direction. The inside of the box contains space, which is formed by adding a third dimension: depth.

The third dimension is just as important as length and width in determining the volume of a space. Calculating in three dimensions is called *cubic measure.* Cubic measurement involves units, such as cubic inches or cubic centimeters, for calculating volume. For instance, the shoe box in Figure 14-3 is 13 inches long and 6 inches wide. The surface area of the bottom of the box is 78 square inches. You simply multiply the length by the width to compute the surface area in square inches.

To calculate the volume, multiply the surface area by the third dimension, depth. The box is 6 inches deep, so the volume is 468 cubic inches. Each surface of this box is a rectangle. The formula for finding the volume of a rectangular container involves multiplying its three dimensions:

$$\text{Volume} = \text{Length} \times \text{Width} \times \text{Depth}$$

EXERCISE 14-2

Read each question carefully before calculating the answer. Round answers to two decimal places.

1. What is the volume of a box 18 inches long, 12 inches wide, and 14 inches deep?

2. What is the volume of a planter 36 inches long, 12 inches wide, and 24 inches deep?

3. What is the volume of a truck bed 6 feet long, 5 feet wide, and 2.5 feet deep?

4. What is the volume of a room 18 feet long, 12 feet wide, and 8 feet high?

5. What is the volume of a briefcase 24 inches long, 20 inches wide, and 4 inches deep?

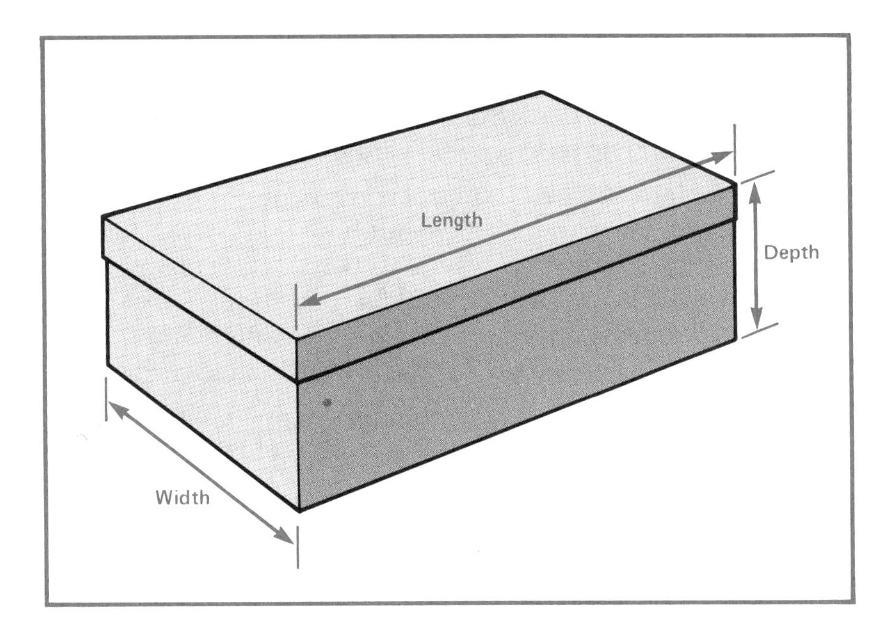

Figure 14-3. A box has three dimensions: length, width, and depth.

Filling Spaces

Mathematics outdoors frequently involves filling spaces with materials, which may be solids, liquids, or gases. In some cases, the materials themselves form the cubic volumes. An example is the construction of a road. The concrete or asphalt itself forms the volume.

Construction materials usually are measured in cubic yards. Think about putting in a new driveway for a private home. Suppose the driveway is 60 feet long and 20 feet wide. To calculate the surface area, multiply the two dimensions:

$$\text{Area} = 60 \times 20 = 1{,}200$$

Thus, the surface area to be repaved is 1,200 square feet. You have to determine how much concrete to order for the driveway. For this volume calculation, you have to include the depth, or thickness, of the pavement. Say that you want the driveway to be 3 inches thick. Knowing this third dimension, you can calculate the volume of concrete required. Three inches is $\frac{1}{4}$ of a foot. Multiply the area by the depth:

$$1{,}200 \times \frac{1}{4} = \frac{1{,}200}{4} = 300$$

Thus, the volume of the driveway pavement will be 300 cubic feet. However, construction materials, such as concrete, are ordered by the cubic yard. So, one more calculation is necessary to convert the cubic feet to cubic yards.

You can look up these types of conversions in Appendix D, but it is useful for you to think about how they are performed. In measuring length, 1 yard equals 3 feet. Calculating area, 1 square yard equals 3 feet times 3 feet, or 9 square feet. Adding the third dimension merely requires a second multiplication: $3 \times 3 \times 3 = 27$. There are 27 cubic feet in 1 cubic yard.

To calculate the cubic yard amount of concrete to order for the driveway, divide the volume in cubic feet by 27:

$$\text{Concrete (cubic yards)} = 300 \div 27 = \frac{300}{27} = 11.1$$

EXERCISE 14-3

Read each question carefully before calculating the answer. Round answers to two decimal places.

1. Antonio is constructing a patio with cement. The dimensions are 20 feet long by 10 feet wide. Antonio wants the patio to be 4 inches thick. How many cubic yards of cement does Antonio need?

2. Oscar built a driveway 15 feet wide, 15 feet long, and 6 inches thick. How many cubic yards of asphalt did he use?

3. Mercedes is building a patio 40 feet long, 30 feet wide, and 3 inches thick. How many cubic yards of cement will she need?

4. Ricki built a patio 35 feet long, 15 feet wide, and 3 inches thick. How many cubic yards of cement did she use?

5. Kent dug a trench 35 feet long, 6 inches wide, and 1 foot deep. How many cubic yards of earth did he remove?

6. Manny dug a hole that is 4 feet long, 4 feet wide, and 8 feet deep. How many cubic yards of earth did he remove?

Large Volumes

The formula for calculating volume is the same for larger spaces. Most buildings and other large structures require foundations. Such foundations usually are constructed below ground level, in an *excavation.* An excavation is a hole in the ground, formed by the removal of earth.

Have you ever looked at a large excavation and wondered where the earth went? Probably, the earth was trucked to another location, where it was needed as fill. The digging, transporting, and dumping of earth and other materials is a large industry. A *borrow site* is the location from which a material is removed. A *dump site* is the location where it is deposited.

For example, think about the construction of an *earthen dam.* An earthen dam is a water barrier constructed of closely compacted earth. Earthen dams usually are low, wide structures designed to hold back water flow in relatively flat areas. For a large earthen dam, several borrow sites may be necessary to provide enough material. The world's largest earthen dam, measured in the volume of material used, is the New Cornelia Tailings Dam in Arizona. The structure contains more than 274 million cubic yards of earth. The dam is only 98 feet high; but it is more than 6.7 miles long. Dump trucks used on smaller construction projects may carry 3 to 5 cubic yards of material. At 5 cubic yards per load, the New Cornelia Tailings Dam would have required 54.8 million truckloads of fill.

The principle is the same, whether trucks are carrying earth, sand, gravel, concrete, or asphalt. For instance, a new two-lane highway may be designed to have 6 inches of base material beneath the paving. If the pavement were to be 27 feet wide, how many cubic yards of material would be needed for each mile of road?

The depth of base material, 6 inches, is $\frac{1}{6}$ yard. The pavement width, 27 feet, is 9 yards. There are 1,760 yards in a mile. The solution is obtained by multiplying the three dimensions:

$$\text{Base material} = 1{,}760 \times 9 \times \frac{1}{6} = \frac{15{,}840}{6} = 2{,}640$$

Thus, 2,640 cubic yards of base material would be needed for every mile of road bed. At 5 cubic yards per truck, 528 truckloads would be delivered to the job.

EXERCISE 14-4

Read each question carefully before calculating the answer. Round answers to two decimal places.

1. The dimensions for the foundation of a house are 40 feet long, 30 feet wide, and 3 feet deep. How many cubic yards of cement are needed?

2. To build a high-rise structure, City Construction must excavate a hole 81 feet deep, 108 feet wide, and 135 feet long. How many cubic yards of earth have to be moved? If a fleet of 5-cubic-yard dump trucks is used to move the earth, how many loads will it take?

3. Country Construction is using cement to make walls for an industrial building. Each wall will be 80 feet long, 20 feet high, and 6 inches thick. How many cubic yards of cement will be needed to make four walls?

4. The dimensions of an earthen dam are 50 feet long, 25 feet high, and 10 feet average thickness. How many cubic yards of earth were used to make the dam?

5. Rafael has a dump truck that can carry 3 cubic yards of earth. He has been hired to fill a hole 25 feet long, 40 feet wide, and 6 feet deep. How many truckloads will the job require?

6. The foundation of Ted's house is 45 feet long, 35 feet wide, and 3.5 feet thick. How many cubic yards of cement were used to pour the foundation?

7. A construction company built a cement dam that was 500 feet high, 750 feet across, and 45 feet thick on the average. How many cubic yards of cement were used?

8. The Road Builders are constructing a four-lane highway that is 64 feet wide. The highway will be 6 miles long and be paved with 2-inch-thick cement. How many cubic yards of cement will the company need to build the highway?

Liquid Volumes

You probably are accustomed to thinking about liquid capacities in everyday terms: fluid ounces, pints, quarts, liters, and gallons. However, none of these volumes is sufficient to describe large bodies of water, such as reservoirs and lakes.

Large bodies of water are measured in terms of *acre-feet.* An *acre* is an area that contains 4,840 square yards, or 43,560 square feet. A square acre has sides measuring 69.57 yards, or 208.71 feet. An *acre-foot* is a volume of water covering one acre, measured to a depth of 1 foot. An easy way to visualize an acre is to think of a football field. An American football field, 100 yards long and 53 $\frac{1}{3}$ yards wide, is 1.1 acres. An acre covers the field from one goal line to just short of the nine-yard line at the opposite end of the field. In baseball, the portion of the infield inside the base paths is just less than one-fifth of an acre.

Acre-feet of water storage capacity can be measured in the same manner as other cubic volumes. For example, suppose a reservoir were two miles long and had an average width of

Reservoirs usually are measured in terms of acre-feet of water.
COURTESY OF TENNESSEE VALLEY AUTHORITY

500 yards. If the reservoir had a uniform depth of 30 feet, what would be its capacity?

The solution is found by determining the surface area of the reservoir and multiplying that by its depth. Two miles equal 3,520 yards. The surface area of the reservoir is 3,520 × 500, or 1,760,000 square yards. Divide by the number of square yards in an acre, 4,840, to find the surface in acres: 363.636. Then, multiply by the depth, 30 feet, to obtain the volume: 10,909.09 acre-feet. As equations, the steps are performed this way:

$$\text{Length in yards} = 1{,}760 \times 2 = 3{,}520$$
$$\text{Area in square yards} = 3{,}520 \times 500 = 1{,}760{,}000$$
$$\text{Area in acres} = 1{,}760{,}000 \div 4{,}840 = \frac{1{,}760{,}000}{4{,}840} = 363.636$$
$$\text{Volume in acre-feet} = 363.636 \times 30 = 10{,}909.09$$

EXERCISE 14-5

Read each question carefully before calculating the answer. Round answers to two decimal places.

1. A lake is 4 miles long with an average width of 1,230 yards and an average depth of 85 feet. How many acre-feet of water does it contain?

2. How many acre-feet of water are in a swimming pool 32 feet long and 16 feet wide, with an average depth of 5 feet?

3. When Pine Reservoir is full, it is 7 miles long and has an average width of $\frac{1}{2}$ mile. Its average depth is 50 yards. How many acre-feet of water does the full reservoir hold?

4. A river has an average depth of 25 feet and an average width of 50 feet. How many acre-feet of water are contained in a 12-mile strip of the river?

5. Lake Mead is the largest man-made lake in the world. It is 115 miles long. On the average, it is 4 miles wide and 103.6 feet deep. How many acre-feet of water does the lake hold?

Agriculture is an industry that relies heavily upon mathematics.
PHOTO BY TIM MCCABE COURTESY OF USDA-SOIL CONSERVATION SERVICE

MATHEMATICS AND AGRICULTURE

Mathematics is involved in every area of decision making in agriculture. The importance of mathematics is universal, from the smallest family farm to giant corporate agribusiness operations.

Agricultural decisions tend to be based on local conditions, such as climate, soil characteristics, and anticipated rainfall levels. In areas with limited amounts of rain, especially the Western states, irrigation can be vital.

Livestock producers also make decisions based on local conditions. Grazing conditions for cattle or sheep provide an example. In dry periods, a feedlot operation may be needed to support livestock.

Fluctuating prices for crops and livestock also play a major role in agricultural decisions. For example, a rancher may expect cattle prices to rise 2% during a two-week period. This anticipated price increase must be weighed against feedlot costs for that two weeks. Such decisions involve calculated risks.

EXERCISE 14-6

Read each question carefully before calculating the answer. Round answers to two decimal places.

1. On the average, a milk cow produces 8 gallons of milk per day. If a dairy owns 550 milk cows, how many gallons of milk can the dairy expect to produce each week?

2. At Helen's Poultry Ranch, it costs an average of 5 cents per day to feed one chicken. The ranch owns 2,000 chickens. Determine how much it costs to feed the chickens for four weeks.

3. The average crop yield for corn is 40 bushels per acre. A bushel is the unit of measurement for many agricultural products. If a farmer plants 1,250 acres of corn, how many bushels can he or she expect to harvest? If the price of corn is $2.68 per bushel, what gross income can the farmer expect to earn from the crop?

4. A farmer has 5,000 bushels of wheat to sell. Currently, the price is $3.45 per bushel. However, the price is expected to increase 2% next week. If the farmer waits and the price does increase, how much more will the farmer earn? What will the total sales price be?

5. A farmer had 10,000 bushels of oats to sell. The going price was $1.66 per bushel. The farmer decided to wait a week to see if prices would increase. Instead, prices dropped 3% and the farmer sold. For what price did the farmer sell the 10,000 bushels? How much did the farmer lose by waiting?

Land and Measurements

Land measurements are important in agriculture. The acre (4,840 square yards) is the most common unit for measuring land area.

For example, suppose you had a 20-acre rectangular field ready for planting. The area is a quarter-mile in length and one-eighth of a mile wide. You decide to plant 10 acres of soybeans.

If you planted the full width of the field, how far would the soybeans extend? First, calculate the number of square yards in 10 acres. Next, divide that number by 220 yards, the width of the field. The result is the length of the area to be planted.

$$\text{Square yards} = 4{,}840 \times 10 = 48{,}400$$

$$\text{Length of field} = 48{,}400 \div 220 = \frac{48{,}400}{220} = 220$$

Thus, the length of the area to be planted in soybeans is 220 yards.

Similar calculations would be used to estimate the amount of materials needed to fence an area, and to estimate the cost. For example, suppose you wanted to install a chain-link fence around

Land measurements can be made using simple mathematical calculations. COURTESY OF USDA-SOIL CONSERVATION SERVICE

the *perimeter* of a 10-acre field. The perimeter is the outer boundary, or edge. If you determine one dimension of a rectangular field, you can calculate the other dimension. Adding the two and multiplying by 2 gives the perimeter of the field.

If the 10-acre field were a quarter-mile long, how many yards of fence would be needed to enclose it? Ten acres equal 48,400 square yards. A quarter mile is 440 yards. Divide 440 into 48,400 to find the width of the field:

$$\text{Width of field} = 48{,}400 \div 440 = \frac{48{,}400}{440} = 110$$

The field is 110 yards wide. The perimeter of the field, then, is 440 + 440 + 110 + 110, or 1,100 yards.

How much chain-link fence would be required if the 10-acre field were square? Each side of a square shape equals the square root of the area. Simply find the square root of 48,400, which is 220. Each side of the square field is 220 yards. Multiply by 4 to find the perimeter, 880 yards.

EXERCISE 14-7

Read each question carefully before calculating the answer. Round answers to two decimal places.

1. How many yards of fence are required to enclose a rectangular field 100 feet long and 55 feet wide?

2. How many yards of fence are required to enclose a rectangular field one-quarter mile wide and one-eighth mile long?

3. A farmer planted wheat in a 5-acre field that is 100 feet wide. How long is the field? How many yards of fence would it take to enclose the field?

4. How many yards of fence are required to enclose a 2-acre field? If the field is 250 yards long, how wide is it?

5. A farmer owns 5 square miles of farmland. How many acres does the farmer own? If the farm is 2 miles wide, how many yards long is it?

Crop dusting for insect control requires careful calculations of the acreage to be covered. COURTESY OF USDA

Covering Areas

Acreage is the basis for many mathematical calculations used in agriculture. For example, crops are planted by the acre. Fertilizer is spread by the acre. Pesticides are sprayed by the acre.

If you were buying seed, you would use a simple formula to calculate the amount needed. Simply multiply the required seed usage per acre by the number of acres you want to plant.

Similarly, fertilizer is spread according to manufacturers' recommendations. The type of fertilizer used and the amount applied per unit of area depend on the chemical content of the soil. All of these calculations involve mathematics.

Spraying pesticides is another operation that requires mathematical computations. Some chemicals are mixed with water. Mathematical formulas guide users in preparing proper mixtures. The formulas are based partly on the type of crop and the insect being eliminated. The amount of mixture, however, depends on the size of the area to be sprayed.

EXERCISE 14-8

Read each question carefully before calculating the answer. Round calculations to two decimal places.

1. It takes 0.75 gallons of pesticide to cover a $\frac{1}{4}$-acre field. How many gallons are required to cover a 150-acre field?

2. A tractor can cover an acre of land with fertilizer in 75 minutes. Using the same tractor, how long would it take to spread fertilizer over a 125-acre field? Give your answer in hours and minutes.

3. A certain pesticide is mixed with water before it is applied to crops. In 1 gallon of the mixture, there is 1 cup of pesticide. It takes 20 gallons of the mixture to cover a 100-acre field. How many cups of pesticide are required to spray a 75-acre field?

4. A farmer owns 350 acres of land. The farmer uses 7,500 pounds of fertilizer to cover 75 acres of soil. How many pounds will the farmer need to fertilize the remaining acreage?

5. A farmer owns a parcel of land that is $\frac{1}{2}$ mile wide and $1\frac{3}{4}$ miles long. The farmer wants to spray pesticide over the entire parcel. It takes 0.25 pounds of pesticide mixed with 1 gallon of water to cover $\frac{1}{2}$ acre. How many pounds of pesticide will the farmer need?

Working for Warmth

There are many parts of the country in which wood remains the major source of household heat. In many rural areas, families can gather their own firewood. In other places, firewood may be purchased from suppliers.

The *cord* is the basic unit of measurement for firewood. A cord is a stack of cut wood that measures 4 feet wide, 4 feet high, and 8 feet long. Packed tightly, a cord of firewood equals 128 cubic feet.

Firewood typically is priced by the cord. Therefore, it is important to be able to estimate bulk quantities. Knowing the dimensions of a cord, you should be able to approach a stack of firewood and estimate its value. It is simple to take quick measurements with a tape measure and calculate how much wood a stack contains.

For example, consider a stack of 4-foot-long logs piled to a height of 6 feet. If the stack were 36 feet long, how many cords would it contain? The solution is to multiply the three dimensions and divide by 128:

$$\text{Cords} = 4 \times 6 \times 36 \div 128 = 864 \div 128 = \frac{864}{128} = 6.75 = 6\,\frac{3}{4}$$

Firewood is cut into short lengths. The cut wood is measured for sale in cords, a mathematical unit of volume. © H. ARMSTRONG ROBERTS

EXERCISE 14-9

Read each question carefully before calculating the answer. Round calculations to two decimal places.

1. How many cords of wood are in a woodpile 8 feet high, 12 feet long, and 6 feet wide?

2. When Carlos finished cutting down a tree, he had a stack of wood 2 feet high, 5 feet long, and 4.5 feet wide. How many cords of wood did he have?

3. Bonnie's Lumber sells a cord of firewood for $95. At that rate, what is the cost for a woodpile 2.5 feet high, 2.5 feet wide, and 4 feet long?

4. For $44, Howard purchased a stack of cut wood that was 2 feet wide, 6 feet long, and 5 feet high. What was the price per cord?

5. Joshua purchased a stack of wood that was 4 feet high, 5 feet wide, and 6 feet long for $95. What was the price per cord?

MATHEMATICS FOR MOVEMENT

Moving freight is a key link in the economic system of getting products from farm or factory to consumers. Four basic methods of transporting cargo are:

- Ship
- Rail
- Truck
- Air.

Three of these four methods involve large quantities of goods. Air freight is rapid but expensive and limited in the weight and bulk that can be handled. For the most part, air freight operations are limited to parcels and small quantities of cargo.

Not surprisingly, ships are at the opposite end of the speed and cargo capacity scales from aircraft. Ocean-going ships and inland waterway barges carry freight in very large quantities. Their cargos are measured in terms of *tonnage* rather than in pounds. Tonnage is the amount of freight expressed in *tons.* Two values of weights are expressed in tons. A *short ton* contains 2,000 pounds. A *long ton* contains 2,240 pounds. In the United States, most cargo is measured in short tons.

Railroads also carry great amounts of freight. The capacity of a typical railway freight car is 110,000 pounds. Think about the amount of cargo in a train of 50 cars:

$$110{,}000 \times 50 = 5{,}500{,}000$$

Thus, the train is carrying 5.5 million pounds of freight. Divide by 2,000 (pounds per ton) to obtain the tonnage: 2,750 tons.

EXERCISE 14-10

Read each question carefully before calculating the answer. Round calculations to two decimal places.

1. A cargo ship can carry 1,250 automobiles across the Pacific Ocean. If each automobile weighs approximately 1,950 pounds, how many tons of automobiles can the ship carry?

2. A train with 45 cargo cars is carrying 2,137.4 short tons of wheat. How many pounds of wheat are in each cargo car?

3. A train is carrying 1,562.5 long tons of grain. Each cargo car contains 100,000 pounds. How many cargo cars are on the train?

4. You are driving a pickup truck full of produce when you approach a bridge. A sign warns that vehicles weighing over 4,500 pounds should not use the bridge. Your truck is loaded with 0.75 short tons of produce. The truck weighs 1.65 tons. Is it safe to drive your truck over the bridge? What is the difference between the loaded weight of your truck and the weight limit for the bridge?

5. A ship carrying 275 short tons of wheat is being unloaded into railroad cars for delivery to warehouses. If one train car can hold 110,000 pounds of wheat, how many cars will be needed to deliver the wheat?

Trucking

While ships and railroads can carry great quantities of freight, they are limited to traveling on waterways or rail lines. It remains for over-the-road and local delivery trucks to bring most cargo to its ultimate consumers.

In most states, highway trucks are limited to 80,000 pounds *gross combined weight (GCW)*. GCW is the total weight of a tractor, the semitrailer or trailers it pulls, and the freight being carried. The less an empty tractor-trailer combination weighs, the heavier the legal load it can carry.

A typical tractor-trailer combination might weigh 27,000 pounds empty. Such a rig could carry 53,000 pounds of cargo. At that rate, how many trucks would be required to transport the cargo carried in the 50-car freight train? The answer is 104 trucks.

$$110{,}000 \times 50 = 5{,}500{,}000$$
$$5{,}500{,}000 \div 53{,}000 = 103.77 = 104 \text{ trucks}$$

Weight saving is an important factor in truck design. The lighter a truck, the heavier the payload it can carry.

EXERCISE 14-11

Read each question carefully before calculating the answer. Round answers to the next whole number.

1. A train carrying 55 short tons of wheat is being unloaded into trucks that can hold 55,000 pounds of grain. How many truckloads will it take to unload the train?

2. A ship carrying 295 short tons of material is being unloaded into trucks that can hold 52,500 pounds of material. How many trucks will be needed to unload the cargo?

3. Burt's tractor-trailer rig weighs 24,000 pounds when empty. The rig's GCW is 80,000 pounds. How many short tons of goods can Burt's rig carry without exceeding the weight limit?

4. Martha's tractor-trailer rig weighs 22,000 pounds when empty and is carrying 24.5 tons of produce. How much more could the rig carry without exceeding the 80,000-pound limit?

5. A train with 55 cars is being unloaded into tractor-trailer combinations that weigh 26,000 pounds when empty. Each railroad car contains 45 tons of wheat. If the GCW limit is 80,000 pounds, how many truckloads will be required to unload the train?

Trucking Costs

There is another difference between the trucking industry and the other forms of cargo hauling: Almost all ships, railroad cars, and aircraft are owned by medium-to-large companies. Many trucks are owned by their operators. The independent owner-operator generally must operate within a narrow margin of profit versus loss. Factors such as the cost of diesel fuel can mean the difference between earning and losing money. Layover time between loads also can be harmful.

Suppose an independent trucker contracted to haul a load of freight from Chicago to New York City. The trip is 830 miles, and the trucker will be paid $1.05 per mile. Gross revenue for the overnight trip will be $871.50.

Also, suppose diesel fuel costs $1.10 per gallon and the truck averages 5.5 miles per gallon (mpg). How much net revenue will the trip earn? Two calculations are needed:

$$\text{Fuel used (gallons)} = 830 \div 5.5 = \frac{830}{5.5} = 150.9$$

$$\text{Fuel cost} = 150.9 \times 1.10 = 165.99$$

Rounding that figure, the trucker will spend $166 for fuel. Subtracting the fuel cost from the gross revenue gives net revenue for the trip:

$$871.50 - 166 = 705.50$$

Thus, the trucker will earn $705.50 on this two-day trip. That seems like a very good income. However, there are several other factors to consider. The trucker probably is paying off a loan for his truck. Other costs include maintenance of the rig, insurance, road use taxes, tolls, and living expenses, such as meals. In addition, the trucker may have to wait for another load. Or, the next load may originate in another city, and the travel costs to get there are the trucker's responsibility. The point is that it is necessary to look beneath the surface to calculate real income and true costs of doing business.

EXERCISE 14-12

Read each question carefully. Then determine each driver's net revenue for the trip. Round calculations to two decimal places. For all questions, fuel costs equal $1.10 per gallon.

1. Connie earned 95 cents per mile to transport a load of produce 250 miles. Her truck averaged 6.3 miles per gallon.

2. Earl's truck averaged 4.5 miles per gallon on a 2,840-mile delivery. He spent $35 dollars on meals and $85 on hotels. He was paid $1.10 per mile.

3. Gene was paid $1.05 per mile to make a long-distance delivery. The trip covered 1,560 miles and the truck averaged 5.1 miles per gallon. In addition, he spent $105 on meals and $155 on hotels.

4. Mary made a long trip that included three loads. The first load paid $1.02 per mile for a 750-mile delivery. The second load paid $1.10 per mile for 1,250 miles. The third load paid $1.08 per mile and the 950 miles brought Mary home. During the trip, Mary was gone for 9 days. She spent $225 on hotels and $150 on meals. In addition, she drove an extra 225 miles to get from job to job. Her truck averaged 5.25 miles per gallon.

5. Jacob made a long trip that included two loads. The first load paid $1 per mile for a 2,500-mile run. The second load paid $1.05 per mile for 3,500 miles. In addition, Jacob drove an extra 375 miles between loads. He spent $175 on meals and $275 on lodging. Jacob's truck averaged 5.75 miles per gallon.

Containerization

In Unit 12, handling was discussed as a cost factor in inventory. Handling also is a key cost factor in moving freight. The development of *containerization* by the transportation industry has minimized handling costs. Containerization is the packaging of large amounts of cargo in large containers. The container usually is designed to carry the cargo from the factory to the final destination or point of use.

Advanced scheduling techniques make it possible to fill a container at a factory and have it delivered to its destination intact. To understand the advantage of reducing handling costs, compare the steps involved in the two processes charted in Figure 14-4.

Traditional Shipping	Containerized Shipping
1. Factory to storage	1. Factory to container
2. To shipping vehicle	2. Road vehicle
3. To terminal	3. Delivered at destination
4. Road vehicle	
5. To terminal	
6. Delivery to customer	

Figure 14-4. Containerization eliminates most of the handling of cargo, thus greatly reducing the cost of shipping.

As the chart illustrates, cargo shipped the old-fashioned way must be handled several times. Each time it is moved, loaded, or unloaded, shipping and handling charges are increased.

Containerization offers another advantage. The containers can be designed for use in two or more methods of transportation. Consider a cargo of automotive parts manufactured in Brazil for an American auto assembly plant. The container might be fitted onto a special semitrailer frame to be trucked from factory to dockside. There, the container is loaded aboard a freighter for the trip to New York. The unloading operation sets the container on another semitrailer frame for its road trip to the assembly plant. Handling has been kept to a minimum.

Another example of containerized movement of cargo is the so-called "piggyback" rail system. It is common to see long freight trains that are made up entirely of flatbed cars carrying semitrailers. Each flatbed rail car can accommodate two semitrailers. From a cost standpoint, railroads offer a most efficient system for moving freight long distances. A 100-car piggyback train takes the place of 200 big trucks, for example, saving fuel as well as handling costs.

EXERCISE 14-13

Read each question carefully before calculating the answer. Round calculations to two decimal places.

1. An automobile parts manufacturer cut shipping costs 25%, or $312,500, through containerization. What were the shipping costs before the company switched?

2. A toy manufacturer cut shipping costs 15%, or $75,000, by changing to containerization. What were the company's costs after the switch?

3. Harriet's company has predicted that containerization methods would cut shipping costs 32%. If the company currently is spending $975,000 per year on shipping, what would shipping costs be using containerization? How much money would be saved?

4. The Trucking Company pays 15 employees $10.50 per hour, 40 hours per week, to load and unload trucks. The company has estimated that, by changing to containerization, the number of employees needed to load and unload trucks would be reduced by two-thirds. How much in wages would the company save each week?

Containerization provides efficient, cost-saving transportation of cargo. COURTESY OF THE PORT AUTHORITY OF NY & NJ

SUMMING UP

- Mathematics plays an important part in supporting life throughout the world. The infrastructure of a nation is the collection of physical installations that support life.
- The national infrastructure includes physical requirements for travel, transportation, water delivery, waste disposal, and power generation and distribution. Mathematics is a basic factor in the design, maintenance, repair, and replacement of infrastructure elements.
- Mathematics is used to calculate volumes. Three-dimensional measurements are required to calculate the amounts of materials that are to be removed from or delivered to construction sites.
- In some cases, such as water reservoirs, decisions are made based on volume measurements. The acre-foot is the basic unit for measuring water supplies.
- Agriculture is another field in which mathematics is vital for decision making. Decisions involving crops, livestock, fertilization, and pest control all involve mathematics. Land area usually is measured in acres. In agriculture, most decisions involving crops are based on acreage.
- Volume measurement is important to winter warmth. Firewood is measured in cords. A cord is a stack of cut firewood measuring 4 feet wide, 4 feet high, and 8 feet long. The volume of a cord of firewood is 128 cubic feet.
- Mathematics is used extensively in the transportation field. For instance, cargo is measured by weight and volume. Freight is moved by four basic methods: air, ship, rail, and truck. Each system has advantages and disadvantages. Efficiency in any form of cargo movement depends upon mathematical calculations.
- Modern movement trends include containerization and combinations of transportation methods. Containerization eliminates most unnecessary handling of bulk cargo, thus lowering freight costs and boosting efficiency.

TESTING YOUR WORKING KNOWLEDGE

Read each question carefully before calculating the answer. Round answers to two decimal places.

1. If two planes land or take off every 5 minutes, how many planes will use an airport's runways in a 24-hour period?

2. What is the volume of a truck trailer 26 feet long, 6 feet wide, and 8 feet deep?

3. Jenny made a two-way freight run. The first delivery covered 1,560 miles and paid $1.05 per mile. The second load paid $1.08 per mile for 2,540 miles. In addition, Jenny drove an extra 425 miles between loads. She spent $185 on meals and $263 on lodging. Fuel cost $1.10 per gallon. If Jenny's truck averaged 5.75 miles per gallon, what was her net revenue?

4. County Construction is using cement to make walls for an industrial building. Each wall will be 60 feet long, 20 feet high, and 6 inches thick. How many cubic yards of cement will be needed to make eight walls?

5. A lake is 6 miles long. It has an average width of 975 yards and an average depth of 105 feet. How many acre-feet of water are contained in the lake?

6. The average crop yield for corn is 45 bushels per acre. If a farmer plants 855 acres of corn, how many bushels can he or she expect to harvest? If the price of corn is $2.71 per bushel, what gross income can the farmer earn from the crop?

7. A farmer owns a strip of land that is $2\frac{1}{2}$ miles long and $\frac{1}{4}$ mile wide. How many yards of fence are required to enclose the field? It takes 0.25 pounds of pesticide mixed with 1 gallon of water to cover $\frac{1}{2}$ acre of land. How many pounds of pesticide will the farmer need to spray the field?

8. A ship carrying 380 short tons of wheat is being unloaded into train cars. If one train car can hold 95,000 pounds of wheat, how many cars will be needed to deliver the wheat?

TERMS THAT COUNT

infrastructure
headwind
volume
dimension
cubic measure
excavation
borrow site
dump site
earthen dam
acre
acre-foot
perimeter
cord
tonnage
ton
short ton
long ton
gross combined weight (GCW)
containerization

PRESENTING INFORMATION

Read each question carefully. Then present the information in the format required. Round calculations to two decimal places.

1. The Redlands Ranch contains 1,000 acres. One-fifth of the ranch is planted with potatoes. Five-tenths is planted in onions. One-tenth is planted in tomatoes. One-fifth has been left unplanted. Determine how many acres are planted in potatoes, onions, and tomatoes, and how many have been left unplanted. Then draw a pie chart that shows how the total acreage is divided.

2. A large diesel truck travels approximately 6 miles on a gallon of fuel. Determine how many gallons of fuel are required to travel 500 miles, 750 miles, 1,000 miles, 1,250 miles, 1,500 miles, and 1,750 miles. Draw a line graph that compares the number of miles traveled with the number of gallons of fuel used. Then, from your graph, determine how many gallons of fuel are required to travel 300 miles, 800 miles, 1,200 miles, and 1,600 miles.

SKILL DRILLS

A. Find the sum in each problem.

1. 543.26 + 26.09 + 832.093 + 78.32 + 510.978 =
2. 45,019 + 8,321.92 + 54.912 + 6,382.028 + 17.734 =
3. 216.436 + 64.7548 + 5,364.65 + 76.986 + 214.536 =
4. 7,142.34 + 43.525 + 231.54 + 98.045 + 42.634 =
5. 325.634 + 657.859 + 343.769 + 3245.98 + 34.733 =

B. Find the difference in each problem.

1. 191.324 − 121.007 =
2. 4,352.657 − 4,232.0187 =
3. 621.231 − 16.7422 =
4. 987.323 − 437.0035 =
5. 4,325.52 − 3,765.99 =

C. Find the product in each problem. Round calculations to three decimal places.

1. 3,875.14 × 0.3541 =
2. 876.01 × 6.547 =
3. 21.423 × 87.34 =
4. 76.32 × 8.4354 =
5. 643.2 × 65.322 =

D. Find the quotient in each problem. Round calculations to one decimal place.

1. 82.218 ÷ 76.32 =
2. 216.43 ÷ 332.28 =
3. 3.1241 ÷ 3.13 =
4. 9.5 ÷ 4.5264 =
5. 12.423 ÷ 29.09 =

P A R T

Data and Information Processing

IN THIS PART

Computer operations are covered in this final part. Computers are expensive, and their costs must be justified. Mathematics is used to calculate the value of computers in business. As you will see, there are two ways to view the advantages of a computer. One is the increased work or business capacity that a computer can provide. The other is the cost savings that can result from increased efficiency. Mathematics helps you see the whole picture, which involves both areas.

Computers can be extremely efficient tools for processing information. Mathematics provides the means to schedule work so as to make the best use of a computer's capacity.

Computer Operations

YOUR LEARNING JOB

When you have completed the assignments and exercises in this unit, you should be able to:

- ☐ Explain the four functions of a computer.
- ☐ Describe how an algorithm can be written to guide the step-by-step processing of a computer program.
- ☐ Calculate the value of a computer based on its tangible benefits.
- ☐ Describe the four parts of an information processing cycle.
- ☐ Schedule a computer based on its capacity and the amount of data to be processed.

WARM-UP

A. Find the sum in each problem.

1. 3,768.99 + 15.234 + 1,721.633 + 3.437 + 836.9 =
2. 347.329 + 328.582 + 37.526 + 29.3872 + 21.214 =
3. 26.1026 + 23.0972 + 1.382 + 52.016 + 264.961 =
4. 9,624.918 + 65.017 + 144.9156 + 7.2391 + 823.92 =
5. 671.9176 + 287.671 + 92.765 + 98.7154 + 20.0168 =

B. Find the difference in each problem.

1. 928.278 – 93.2826 =
2. 5,432.273 – 2,187.0019 =
3. 23.2371 – 1.00293 =
4. 2.9473 – 2.039275 =
5. 341.425 – 54.79373 =

C. Find the product in each problem. Round calculations to three decimal places.

1. 5.21 × 92.927 =
2. 265.41 × 19.9267 =
3. 93.039 × 2.934 =
4. 463.187 × 58.29 =
5. 918.5725 × 9.25 =

D. Find the quotient in each problem. Round calculations to one decimal place.

1. 650.273 ÷ 83.8 =
2. 82.197 ÷ 34.64 =
3. 95.324 ÷ 45.98 =
4. 75.25 ÷ 65.38 =
5. 7,652.1 ÷ 634.93 =

COMPUTERS AS TOOLS

Computers are tools used by people to change *data* into useful *information*. Data are raw facts and figures that, by themselves, have no meaning. Examples of data might be the letters A and L and the numbers 1 and 5. Information is data that have been given meaning by *processing*. A *process* is a series of steps that are followed to reach a known result. Processing of data is the performance of certain functions, or steps, to change meaningless data into information, or knowledge. The data examples above, after processing, might read AL and 15—someone's first name and age.

Data processing, then, is something that people do all the time. Computers are used to increase people's capacity for processing data.

As discussed in Unit 1, one important difference between computers and calculators lies in the way processing operations occur: Calculators must receive all data and instructions, one at a time, from humans. Computers have the ability to store and follow sets of instructions called programs.

Programs are written by people to tell computers what steps to follow in processing data. To qualify as a computer, a device or set of equipment must have several capabilities. A computer must be able to:

- Accept data items as inputs
- Process data to produce information
- Deliver results in usable form, called outputs
- Store data, information, and programs.

Because of these capabilities, computers are different from other tools that people have used in the past. Most tools are under total control of people. Thus, if you are driving a nail, the hammer depends on you for the movement necessary to get the job done. If you are tightening a screw, you don't expect the screwdriver to turn itself. But, with a computer, you can expect the tool to do the work on its own. All you have to do is figure out what is to be done. You do this in advance. Then you enter instructions that control the computer, which goes ahead and does the work. These sets of instructions are called *programs*.

Computer systems are tools for solving problems. COURTESY OF NORTHERN TELECOM, INC.

In solving problems with computers, you use some special principles of mathematics. One of these principles is that sets of numbers are separate from problems to be solved. The basic numbers are separate from the math to be performed upon them. For example, consider the mathematics of paying people for the work they do. People who work for a company deal in only one part of a payroll problem: time on the job. Workers report the length of time they work, sometimes on individual, separate assignments.

As far as the information system is concerned, data to be input are worker number and hours. This approach to solving problems is different from what you do with a pencil and paper. If you wanted to figure your own earnings, you would need more than hours worked. You would have to know your pay rate. You would also need information on income tax and other deductions. You wouldn't need your employee number because you know who you are. The computer doesn't know you. But it is built to do some amazing things when it comes to storing and processing numbers.

Data storage is a basic part of any computer system. PHOTO BY LISA SCHWABER-BARZILAY

For a computer system, the data needed to pay you consist only of your employee number and hours worked. In computer terms, these are *data items.* A data item is any unit of information that can be used in describing a person, place, thing, or event. Employee number and hours worked are data items. Where you are concerned, other data items might be name, color of eyes, height, weight, and other descriptive facts. In other words, data items are basic facts that can be processed to develop greater meaning. The processed data, in turn, produce information. All of the data items about a person combine to produce information in the form of identity. One collection of data items represents Harvey Heinz. Others represent Julio Gomez, Shizu Ito, Anita McShane, and so on.

Now, return to your payroll calculation. From the input about your number and hours worked, the computer needs to add other data items. These items include your pay rate, deductions, tax status, etc. These other items exist in a device that also helps determine that a collection of equipment is a computer. The device is a *storage* unit. The data about you are recorded on tape or round, flat disks that are coated with magnetic materials. A computer tape or disk may store thousands, even millions, of data items. The trick is to find the data about you, and to find it in perhaps hundredths of a second. To do this, the computer uses a special kind of mathematical function, *compare.*

Computers are sometimes called electronic brains. Of course, computers don't think. They are collections of equipment, wires, and electronic parts. The closest a computer comes to thinking is simple comparison. That is, the computer can compare two values. For example, the computer can tell whether one number is equal to, smaller than, or greater than another. This kind of math is important for such jobs as payrolls.

The computer can be given a value, such as 40. Then, the computer can be told to compare the number of hours you worked with 40. If your hours worked value is greater than 40, you earn overtime. If hours worked are equal to or less than 40, there is no overtime. Making comparisons in mathematics is known as *comparison logic* or *computer logic.* Comparison logic is a special capability of computers.

You use comparison logic in your own decision making. For example, you may look at your watch and learn it is time to go home. When you do this, you are applying judgment. You know the meaning of home and the value of time. The computer isn't that smart. Every processing function performed by a computer requires advance thought—and instructions—from a person.

To use computers, people must learn to think in the step-by-step processing terms that are required. One way to do this is to prepare an algorithm as a way of thinking through needed processing. The algorithm below gives some of the steps in payroll processing. This is not a complete algorithm, but a set of steps designed to illustrate comparison logic. Note the terms IF, THEN, and ELSE. These are basic instructions for making comparisons. The algorithm:

- ENTER employee number and hours worked.
- COMPARE hours worked with 40.
- IF hours worked >40 (are greater than 40),
- THEN subtract 40 from hours worked.
- MULTIPLY difference by 1.5.
- ADD product to 40, increasing hours worked.
- ELSE use actual hours worked.
- MULTIPLY hours by rate.
- STORE product as gross pay.

To demonstrate how this works, assume you earn $5.25 per hour. This week, you worked 46 hours. The data items 46 and your employee number are entered into the computer. The computer looks up the data on you, using your number as a *record key*. In information processing, a key is a basis for finding information. At work, your employee number would be a key. At the bank, your account number might be a key. Other keys you will use frequently include your social security number and driver's license number.

A record key is used to find a collection of data items known as a *record*. Records, simply, are collections of data about the same person, place, thing, or event. All the information a company has about you would be in your employee record.

For a week in which you work 46 hours, your payroll calculations would be as follows:

$$46 > 40$$
$$46 - 40 = 6$$
$$6 \times 1.5 = 9$$
$$40 + 9 = 49$$
$$49 \times \$5.25 = \$257.25$$

EXERCISE 15-1

Read each question carefully before calculating the answer. Round calculations to the nearest cent.

1. From the weekly payroll data presented below, determine the number of earned hours and calculate gross pay. Employees who work more than 40 hours are entitled to overtime pay (time-and-a-half).
   - **a.** Rick Mitchell earned $8.50 per hour and worked 48 hours.
   - **b.** Jose Cardinas earned $12.50 per hour and worked 52 hours.
   - **c.** David Weinstein earns $4.50 per hour. He has worked 39 hours.
   - **d.** Shizo Ito worked 44 hours. Her pay rate was $6.50 per hour.
   - **e.** Trudi Marshall worked 52 hours. Her pay rate was $11.50 per hour.

2. From the data given below, determine the interest earned by each account for one year. Interest is compounded annually. Accounts with a balance under $2,000 earn 5.5% interest. Accounts with a balance between $2,000 and $10,000 earn 8.5% interest. Accounts with a balance over $10,000 earn 10% interest.

a. Marlene has $1,500 in her account.

b. Ken has $15,000 in his account.

c. Heidi has $3,875.35 in her account.

d. Ignacio has $8,358.64 in his account.

e. Sarah has $11,479.94 in her account.

3. You work for a distributor. From the data presented below, determine the amount the customer should be charged for the order. Orders that exceed $2,000 before sales tax is charged receive a 10% discount. Orders that are more than $5,000 before sales tax is charged receive a 15% discount. The sales tax rate is 6.5%.

a. O.K. Auto Parts orders 150 rubber mats that sell for $20 each.

b. Fred's Furniture orders 250 office chairs. The price of each chair is $24.95.

c. Mary's Fast Food orders 2,000 packages of hamburger buns that sell for 97 cents per package.

d. Plastimold orders 7,500 pounds of plastic pellets. The price is $1.15 per pound.

e. Canyon Grocery Supply orders 500 cases of soft drinks. Each case sells for $4.05.

COMPUTERS AND COSTS

Computers are tools for performing mathematical calculations. However, you can turn that statement around and say that mathematics is a tool for determining the value of computers.

It is important to measure value because computers can be extremely expensive. Measuring values mathematically helps to determine if the computers are worth what they cost.

For example, think about buying a desk-top computer for use as a *word processor* in a small business office. A small computer loaded with special word processing programs can capture and store text. Word processing computers can be used for editing, revising, formatting, and printing of text files. Such files may consist of letters, reports, and other documents.

The executive who considers buying a computer must compare the costs of the equipment with its anticipated benefits. The purchase is justified if the benefits outweigh the costs within a reasonable time.

Suppose that the computer and the word processing *software* (programs and operating instructions) cost $2,500. There is an equation to determine if and when the equipment will pay for itself:

$$\text{Investment} = \text{Unit savings} \times \text{Units}$$

Now, assume that an office worker retypes a typical letter four to six times at a cost of $2 per copy. With word processing equipment, the body of the letter would be stored in memory. To prepare additional copies for mailing, only the heading and salutation would have to be typed. The ability of the computer to reproduce the body of the letter would save $1.25 per page. How many pages would have to be produced in this manner to justify the purchase price of the computer? Use the same equation:

$$\text{Investment} = \text{Unit savings} \times \text{Units}$$

$$\text{Units} = \frac{\text{Investment}}{\text{Unit savings}}$$

$$\text{Units} = \frac{2{,}500}{1.25} = 2{,}000$$

Thus, at a saving of $1.25 per page, the computer's price will be matched when 2,000 pages have been produced. With this information, the executive can determine the length of time it will take for the computer to pay for itself.

To calculate the time, think about the production level of the typist. Suppose the typist had been producing 40 copies of letters per day without the computer. Divide 2,000 by 40 and the result is 50 working days.

Actually, the real advantage of the computer is the increased production it permits. The typist may still work an eight-hour day. With the word processor, however, he or she can produce two and two-thirds times as many letters in a day. This produces a rate of more than 100 letters a day. Thus, the price of the computer could be realized in less than 20 working days.

The point is that, once a computer has saved its initial cost, it continues to increase efficiency and speed. Also, the savings that pay for the computer keep mounting after the break-even point is reached.

EXERCISE 15-2

Read each question carefully before calculating the answer. Round calculations to two decimal places, or to the nearest cent.

1. Caroline takes orders over the phone. She can fill out one order form every 2.5 minutes by hand. Using a computer, she could fill out three order forms every 2.5 minutes. By using a computer, how many more orders could Caroline take during an eight-hour period?

2. Carl is a manager. He spends an average of six hours every day calculating budget amounts and constructing budget reports. With a computer, Carl could perform the same function in one-third the time. By purchasing a computer, how many more hours, in a five-day week, could Carl spend managing employees?

3. Dana is a writer. On a typical day, she spends five hours retyping manuscripts. With a word processor, Dana could reduce her retyping time by three-fourths. By purchasing a word processor, how many more hours, per day, could Dana devote to writing? If her average income was $15 per hour, and the computer cost $2,500, how many days would it take to recover the cost?

4. The Mail House employs three secretaries to type labels for envelopes. Each secretary works six hours per day and earns $6.50 per hour. With a computer, names and addresses would have to be typed only once. The computer then could generate the mailing labels as needed. Management believes that, by purchasing a computer, the company could reduce secretarial costs by two-thirds. How much money would the company save each day by purchasing a computer? If the computer cost $3,000, how many work days would it take to recover the purchase price?

5. At the end of each month, Primer Printers sends bills to its customers. It takes two secretaries to add up charges, subtract payments, and type the bills, or statements. Each secretary works for 42 hours and earns $7.50 per hour. With a computer, the time required to prepare and mail billing statements would be cut in half. If the computer system cost $2,250, how many months would it take to recover the expense?

Word-processing computers greatly improve business efficiency.

COMPUTERS AND INCREASED BUSINESS

Think about another use of computers, one with which you probably come in contact frequently: supermarket check-out counter equipment. Most supermarkets have scanning equipment that "reads" code labels on packaged products. The product code data are processed by a computer, which adds the price of the item to the running sales total automatically.

Computerized check-out systems have a big job to do. A typical supermarket probably stocks 80,000 to 100,000 items. The tasks performed by computer check-out systems can include:

- Finding data on products in computer files and displaying the information at the check-out counter in *milliseconds* (thousandths of a second)
- Accumulating data on all items sold (typically 100,000 a week or more)
- Keeping track of inventory and reordering merchandise to replace that which has been sold
- Analyzing sales
- Analyzing the use of shelf space (based on sales volumes).

Clearly, this volume of work requires a computer. The efficiency of operation in modern supermarkets could not be attained by people operating mechanical devices. This efficiency level is an *intangible* benefit of computers. Intangible means unable to be assigned a measurable value.

However, such computer systems do offer *tangible,* or measurable, benefits. Think about how the cost of a supermarket check-out computer system can be justified. First, look at the approximate cost of such a system for a single store. Assume the following costs:

Equipment	Cost
Check-out terminals (15 @ $10,000)	$150,000
Computer	150,000
Installation and communication equipment	25,000

The supermarket's management has committed $325,000 per store for a computer system. The tangible benefits of such a

Computerized check-out counters with optical scanning readers have increased the efficiency of retail store employees.
PHOTO BY LISA SCHWABER-BARZILAY

system can be measured in different ways. One way is to compare check-out counter operating costs. Suppose that the supermarket has a retail clerk and a bagger at each counter. The hourly cost of the two employees might be $17.

If the computer system increases check-out efficiency by 50%, the effect is the same as cutting those salaries in half. Thus, the computer's value can be measured at $8.50 per hour, per check-out counter. If all 15 counters were operating continuously during a 12-hour store day, the savings would be calculated this way:

$$\text{Check-out savings} = \$8.50 \times 15 \times 12 = \$1{,}530$$

At a savings of $1,530 a day, the computer would pay for itself after 213 days of use.

Notice, however, that the major advantage of the computer is identified above as increased efficiency. The computer does not cause half of the check-out employees to be dismissed. The idea of having a computer is not so much to save costs as to increase business volume and income. The computer permits employees to serve customers more rapidly. This means that money is collected faster than before.

The computer's ability to analyze sales and use of shelf space also enables management to react more quickly to customer demand. The store is less likely to run out of items that customers want to buy. On the other hand, the store also is less likely to overstock items that sell slowly. It all translates into greater profit levels through increased efficiency.

EXERCISE 15-3

Read each question carefully before calculating the answer. Round answers to two decimal places, or to the nearest cent.

1. Mountain Market has four check-out counters. It costs $15 per hour to operate each counter. The market is considering the purchase of a computer system for $50,000. The computer would reduce the cost of operating check-out counters by 75%. If the market is open 10 hours per day, how much money would be saved each day? How many days would it take to recover the cost of the computer system?

2. Joe's Market can process an average of 25 customers per hour through three check-out counters. With a computer system, the number of customers the market could process would be increased 2.5 times. How many more customers could the market process in a 10-hour period by using computerized check-out counters?

3. A large supermarket employed five workers to monitor inventory levels. Each employee worked 35 hours per week and earned $8.50 per hour. When the store set up a computerized system, the cost of monitoring inventory was cut in half. How much money did the store save per week on inventory cost?

4. With a manual system, a large supermarket lost $250 in a typical week due to cashier errors. When a computerized check-out system was installed, this loss was reduced by 75%. How much money was saved each week? If the store was open 52 weeks per year, how much did the computer save the company during the first year?

5. Management at a large supermarket predicts that a computerized system will cut the cost of inventory management by 85%. In addition, the cost of operating check-out counters will be reduced by 30%. Currently, the company spends $175,000 per month on management of inventory and $145,000 on operating check-out counters. How much money will the supermarket save per month? If the computer system costs $350,000, how many months will it take to recover the cost?

MATHEMATICS TO BUILD COMPUTER SYSTEMS

Computers are tools for solving problems. In business, computers are used to process data into usable information. This function is called *information processing*. Information processing is not a new idea. People always have processed information. Computers are modern tools that have made it possible for people to improve data processing methods and capabilities.

With or without computers, all *information processing systems* follow a series of steps, in order. The system enters and processes data to produce information. The series of steps is known as the *information processing cycle.* An information processing cycle, illustrated in Figure 15-1, consists of four parts:

- Input
- Processing
- Output
- Storage.

These four steps are basic to any information processing system, whether data are processed by hand or by computer. Computers make possible the processing, delivery, and storage of large amounts of data. Mathematics is used extensively throughout this cycle to assure accuracy of processing.

Input. Before data are input, controls are established. A typical input control technique is to total values of input items prior to

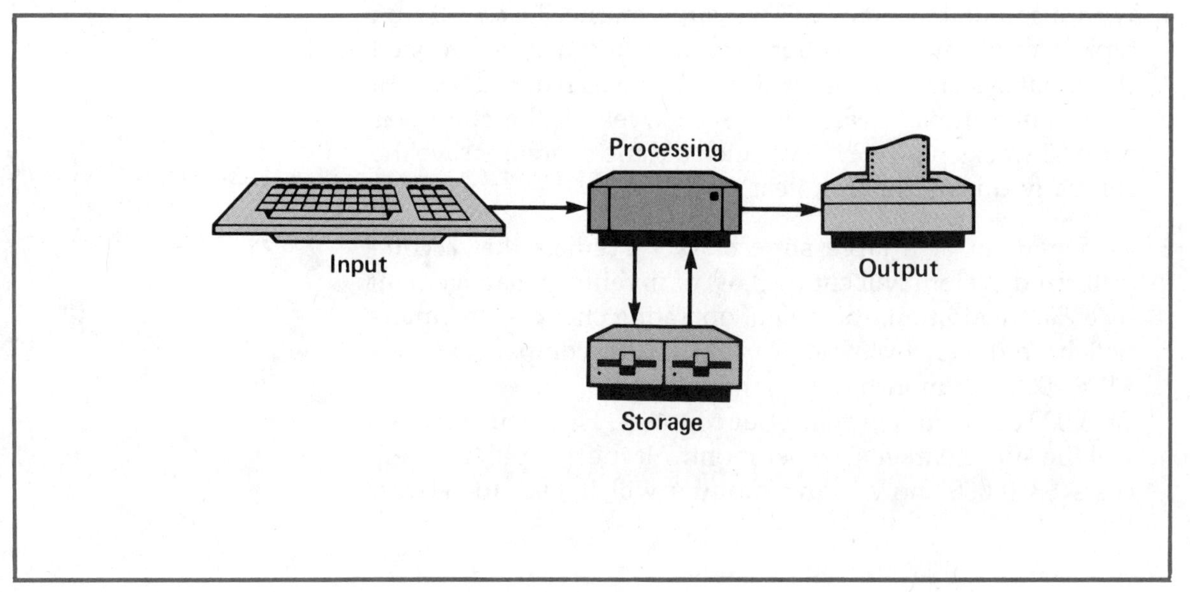

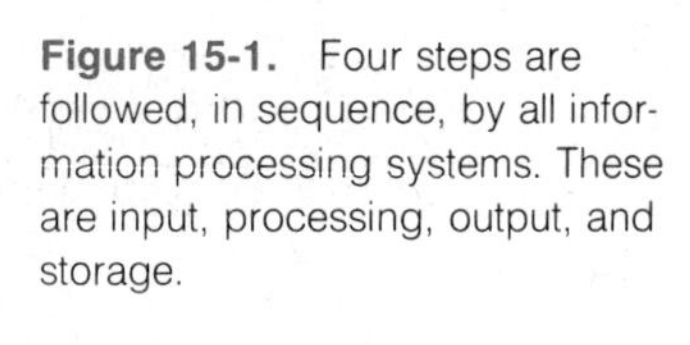
Figure 15-1. Four steps are followed, in sequence, by all information processing systems. These are input, processing, output, and storage.

processing. For example, when you make a bank deposit, you fill out a deposit slip. On this slip, you list all checks to be deposited. If you have cash to deposit, the amount of money is entered also.

In the bank, the amount of your deposit is checked by a teller. Then the teller groups your deposit with perhaps 50 or 100 others. The teller adds all deposited items to produce a *control total.* A control total is an amount that represents a value for all data items to be input. An example would be the total value of all checks in a deposited group.

Processing. A group, or *batch,* of data items then is entered into a computer for processing. The computer, in turn, totals the amounts of all the items that are entered. The total developed by the computer is then compared with the batch control amount. These two items must be equal, or must match, before processing takes place.

Output. A special output operation is used to establish control over an information system. This is a listing of all items that are input and processed as a batch. Again, the output must be balanced to the input to be sure of accuracy.

Storage. When data are stored, controls are recorded along with them. Groups of data records are kept on tape or magnetic disk.

Input to computer of patients' menu selections is done by use of a special optical mark reader. This computerized system reduces menu handling and frees hospital employees for other duties.
COURTESY OF SCAN-TRON CORPORATION

For each set of records, or *file,* a control record is created. This special record contains a count of all records in the file as well as a control total for value. Thus, in records of bank accounts, the control record would show the number of accounts stored. Also shown would be a total for the value of deposits in all the accounts.

When new deposits are added to a file of account records, the total is changed. The new control total, then, must be equal to the previous total, plus the new deposits.

Computer files are sets of records. COURTESY OF WANG LABORATORIES, INC.

To see how the information processing cycle works, consider this example:

- Deposit 100 checks with a total value of $4,534.98.
- Send checks to input section of computer center. Check the total. Produce an input record containing the control total.
- Enter and process deposit information. In processing, the computer develops a total.
- Check computer total against input control total. Both must be equal to $4,534.98. If not, recheck inputs. If totals are equal, use deposit records to update customer account files.
- When deposits are processed, develop a total for value of all records in file. Add deposit control total to previous total for file. The new total must be equal to the old total plus $4,534.98. Thus, output and storage operations are checked back to input controls.

EXERCISE 15-4

For each question below, think of yourself as a bank teller. You are given a list of deposit amounts. From these data, produce a control total.

1. $1,243.54; $500; $350; $75.45; $104.50; $750; $1,500; and $356.84.

2. $10.45; $305; $5,690; $212.98; $3,450; $23.75; and $365.47.

3. $5,000; $1,200; $3,287.98; $675.28; $1,283.76; $350; $750; $313.33; and $980.

4. $3,600; $250; $450; $2,093.65; $560; $1,200; $410; $569.87; and $3,250.

5. $5,500; $1,290.65; $335.55; $275.99; $2,374.13; $560; $13,209.46; $25,000; and $1,357.99.

6. $4,574.81; $567.32; $51.25; $815.35; $8,100; $3,598.29; $579.75; $23,456.84; $4,238.87; and $300.

Scheduling is an important part of computer operations. Understanding the capacity of equipment is necessary to make good use of computers.

SCHEDULING COMPUTER JOBS

As fast as they are, computers still have limits on their *capacities*, or rates at which they do work. For example, the text of this book was prepared on a computer. For checking purposes, the text would be printed out. The printer used has a capacity of 55 characters per second. Each page has 27 lines with up to 78 characters per line. Thus, the maximum time required to print one full page of copy (material to be printed) is:

$$78 \times 27 \div 55 = 78 \times \frac{27}{55} = \frac{2{,}106}{55} = 38.29$$

Rounding this number gives a printing speed of 38 seconds per page. Suppose that the book text runs 300 pages of single-spaced copy. At 38 seconds per page, the printer can produce the full text in a little over three hours. The calculation is performed by multiplying the number of pages by the time required to print each page. Dividing the product by 60 gives the time in minutes.

$$300 \times 38 = 11{,}400$$
$$11{,}400 \div 60 = \frac{11{,}400}{60} = 190$$

Since there are 60 minutes in an hour, the job would require 3 hours 10 minutes.

Once this text had been checked and corrected, it was transmitted to another computer at the typesetting company. This transmission was done over telephone lines through the use of a *modem*. A modem is a device that changes data into a form that can be transmitted over telephone lines. The signals are changed back into usable form by another modem at the receiving end.

The computer can transmit data in this manner at a rate of 1,200 *bits* per second. A bit is a single electrical pulse. In this computer system, nine bits equal one *byte*. A byte is a basic unit of data. In this system, 256 bytes form a *block* of text, which roughly equals 50 typewritten words.

Say that a book text contains 2,856 blocks. You want to know how long it will take to transmit the text to the typesetter.

First, convert blocks to bits. Then, divide the total number of bits by the transmission rate of 1,200 bits per second. Divide that answer by 60 to determine the time in minutes.

$$2{,}856 \times 256 \times 9 = 6{,}580{,}224$$
$$6{,}580{,}224 \div 1{,}200 = \frac{6{,}580{,}224}{1{,}200} = 5{,}483.52$$
$$5{,}483.52 \div 60 = \frac{5{,}483.52}{60} = 91.39$$

Thus, total transmission time would be 91.39 minutes, or approximately 1 hour 31 minutes. The computer's capacity on a single transmission might be about 500 blocks. Therefore, six transmissions would be necessary. You might have to block out an additional 4 minutes per transmission. This added time would be used to load data to be sent and to check data received.

Computer scheduling requires knowledge of a computer's capacity and the amount of data to be processed.

SUMMING UP

- People process data to change it into useful information, or knowledge. Computers are special tools used to increase people's capacity for processing data.
- A computer must be able to do four things. First, it must accept data items as inputs. Second, it must process data items to produce information. Third, it must deliver results in usable form, called outputs. Fourth, it must store data, information, and programs.
- Computers differ from other tools in that they can operate on their own, once people give them instructions.
- Special principles of mathematics are used in working with computers. It is important to understand these concepts: Numbers are different from problems to be solved, and numbers are different from the mathematics performed on them.
- To work with computers, people must learn to think in the step-by-step processing terms required by these devices. Algorithms help you in organizing the steps by which a computer can be used to solve problems.
- Comparison is a special mathematical function that is the closest a computer comes to thinking. A computer can tell whether one number is larger than, equal to, or smaller than another number. This ability to compare numbers is called computer logic.
- Mathematics is used to determine the value, or usefulness, of computers. Computers can be extremely expensive. Therefore, it is important to be able to measure their value in mathematical terms. Computers increase efficiency and the capacity for doing business. Some computer benefits are tangible, or measurable; others are intangible.
- An information processing cycle has four parts: input, processing, output, and storage.
- Scheduling work for a computer requires knowledge of the computer's capacity and of the amount of data to be processed.

TESTING YOUR WORKING KNOWLEDGE

Read each question carefully before calculating the answer. Round calculations to the nearest cent or to two decimal places.

1. Your job is to process orders for a lumber yard. Orders over $1,500 receive a 10% discount. Orders over $5,000 receive a 15% discount. The sales tax rate is 5.5%. How much would an order for 500 sheets of 4-by-8 plywood that is $\frac{1}{2}$ inch thick cost if one sheet sells for $7.99?

2. Mable earns $9.75 per hour and works 46.5 hours during a week. Calculate her gross pay. (Note: Use the earned-hours method for overtime hours.)

3. At the end of each month, Eagle Machining sends bills to customers. It takes 124 hours per month to add up charges, subtract payments, and type bills, or statements. Secretaries are paid $6.75 per hour to produce the statements. With a computer, the time required to prepare and mail billing statements would be reduced 60%. If the computer system cost $2,300, how many months would it take to recover the initial cost?

4. A large supermarket employed eight workers to monitor inventory levels. Each employee worked 40 hours per week and earned $6.50 per hour. When the store implemented a computerized system, the costs of monitoring inventory were reduced 65%. How much money did the store save per week on inventory costs?

5. A large market purchased a computer system to help reduce the amount of money being lost due to cashier errors. By implementing the system, weekly losses were reduced 75%, to $87.50. How much were weekly losses before implementation of the computer system? If the store was open 51 weeks per year, how much did the company save during the first year?

6. You are a bank teller. The following deposits were made at your window: $3,546; $4,211.98; $1,765.43; $343.68; $2,123.12; $213; $330; $343.09; and $6,980.92. From these data, produce a control total that can be used to verify the accuracy of computer inputs.

TERMS THAT COUNT

data
information
process
processing
program
data item
storage
compare
comparison logic
computer logic
record key
record
word processor
software
millisecond
intangible
tangible
information processing
information processing system
information processing cycle
control total
batch
file
capacity
modem
bit
byte
block

PRESENTING INFORMATION

Read each question carefully. Present the information in the format required. Round calculations to the nearest cent or to two decimal places.

1. Hal bought a computer system for $3,500. The computer itself cost only $2,200. To complete the system, Hal also purchased a printer for $500 and software (programs) for $800. Draw a pie chart that shows how the cost of the computer, printer, and software compare with the total cost of the system.

2. On Monday, salespeople at Palmer Plastics took 150 orders over the phone. On Tuesday, they took 165 orders. On Wednesday, 155 were received. On Thursday, 145 were received. On Friday, 185 orders came in. Management believes that a computer would allow salespeople to shorten the time it takes to fill out order forms. As a result, the number of orders taken per day would be increased 25%. You have been assigned to present this information in a clear format. To perform your job, draw a two-line graph. One line should show the number of orders salespeople received each day. The second line should show a 25% increase. Then, shade the area on the graph that represents the increase in sales.

SKILL DRILLS

A. Find the sum in each problem.

1. 5,323.09 + 24.54 + 2,412.56 + 6.33 + 887.90 =
2. 315.65 + 621.41 + 19.45 + 3,429.72 + 2,761.14 =
3. 29.26 + 4,323.97 + 198.80 + 947.55 + 375.34 =
4. 19,324.76 + 325.45 + 1,943.96 + 7,325.91 + 801.65 =
5. 176.16 + 823.78 + 835.25 + 5,695.21 + 2,473.68 =

B. Find the difference in each problem.

1. 45,932.78 − 32,393.26 =
2. 75,213.75 − 63,231.19 =
3. 1,209,472.31 − 1,199,421.93 =
4. 2,432,534.43 − 2,116,322.25 =
5. 34,736,266.29 − 25,918,312.33 =

C. Find the product in each problem. Round calculations to three decimal places.

1. 1,235.45 × 2.27 =
2. 25.91 × 356.43 =
3. 75.39 × 9.154 =
4. 4,783.7 × 39.96 =
5. 63.3221 × 19.73 =

D. Find the quotient in each problem. Round calculations to one decimal place.

1. 917.73 ÷ 3.85 =
2. 872.97 ÷ 65.32 =
3. 83.25 ÷ 39.12 =
4. 778.18 ÷ 18.18 =
5. 9,321.2 ÷ 432.03 =

Rules for Fractions and Decimals

ADDING FRACTIONS

1. If you have mixed numbers, add the fractions first, then add whole numbers separately.
2. If necessary, convert unlike fractions to fractions with least common denominators.
3. Add the numerators and write the sum over the least common denominator.
4. If the addition of fractions produces an improper fraction, convert it to a mixed number and carry the whole number into the units column.
5. Express the fraction in the sum in lowest terms.
6. Add the whole numbers.

MULTIPLYING FRACTIONS

1. Write out the problem horizontally.
2. Convert mixed or whole numbers to improper fractions.
3. Cross-cancel where possible.
4. Multiply numerators by each other.
5. Multiply denominators by each other.
6. Convert improper fractions to mixed numbers.
7. Express the product in lowest terms.

SUBTRACTING FRACTIONS

1. Convert unlike fractions to fractions with least common denominators.
2. If you have mixed numbers, work with the fractions first, borrowing from the whole unit if necessary.
3. Subtract the numerators and write the difference over the common denominator.
4. Express fraction in difference in lowest terms.
5. Subtract whole numbers.

DIVIDING FRACTIONS

1. Write the problem out horizontally with the divisor in the second position, after the division sign.
2. Convert any mixed numbers to improper fractions.
3. Invert the divisor and change the division sign to multiplication.
4. Multiply following the rules for multiplication of fractions.
5. Convert any improper fractions to mixed numbers.
6. Express the answer in lowest terms.

VALUES OF WHOLE NUMBERS AND DECIMALS

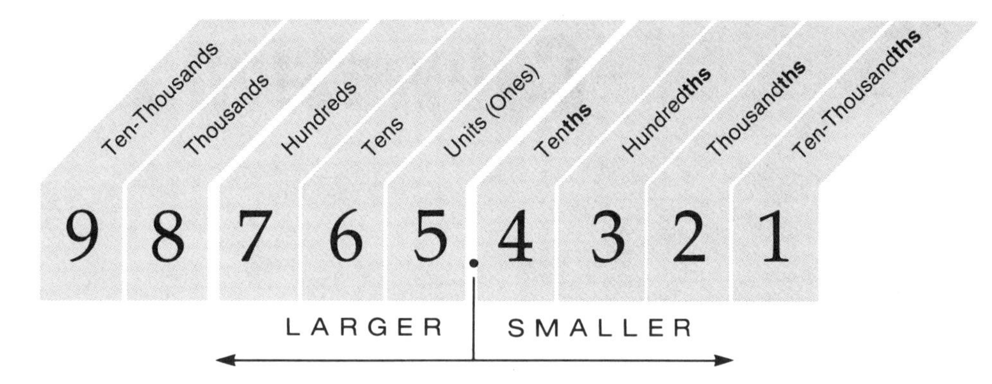

ADDING DECIMALS

1. Line up the decimal points in all of the addendends.
2. Add as you do with whole numbers.
3. If whole numbers are included, the decimal point is at the right. If some decimal numbers have fewer places than others, add zeroes to help you keep track of the places. Adding zeroes to the right does not change the value of decimal numbers.
4. If the addition of a decimal produces a whole number, carry the whole unit over to the units column.

MULTIPLYING DECIMALS

1. Solve the problem as though the decimals were not there, lining up the factors (the multiplicand and the multiplier) so their digits are under each other from right to left.
2. Count the total number of digits **to** the right of the decimal points in both the multiplicand and the multiplier.
3. Count the same total number of digits **from** the right in the product and insert the decimal point.

SUBTRACTING DECIMALS

1. Line up the decimal points of the minuend and the subtrahend.
2. Subtract as you do with whole numbers.
3. If whole numbers are included, the decimal point goes to the right. You can add zeroes to the right without changing value.
4. If necessary, you can borrow from a whole unit by adding the whole equivalent of the decimal value.

DIVIDING DECIMALS

1. Make the divisor a whole number by moving the decimal point to the right if necessary.
2. If you move the decimal point in the divisor, you must also move the decimal point the same number of digits or places to the right in the dividend.
3. Put a decimal point in the quotient directly above the one placed in the dividend.
4. Add zeroes to the right in the dividend if necessary.
5. Solve the problem as though you were dividing whole numbers. Line the quotient up with the dividend.

Units of Measurement: Conversion Factors

All of the conversion factors below indicate multiplication; however, it is possible to convert by dividing. To divide, use the reciprocal of any indicated multiplier as divisor. For example, to convert from inches to feet, refer to the table headed "To Convert from Feet" and use the factor listed at "inches" (12) as a divisor. Similarly, to convert from centimeters to inches by division, use the table headed "To Convert from Inches" and use the factor listed at "centimeters" (2.54) as divisor.

UNITS OF MASS

TO CONVERT FROM SHORT TONS

TO	MULTIPLY BY
Long Tons	0.892 857 1
Kilograms	907.184 74
Metric Tons	0.907 184 74

TO CONVERT FROM LONG TONS

TO	MULTIPLY BY
Short Tons	1.12
Kilograms	1 016.046 908 8
Metric Tons	1.016 046 908 8

TO CONVERT FROM METRIC TONS

TO	MULTIPLY BY
Short Tons	1.102 311 3
Long Tons	0.984 206 5
Kilograms	1 000

TO CONVERT FROM KILOGRAMS

TO	MULTIPLY BY
Grams	1 000
Short Tons	0.001 102 31
Long Tons	0.000 984 2
Metric Tons	0.001

UNITS OF LENGTH

TO CONVERT FROM METERS TO	MULTIPLY BY
Inches	39.370
Feet	3.280
Yards	1.093
Miles	0.000 62
Millimeters	1 000
Centimeters	100
Kilometers	0.001

TO CONVERT FROM FEET TO	MULTIPLY BY
Inches	12
Yards	0.333 333 3
Miles	0.000 189 39
Centimeters	30.48
Meters	0.304 8
Kilometers	0.000 304 8

TO CONVERT FROM CENTIMETERS TO	MULTIPLY BY
Inches	0.393
Feet	0.032 8
Yards	0.010 9
Meters	0.01

TO CONVERT FROM YARDS TO	MULTIPLY BY
Inches	36
Feet	3
Centimeters	91.44
Meters	0.914

TO CONVERT FROM INCHES TO	MULTIPLY BY
Feet	0.083
Yards	0.028
Centimeters	2.54
Meters	0.025 4

TO CONVERT FROM MILES TO	MULTIPLY BY
Inches	63 360
Feet	5 280
Yards	1 760
Centimeters	160 934.4
Meters	1 609
Kilometers	1.609

UNITS OF LIQUID VOLUME

TO CONVERT FROM LITERS TO	MULTIPLY BY
Liquid Ounces	33.814 02
Liquid Pints	2.113 376
Liquid Quarts	1.056 688
Gallons	0.264 172 05
Cubic Inches	61.023 74
Cubic Feet	0.035 314 67
Milliliters	1 000
Cubic Meters	0.001
Cubic Yards	0.001 307 95

TO CONVERT FROM CUBIC INCHES TO	MULTIPLY BY
Cubic Feet	0.000 578 7
Milliliters	16.387 064
Liters	0.016 387 064
Cubic Meters	0.000 016 387 064
Cubic Yards	0.000 021 43

TO CONVERT FROM LIQUID OUNCES TO	MULTIPLY BY
Liquid Pints	0.062 5
Liquid Quarts	0.031 25
Gallons	0.007 812 5

TO CONVERT FROM CUBIC FEET TO	MULTIPLY BY
Cubic Inches	1 728
Liters	28.316 846 592
Cubic Meters	0.028 316 846 592
Cubic Yards	0.037 037 04

TO CONVERT FROM LIQUID PINTS TO	MULTIPLY BY
Liquid Ounces	16
Liquid Quarts	0.5
Gallons	0.125
Milliliters	473.176 473
Liters	0.473 176 473

TO CONVERT FROM CUBIC YARDS TO	MULTIPLY BY
Cubic Inches	46 656
Cubic Feet	27
Liters	764.554 857 984
Cubic Meters	0.764 554 857 984

TO CONVERT FROM

CUBIC METERS

TO	MULTIPLY BY
Gallons	264.172 05
Cubic Inches	61 023.74
Cubic Feet	35.314 67
Liters	1 000
Cubic Yards	1.307 950 6

TO CONVERT FROM

LIQUID QUARTS

TO	MULTIPLY BY
Liquid Ounces	32
Liquid Pints	2
Gallons	0.25
Cubic Inches	57.75
Cubic Feet	0.033 420 14
Milliliters	946.352 946
Liters	0.946 352 946

TO CONVERT FROM

GALLONS

TO	MULTIPLY BY
Liquid Ounces	128
Liquid Pints	8
Liquid Quarts	4
Cubic Inches	231
Cubic Feet	0.133 680 6
Milliliters	3 785.411 784
Liters	3.785 411 784
Cubic Meters	0.003 785 411 784
Cubic Yards	0.004 951 13

UNITS OF AREA

TO CONVERT FROM SQUARE CENTIMETERS

TO	MULTIPLY BY
Square Inches	0.155 000 3
Square Feet	0.001 076 39
Square Yards	0.000 119 599
Square Meters	0.000 1

TO CONVERT FROM SQUARE INCHES

TO	MULTIPLY BY
Square Feet	0.006 944 44
Square Yards	0.000 771 605
Square Centimeters	6.451 6
Square Meters	0.000 645 16

TO CONVERT FROM SQUARE METERS

TO	MULTIPLY BY
Square Inches	1 550.003
Square Feet	10.763 91
Square Yards	1.195 990
Acres	0.000 247 105
Square Centimeters	10 000

TO CONVERT FROM SQUARE FEET

TO	MULTIPLY BY
Square Inches	144
Square Yards	0.111 111 1
Acres	0.000 022 957
Square Centimeters	929.030 4
Square Meters	0.092 903 04

TO CONVERT FROM SQUARE YARDS

TO	MULTIPLY BY
Square Inches	1 296
Square Feet	9
Acres	0.000 206 611 6
Square Miles	0.000 000 322 830 6
Square Centimeters	8 361.273 6
Square Meters	0.836 127 36

TO CONVERT FROM ACRES	
TO	MULTIPLY BY
Square Feet	43 560
Square Yards	4 840
Square Miles	0.001 562 5
Square Meters	4 046.856 422 4

TO CONVERT FROM SQUARE MILES	
TO	MULTIPLY BY
Square Feet	27 878 400
Square Yards	3 097 600
Acres	640
Square Meters	2 589 988.110 336

Squares, Cubes, and Square Roots

The square of a number is the number multiplied by itself. Thus, 4^2 means 4×4 which equals 16. The cube of a number is the number multiplied by itself twice. Thus, 4^3 means $4 \times 4 \times 4$ which equals 64.

To find the square root of a number, it is necessary to find the number, or root, which when raised to the second power (or squared) will produce the square root number. Thus $\sqrt{16} = 4$ because $4^2 = 16$.

The table below gives the square, cube, and square root of numbers 1 through 100. The ''n'' represents ''number.''

n	n^2	n^3	$\sqrt{n}$	n	n^2	n^3	$\sqrt{n}$
1	1	1	1.00	15	225	3 375	3.873
2	4	8	1.414	16	256	4 096	4.000
3	9	27	1.732	17	289	4 913	4.123
4	16	64	2.000	18	324	5 832	4.243
5	25	125	2.236	19	361	6 859	4.359
6	36	216	2.449	20	400	8 000	4.472
7	49	343	2.646	21	441	9 261	4.583
8	64	512	2.828	22	484	10 648	4.690
9	81	729	3.000	23	529	12 167	4.796
10	100	1 000	3.162	24	576	13 824	4.899
11	121	1 331	3.317	25	625	15 625	5.000
12	144	1 728	3.464	26	676	17 576	5.099
13	169	2 197	3.606	27	729	19 683	5.196
14	196	2 744	3.742	28	784	21 952	5.292
				29	841	24 389	5.385

n	n^2	n^3	$\sqrt{n}$
30	900	27 000	5.477
31	961	29 791	5.568
32	1 024	32 768	5.657
33	1 089	35 937	5.745
34	1 156	39 304	5.831
35	1 225	42 875	5.916
36	1 296	46 656	6.000
37	1 369	50 653	6.083
38	1 444	54 872	6.164
39	1 521	59 319	6.245
40	1 600	64 000	6.325
41	1 681	68 921	6.403
42	1 764	74 088	6.481
43	1 849	79 507	6.557
44	1 936	85 184	6.633
45	2 025	91 125	6.708
46	2 116	97 336	6.782
47	2 209	193 823	6.856
48	2 304	110 592	6.928
49	2 401	117 649	7.000
50	2 500	125 000	7.071
51	2 601	132 651	7.141
52	2 704	140 608	7.211
53	2 809	148 877	7.280
54	2 916	157 464	7.347
55	3 025	166 375	7.416
56	3 136	175 616	7.483
57	3 249	185 193	7.550
58	3 364	195 112	7.616
59	3 481	205 379	7.681
60	3 600	216 000	7.746
61	3 721	226 981	7.810
62	3 844	238 328	7.874
63	3 969	250 047	7.937
64	4 096	262 144	8.000
65	4 225	274 625	8.062
66	4 356	287 496	8.124
67	4 489	300 763	8.185
68	4 624	314 432	8.246
69	4 761	328 509	8.307
70	4 900	343 000	8.367
71	5 041	357 911	8.426
72	5 184	373 248	8.485
73	5 329	389 017	8.544
74	5 476	405 224	8.602
75	5 625	421 875	8.660
76	5 776	438 976	8.718
77	5 929	456 533	8.775
78	6 084	474 552	8.832
79	6 241	493 039	8.888
80	6 400	512 000	8.944
81	6 561	531 441	9.000
82	6 724	551 368	9.055
83	6 889	571 787	9.110
84	7 056	592 704	9.165
85	7 225	614 125	9.220
86	7 396	636 056	9.274
87	7 569	658 503	9.327
88	7 744	681 472	9.381
89	7 921	704 969	9.434
90	8 100	729 000	9.487
91	8 281	753 571	9.539
92	8 464	778 688	9.592
93	8 649	804 357	9.644
94	8 836	830 584	9.695
95	9 025	857 375	9.747
96	9 216	884 736	9.798
97	9 409	912 673	9.849
98	9 604	941 192	9.899
99	9 801	970 299	9.950
100	10 000	1 000 000	10.000

Index

Page numbers in *italics* refer to figures in the text.

A

B

C

D

E

F

G

H

I

N

O

P

S

T

U

V

W